S0-BVO-685

ASTRONOMY OF STAR POSITIONS

ASTRONOMY OF STAR POSITIONS

A critical investigation of star catalogues, the methods of their construction, and their purpose

HEINRICH EICHHORN

Frederick Ungar Publishing Co., New York

Printed in the United States of America
Library of Congress Catalog Card No. 73-81764
Designed by Francis Reisz
ISBN: 0-8044-4187-1

In memory of my teachers

ADALBERT PREY
(1873–1949)

and

JOHANN RADON
(1887–1952)

Preface

There has never been any decade in which such a mass of exciting astronomical discoveries has been made as in the last one. We have learned about the existence of quasars, pulsars, the 3°K background radiation, and other phenomena that have brought about a great deal of new knowledge of things in the universe.

Under these circumstances, a field like positional astronomy is easily neglected. Most of the best young minds in astronomy turn their attention to these new, exciting fields where a reputation may be established based on observations made during but a few nights. With such possibilities existing, what bright young astronomer is going to devote his efforts to positional astronomy where spectacular discoveries are hardly to be expected, and whenever they should be made, will be based on years and decades of painstaking work?

And yet, positional astronomy has not lost any of its importance for astronomy as a whole. The empirical realization of an inertial system has not yet been accomplished with the desired accuracy. The distance scale of the universe ultimately depends on astrometric investigations which yield secular, statistical and trigonometric parallaxes of the stars. The investigation of the systematic trends in the stars' proper motions (through galactic rotation) allows one to draw conclusions to the dimensions of our galaxy. The continued verification of Newton's postulates of motion needs a self-consistent system of reference positions.

The emergence of electronic computers has—at least potentially—revolutionized the techniques used in positional astronomy. Tasks which only a decade ago would have required years of an individual working at a desk calculator can now be performed in a few days of programming and a few minutes at the computer, and problems whose solution was impossible a couple of decades ago because of the huge amount of numerical work involved, can now be tackled successfully.

It has thus become possible to not only solve the problems of positional astronomy by some (not always elegant) method of approximation but to apply rigorous methods to them, and to drive the sophistication of astrometric investigations to a level which, without computers, would have been completely impossible. An enormous field has thus opened up for intelligent investigators, and any knowledgable astronomer will appreciate the impact of improved astrometric data on astronomy as a whole.

Unfortunately, the present generation of young doctorate-level astronomers does not know enough about the problems in astrometry to meaningfully work on them. There is just so much any individual can know, and with all the excitement that surrounds the new pioneer fields in astrophysics it is understandable that there is not much room in their education for a thorough acquaintance with astrometry and its problems, however important this may be even for the fields in which they are interested. And thus, the exact meaning of the information in the star catalogues which has accumulated in the last two centuries is for many locked away in a book with seven seals.

This book was written in an attempt to unlock these seals. It gives, I hope, some insight into the meaning of the concepts of system, of systematic errors, and so on. It tries to outline in principle the methods which are used for getting the data one usually finds in star catalogues, namely, right ascensions and declinations. Clear instructions are given for the transformation of the data found in the catalogues into different form, that is, to different orientations and epochs and to different systems, and for the derivation of secondary information, such as proper motions, from these data.

Special attention is given to some important catalogues, fundamental ones and others. They are described in detail, together with a discussion of the processes by which they were formed. The section on the Astrographic Catalogues is intended to help in the lifting of the treasure that is buried in its many volumes.

The book does not pretend to be, nor is it, a text on spherical astronomy. I have rather assumed throughout that the reader is familiar with this field and only when I found it desirable have I reminded the reader of some facts of spherical astronomy, or elaborated on a point which in my judgment is not covered in sufficient detail in the well-known texts. Throughout, I have used matrix and vector notation, and I do not feel any need to apologize for it. Although I am indebted to many other authors for the information presented in this book, much of it is based on my own, previously unpublished, work. The expert will recognize these passages.

Originally, an investigation covering the scope of this book was commissioned by the RCA Service Co. under contract No. AF 08(606)-3413 with the U.S. Air Force. Later, the manuscript was expanded to its present form. Both the RCA Service Co. and my present publisher, Mr. Frederick Ungar, have earned my gratitude for the angelic patience they have displayed in waiting for the manuscript to be finished.

A number of people have read the manuscript and given me valuable hints. Foremost of these is Mr. Francis Patrick Scott, formerly of the U.S. Naval Observatory, who has not only carefully read the entire manuscript, but also made many suggestions for the improvement of the book, to the point of rewriting whole paragraphs. For this, I want to thank him most profoundly. I have, however, not in all cases accepted his suggestions, and the responsibility for any inaccuracies. omissions, mistakes or any other shortcomings is entirely mine.

Dr. Hermann v. Socher (Vienna) also read the first four chapters, and I owe some good advice to him. Other colleagues who gave me the benefit of their judgment by reading parts of the book and commenting on it are P. Brosche (Heidelberg), W. Dieckvoss (Hamburg), J. Gibson (New Haven), W. Gliese (Heidelberg), D. Hoffleit (New Haven), G. C. McVittie (Urbana), D. Mulholland (Austin), G. v. Schrutka-Rechtenstamm (Vienna), T. R. van Flandern (Washington) and C. A. Williams (Tampa). Mr. Duane Brown of Melbourne, Florida was the first person who suggested that this monograph be written. The editor, Miss Calliope Scumas, has diligently ploughed through many pages of what cannot by any stretch of the imagination be considered breathtaking material, and has substantially contributed to the clarity and unambiguity of the presentation by catching and correcting many shortcomings of grammar and punctuation in the unedited manuscript. Miss Jane Russell and Messrs. J. A. Carr and F. W. Fallon have assisted substantially with proofreading, and pointed out several errors which otherwise would have been overlooked. Finally, my secretaries, in particular Mrs. Mary Haney and Mrs. Iris Rose, have spent many arduous hours in typing the manuscript from heavily marked up, and often handwritten, material. To all these individuals I express my sincere thanks.

HEINRICH EICHHORN

Tampa
May 1973

Contents

Chapter 1
ASTRONOMICAL COORDINATE SYSTEMS

Chapter 2 THE ACQUISITION OF ASTRONOMICAL DATA

Chapter 4
COMPILATION CATALOGUES

Chapter 1

ASTRONOMICAL COORDINATE SYSTEMS

1.1 Introduction

1.1.1 DEFINITION OF A STAR CATALOGUE. In the usage of astrometry, that is, the science dealing with the positions and position changes of celestial bodies, a "Star Catalogue" is defined as *a list of spherical coordinates of stars or their equivalent, such as standard coordinates, that were obtained by the evaluation and reduction of observations made expressly for the accurate determination of stellar positions.*[1] This definition therefore includes catalogues in which accurate values are listed for only one coordinate while the second coordinate, if listed at all, is given only approximately for identification purposes; for example, catalogues of right ascensions determined at a transit instrument and lists of declinations determined with a vertical circle would be included under this definition. Compiled catalogues, such as the "General Catalogue" by Benjamin Boss (see Section 4.2.4) would also be included under this definition because the positions of the stars in such catalogues are obtained from observations made for the purpose of determining positions, even though the positions given in the catalogue were derived from more involved reductions than those used in the preparation of a catalogue from a single series of more or less homogeneous observations. On the other hand, the "Bonner Durchmusterung," various catalogues of spectral types, and lists of other data characterizing the stars, would be excluded by the definition because, in general, only approximate positions of the stars are given in them for identification purposes.

Accurate positions are positions determined from observations made on instruments, such as transit circles, transit instruments and certain photographic tele-

1. A star's *position* is equivalent to the direction in which it is located and is therefore completely *given by a pair of coordinates.* The object's *location*, however, describes the point in space at which the star is actually located and is therefore *uniquely described by three coordinates*, for instance the star's position and its distance from the observer.

scopes, whose primary function is to serve in the gathering of observations for the determination of high precision star positions. For this purpose, these instruments must be used in a mode which will ultimately lead to positions whose precision is reasonably close to the best the instruments are capable of producing. This precision varies from one instrument type to another and even from one instrument of the same type to another. It has also increased in the course of time. While 100 years ago any positions regarded as accurate had rms. errors of 1.″0, we nowadays require positions to have rms. errors of no more than about $\frac{1}{3}''$ before we regard them as "accurate."

1.1.2 DEFINITION OF A SYSTEM OF SPHERICAL POLAR COORDINATES. The *location* of the stars and other celestial objects in space would be defined by

a) the definition and realization of a coordinate system;

b) their coordinates in this system.

The establishment of coordinate systems and the measurement of the object's coordinates with respect to them will be discussed in Chapter II.

In a meaningful coordinate system, there must be a reversibly unique relationship between the location of a point and its coordinates, that is, every point must be uniquely described by only one set of coordinates, and every set of coordinates must define the location of only one point.

Cartesian coordinates have this property but, since knowing a point's Cartesian coordinates implies knowing its distance from the system's origin, they are of limited use for astronomy, for the following reason. While in recent years the direct determination of topocentric *distances*[2] of other members of our solar system, notably of the Moon and of Venus, has been achieved by radar with high accuracy, the distances to bodies (stars) outside our solar system cannot (yet) be measured directly but must be computed or rather estimated. In no case is the relative accuracy of this distance estimation comparable to that of the direction, which can be measured directly and accurately. We may take advantage of the properties of a system of spherical polar coordinates which defines a point's location, by giving its distance and its direction.

The two quantities defining the direction may be regarded as curvilinear coordinates on the surface of the two dimensional unit sphere. Traditionally, the unit sphere is replaced by the (actually nonexistent) *celestial sphere*, on which the celestial objects seem to be located. The *topocentric direction* of an object, that is, its direction from the observer, therefore specifies its position on the celestial sphere. This position is given by a pair of spherical coordinates, *longitude angle and latitude*

---

2. The author regrets the intrusion of the ambiguous expression "*range*" to denote the unique "*distance*." This term used to be restricted to ballistics, and in ordinary usage may mean a place where cattle live, something to cook one's food on, an interval within which a quantity may vary, and so forth. Must the technical man use expressions which are meaningless to the layman in order to preserve his nimbus?

angle as defined below. On the celestial sphere a point is chosen according to certain principles and denoted as the *pole* or the *first pole*; the point opposite of the first pole on the sphere is called the *second pole.* The great circle on the sphere, which runs between the two poles and is perpendicular to the line connecting them, is called the *fundamental circle.* A zero point on the fundamental circle is chosen arbitrarily or by some convention; half the great circle through it between the poles is called the *zero vertical.* The position of a point (star) on the sphere is then defined by its latitude and longitude angles. The latitude angle is the angle which a line through the object and the center of the sphere makes with the plane of the fundamental circle; it is customarily restricted to vary between $-90°$ (second pole) and $+90°$ (first pole). It will be negative on the "second" hemisphere, that is the hemisphere on that side of the fundamental circle which contains the second pole, and positive in the first hemisphere which is defined analogously. The longitude angle is the angle by which the zero vertical must be turned in the positive direction on an axis through the poles in order to make it coincide with the vertical through the object, which is the half of the great circle that connects the poles and runs through the star.

The *"positive" direction*, looking at the sphere from outside from the direction of the first pole, is *counterclockwise* if the associated rectangular Cartesian coordinate system is a right-handed one, and *clockwise* if the associated system is left-handed. This system, which has units of equal length on each axis, is defined as follows: the z-axis points toward the first pole, the x-axis towards the intersection point of the zero vertical with the fundamental circle, and the y-axis is in the plane of the fundamental circle so that a rotation by $+90°$ about the z-axis will bring the x-axis to where the y-axis was. It is generally advantageous to replace the latitude angle b by its complement, the colatitude angle $c = 90° - b$. The relationship between the coordinates (x, y, z) of a point in the associated Cartesian coordinate system, and its spherical polar coordinates (r, l, c) where r is the point's distance from the origin, is as follows:

$$\boldsymbol{x} = r\hat{\boldsymbol{x}}(l, c)$$

where

$$\boldsymbol{x}^{\mathrm{T}} = (x, y, z)$$

and the position unit vector

$$\hat{\boldsymbol{x}}(l, c) = \begin{pmatrix} \sin c \cos l \\ \sin c \sin l \\ \cos c \end{pmatrix} \tag{1.1}$$

A reversibly unique relationship between Cartesian and spherical coordinates is preserved by restricting $|b| \leqq 90°$, equivalent to $0° \leqq c \leqq 180°$ and $0° \leqq l < 360°$. c and l can, of course, be very easily calculated without ambiguity from the components of $\hat{\boldsymbol{x}}$.

1.1.3 THE CHOICE OF SUITABLE COORDINATE SYSTEMS. The question of choosing a system in which to record star positions has by no means an obvious, and not even a simple, answer. So that it can serve as a basis for investigations in mechanics, it should satisfy the condition that it can be realized by observations and that it is either *inertial*, or that a rigorous mathematical expression is available describing how the relationship of this system to an inertial one changes with time.

The definition and empirical establishment of the relationship of a system used for the recording of star positions to an inertial system is indeed the cardinal problem of catalogue astronomy, and one of the most important tasks of spherical astronomy. One of the reasons for this is that the observed motions of the members of our solar system will also be related to this system. These motions will be theoretically explained on the basis of Newtonian mechanics, which assumes that the positions are referred to an inertial system. If the system in which the bodies' positions are recorded is not an inertial one, contradictions will appear between the observed positions and those calculated from the theory. Such differences are by no means proof that "all is not well" with Newtonian mechanics; they do, however, provide an important tool for the empirical establishment of the relationship between the system in which the positions are recorded and an inertial one. The transformation which describes this relationship mathematically depends of course on time. This does not mean that the system in which the star positions are directly observed need be even an approximately inertial one. It must be realized that, if all observed relative positions are kept invariable, that is, if all celestial objects retain their directions with respect to each other, all systems, except for a possible reversal in the direction of the longitude angle, can be transformed into each other by a single rotation which brings their first poles and zero verticals (and thereby also automatically the fundamental circle and second poles) into coincidence. In case the relationship between two systems varies with time, the transformation between the two systems can be established as a function of time, only if the ways in which the systems change their relative orientations are known. The detailed transformations which describe the relationships between the various systems in use, including an inertial one, are a subject of spherical astronomy. More will be said about this in Section 1.6.5.

Furthermore, a usable coordinate system must be defined in such a way that it can be realized, that is, that it is both *producible* and *reproducible*. This means that two observers working independently and using the same procedures should arrive at the same conclusion as to the coordinates of the stars in a particular system of reference.

This requirement has led astronomers to seek out directions and planes defined by natural phenomena to serve as axes and principal planes for their coordinate systems, such as the direction of gravity and the direction of the celestial poles. The manner in which natural phenomena contribute to the establishment of a celestial coordinate system will be treated in Sections 1.2 and 1.3.

1.2 The System of the Horizon

1.2.1 THE ZENITH. A system, the first pole of which can at all times be realized by considering the Earth's statics, that is, the Earth's gravitation field only, is the so-called *horizon system.* This *first pole* is the *zenith* of the observer; its *fundamental circle,* the observer's ideal horizon. The direction towards the zenith of the observer or, more accurately, the zenith of the instrument which the observer uses for the realization of the system, is given as the direction of gravitation at the telescope, or, more precisely *the direction of the upward tangent at the center of the crosshairs of the observing instrument to the orthogonal trajectory to the equipotential surface of the Earth's gravitation which passes through the center of the observing instrument.* (Note here that this is not the same as the normal to the surface through the observer parallel to the geoid.)

This definition of the zenith follows the actual process by which the direction to the zenith is established. The second pole in the horizon system is called the *nadir.* If the equipotential surfaces of the Earth's gravitation field retained an unchanging shape and if the position of the observer within the system defined by the equipotential surfaces remained constant, a zenith could be defined for every point where position within the field of the Earth's gravitation can be established, and the direction to this zenith would be constant with respect to a coordinate system that is anchored within the body of the Earth.

The shape of the equipotential surfaces, however, undergoes periodic and secular changes due to tectonic processes, the seasonal melting of the Earth's polar ice caps, changes in the Earth's speed of rotation, the varying positions of the tide-generating bodies, and so on. This renders the geocentric direction to the zenith for every point on the Earth's surface very slowly and, where short period changes are involved, very slightly variable. The records of observing astronomy cover too brief a period to show a secular change in the zenith of observing stations; however, short period zenith fluctuations have been detected by astrometric observations. At the (old) Greenwich Observatory, for instance, the tides in the Thames River apparently caused observable fluctuations of the direction to the zenith. (Danjon [1959], p. 57.)

1.2.2 THE ZERO VERTICAL. The *zero vertical* is that half great circle connecting zenith and nadir which passes through the north celestial pole (defined below). This vertical intersects the horizon at the *north point.* The colatitude angle is called *zenith distance,* and is usually denoted by z. Its complement, the latitude angle, is called *altitude*[3] and is usually denoted by h. (At the Earth's astronomical

3. Recently the altitude has often been termed *elevation* in publications by nonastronomers writing on the subject. This term, which originated in ballistics, means something else in geodesy and astronomy and ought to be avoided. Unfortunately, its use is already so widespread that attempts to purge it from the literature have probably no chance of success.

poles, the horizon system is not defined because at these points the celestial poles coincide with zenith and nadir, respectively. This difficulty can be overcome by arbitrarily assigning a zero point on the horizon.) The name of the longitude angle in the horizon system is *azimuth* and it is usually denoted by a. It is counted positive from north through east, which makes the associated Cartesian system left-handed. In older publications and in the Continental European literature, the azimuth is counted from the south point rather than the north point. The great circle, of which this system's zero vertical is one part, is called the *meridian*. The great circle described by $a = 90°$ and $a = 270°$ is called the *prime vertical*. It intersects horizon and meridian at right angles and passes through the zenith of the observer.

The usual definition of the celestial north pole as the point where the celestial sphere is penetrated by the continuation of the Earth's axis through the Earth's north pole is suitable for all practical purposes and adequate for the position accuracy which the best astronomical measuring instruments at this time are capable of yielding. In the corresponding Cartesian system, the direction to the celestial north pole is identical with the direction of the Earth's axis, taken positive toward the terrestrial north pole.

The "Earth's axis" is not a static, but a kinematical, concept. Therefore it is clear that not even the simplest coordinate system in astronomy can be set up without using concepts that are essentially mechanical. As we shall see later, some astronomical coordinate systems even require recourse to dynamical concepts. If we use the term "Earth's axis," we are tacitly assuming that the Earth is a rigid body. From a very critical standpoint this is not true. There are, for example, tectonic processes which regularly change the relative position of points within the Earth's body and at the Earth's surface with respect to each other; furthermore, there are the influences of extraterrestrial bodies which produce the *continental tides*. This means that the Earth's axis, strictly speaking, is not a well-defined concept.

To form an entirely rigorous definition of the north pole, which would be valid even for a nonrigid Earth, let us consider the geocentric[4] space curve in a coordinate system whose axes always remain parallel to those of an inertial system which is described by a point physically connected with the Earth, for example, a point on the Earth's surface. If the Earth were strictly rigid, this curve would be a perfect circle.

At the point under consideration, the direction to the north pole is defined as the direction of the binormal to the curve just described. This is, of course, parallel to the direction of the axis of the osculating circle to the curve in the point considered.

This definition is a generalization of the previous one and would still apply even if only one "Earth's axis" could be defined. It takes into account the fact that,

---

4. Geocentric means referred to the Earth's center, in this case the center of the Earth's mass.

strictly speaking, every point on the Earth's surface has its own north polar direction.

This definition also takes account of all the changes of the position of the Earth's axis in space as caused by precession and nutation, and the shifts of the Earth's axis within the Earth, such as polar motion. Thus, when we then speak of the Earth's axis, we imply the mean of all "local Earth's axes." However, we should bear in mind that actual observations at any one point which is physically connected to the earth will always yield the direction to the local pole.

1.3 The Equator System

1.3.1 DECLINATION AND HOUR ANGLE. The system of the horizon suffers from two disadvantages which make it poorly suited to serve as the basis for a star catalogue. The coordinates of the stars in this system vary rapidly and nonlinearly with time and they are different for all points which have different zeniths. An ideal coordinate system, for the purposes of a star catalogue, would be one in which the positions of these points that do not change their coordinates in an appropriately chosen inertial system would also remain unchanged. In an inertial system whose center moves, say, parallel with the center of our galaxy, the stars would change their positions only because they move with respect to the inertial system. However, sooner or later it will be shown that the center of our galaxy is also accelerated. Indeed, the ultimate aim of astrometry is the realization of an inertial system. This is very difficult. A rigorous realization of an inertial system by astrometric measurements is essentially impossible because of the statistic character of the observations. The quantities describing the relationship defined by observations and an inertial system will always have a covariance matrix associated to them, and all we can hope to do is to reduce the values of the variances.

The desiderata for a system in which star positions are to be recorded are as follows:

(1) it should be fairly easy to make direct and accurate measurements of the coordinates of the stars in this system;

(2) the stars' coordinates in this system should be as independent from the place of observation as possible;

(3) the transformation of this system to an inertial system should be known as accurately as possible and should depend only slightly on time.

If (3) is satisfied, the stars' coordinates in the system will vary only slowly with time.

Direct observations are easiest in a system which is most directly defined and can be most simply realized. This is the system of the horizon, because the only observations which can be made quite directly are zenith distance or altitude observations, and we shall see that these play an important role in the derivation of star positions.

In a system of coordinates in which the *latitude angle* is the *declination*, at least this one coordinate is practically independent of the place of observation. This is because the declination is defined as the complement of the north polar distance which is the angle between the position unit vectors to the celestial north pole and the object, respectively. As we have stated above, local variations of the declination could therefore only be caused by local variations of the direction to the celestial poles, but since the differences in the directions to the local celestial poles are caused only by the deviations of the physical earth from strict rigidity, they are too small to cause local variations of declination that are detectable by present-day astronomical methods. Due to the finite distance of the celestial object, the declination still depends on the observer's position. For bodies outside our solar system, the origin of the coordinate system is considered to be in the sun, and the difference between the ideal heliocentric and the more directly available geocentric direction of an object outside the solar system is its annual parallax. The *daily parallax*, being the difference between an object's geocentric and its more directly available topocentric direction, is, with the presently available accuracy, measurable only for objects within our solar system. If not expressly stated otherwise, all coordinate systems discussed below should be understood to be heliocentric, that is, with their origins in the Sun's center, or, more accurately, the barycenter of the solar system.

The fundamental circle of the equator system, the celestial equator, is the locus of all points having a declination of 0°, or a north polar distance of 90°. With the reservations outlined above in Section 1.2.2, we may regard it as common to all points on the earth for a given moment of time. The declination, usually denoted by δ, satisfies the inequality $-90° \leqq \delta \leqq 90°$. Positive declinations define the "northern hemisphere" of the sky. The connection between the equatorial systems and the horizon system is given by the *geographic* or *astronomical latitude*, usually denoted by ϕ, which is the *altitude of the north celestial pole*.

In order to completely define the system, we must define a longitude angle. If we choose the associated Cartesian coordinate system to be left-handed and the observer's meridian to be the zero vertical, the *longitude angle* in this system is called the *hour angle*, usually denoted by t, and is counted from 0^h to 24^h ($1^h = 15°$). The system itself is usually called the *hour angle system*. We may also call it the first equatorial system. (We use the word "first" to distinguish it from the equatorial system which will be defined below in Section 1.3.2).

Hour angle and declination are the coordinates by which a telescope is set to point to a certain region of the sky. Observers avoid hour angles larger than 12^h and, instead, count the hour angle negative from the meridian towards east so that -1^h replaces 23^h, -2^h replaces 22^h, and so forth.

A celestial object has the same hour angle at all observing stations which lie on the same *terrestrial meridian*.[5] The hour angle is still rapidly variable with time,

5. This statement is precise only if one considers the positions unaffected by diurnal aberration, see Section 1.6.4.

mostly due to the Earth's rotation on its axis so that it increases by 24^h or $360°$ in one sidereal day. For any given location on Earth, the first equatorial system and the horizon system are rigidly connected to each other. Since both of them are Earth anchored, they move with respect to the stars.

1.3.2 RIGHT ASCENSION. The strong dependence of the longitude angle on the time in the first equatorial system (that is, the hour angle) makes this system very poorly suited for catalogue purposes. This strong dependence was therefore eliminated by introducing a new longitude angle which is counted from a point which participates in the daily rotation of the sky, that is, by regarding as a point's longitude angle in a second equator system the difference between the hour angles of the zero point and the object itself. This reference or zero point might have been chosen quite arbitrarily. Any easily identifiable star might have served. The choice which was actually made is best appreciated when one considers the historical development of astronomy. In the days when the coordinate systems were defined, the primary interest of astronomy was the explanation and prediction of the motions of the bodies in the solar system. It is natural that the fundamental circles of the coordinate systems were associated with and derived from the observation of the motion of the Earth. One, the celestial equator, is parallel to the plane of the Earth's equator, the other is in the plane of the Earth's orbit around the Sun. This latter circle almost coincides with the ecliptic which will be rigorously defined in the next section.

1.3.3 THE ECLIPTIC. The ecliptic is a great circle, and very nearly the apparent path of the Sun's center in the sky. Its plane is very close to that of the Earth's heliocentric orbit. Ecliptic and equator intersect in two diametrically opposite points in the sky. One, the *vernal equinox*, commonly called the *First Point of Aries*, is, essentially, the point where the Sun passes from the southern to the northern celestial hemisphere; the other one is called the *autumnal equinox*. The hour angle of the vernal equinox is called *sidereal time*, and is frequently denoted by θ. The great circle passing through the celestial poles and the equinoxes is called the *equinoctial colure*. That half of the equinoctial colure which passes through the vernal equinox is the zero vertical of the second equator(ial) system or the equator(ial) system. In this system the *longitude angle* is the *right ascension*, normally denoted by α, and the latitude angle is the declination. It is right-handed in contrast to the first equatorial system and the horizon system. The equator system changes its relationship to an inertial system only as a consequence of a change of the orientation of the Earth's axis and the ecliptic with respect to an inertial system. These changes originate in attractions on the equatorial bulge and perturbations of the orbital plane of the Earth by the actions of the Sun, the Moon and the planets, and are known as *precession* and *nutation*.

The rigorous definition of the ecliptic is more complicated than that of the celestial equator. While the equator can be realized instantly, at least in principle, from only the kinematics of the point of observation, no such instant realization is even theoretically possible for the ecliptic. One might define the plane of the "true ecliptic" as that of the osculating circle to the space curve which the center of gravity of the Earth-Moon system describes with respect to the Sun. The ecliptic is, however, not defined in this way, mainly for historical reasons. The ecliptic, so close to the apparent paths of all major bodies of the solar system, was a concept used even by the ancient astronomers. Until the influence of the other bodies of the planetary system on the ecliptic's orientation was appreciated, the ecliptic was defined as the apparent path of the center of the sun as seen from the barycenter of the Earth-Moon system. It was thus natural that the plane defined by the solar center, the Earth-Moon system's barycenter and this barycenter's velocity vector, was taken as the basis for the rigorous definition of the ecliptic. The orientation of this plane with respect to an inertial system will be given by two Eulerian angles, ϵ^* and Ω^*. These will be functions of time, say

$$\epsilon^* = \epsilon_0 + \epsilon_1 t + \epsilon_2 t^2 + \epsilon_3 t^3 + P_\epsilon$$

$$\Omega^* = \Omega_0 + \Omega_1 t + \Omega_2 t^2 + \Omega_3 t^3 + P_\Omega,$$

where t is the time. The coefficients $\epsilon_0, \epsilon_1, \ldots, \Omega_3$ as well as P_ϵ and P_Ω are calculated from the theory of the motion of the planets. P_ϵ and P_Ω are sums of periodic functions of the time, and originate from periodic perturbations of the orientation of the earth's orbital plane by the planets, chiefly Venus and Jupiter. The plane whose orientation is defined by the angles $\epsilon = \epsilon^* - P_\epsilon$ and $\Omega = \Omega^* - P_\Omega$ is rigorously regarded as the plane of the (true) ecliptic. This means that the plane of the ecliptic changes its orientation with respect to an inertial system secularly, but not periodically.

From this definition it is seen that the ecliptic does not coincide with the apparent path of the center of the Sun for two reasons:

(1) The barycenter of the Earth-Moon system, not the center of gravity of the Earth, is used in the definition. The Sun's ecliptic latitude will be negative or positive depending on whether the center of the Earth is above or below the plane of the ecliptic.

(2) The Sun will not be in the ecliptic even as viewed from the barycenter because of the eliminated periodic terms P_ϵ and P_Ω.

If the ecliptic is chosen as the fundamental circle and the vernal equinox as zero point in a right-handed system, we get as corresponding spherical coordinates *ecliptic latitude*, usually denoted by β, and *ecliptic longitude*, mostly denoted by λ.

Since only the barycenter of the Earth-Moon system is used in the definition of the ecliptic, this definition—unlike that of the equator—does not depend on the additional assumption that the Earth is a rigid body. The ecliptic is therefore truly independent of the point of observation, and so is the ecliptic latitude. Unfortunately, this independence of the ecliptic latitude from the point of observation is not preserved for the ecliptic longitude because its zero point (vernal equinox) requires the equator, which is dependent on the observation point for its definition.

It is worth noting that the realization of the ecliptic requires either the application of a dynamical theory for the computing of the neglected periodic terms, or observations over a long period of time to separate the secular changes of the position of the plane from the periodic ones and thus establish a good estimate of the plane which defines the ecliptic. At the present state, the computation of the periodic terms from the theory of the motion of the bodies in the solar system is far more accurate than from an analysis of the observation material. Fortunately, the geocentric ecliptic latitude of the Sun is never larger than 1.″2, of which at the most 0.″6 may be due to the perturbing action of the planets. Essentially dynamical considerations concern the definition of the vernal equinox to only a very slight degree; the main contribution to the definition of the ecliptic is still made by a concept that is purely kinematic in nature, namely, the Sun's apparent path in the sky.

1.4 Tables and Formulas

Table 1.1 summarizes the information pertinent to the coordinate systems that are principally used in positional astronomy.

The formulas for the transformation of the positions of celestial objects from one coordinate system to another are most concisely written in the form of the transformation of position unit vectors in the associated Cartesian coordinate systems.

If $\mathbf{R}_1(\alpha)$ denotes the orthogonal matrix associated with a rotation by α on the x-axis, that is,

$$\mathbf{R}_1(\alpha) = \begin{pmatrix} 1 & 0 & 0 \\ 0 & \cos\alpha & \sin\alpha \\ 0 & -\sin\alpha & \cos\alpha \end{pmatrix}$$

with 2 corresponding to the y-, and 3 to the z-axis so that $\mathbf{R}_2(\alpha)$ and $\mathbf{R}_3(\alpha)$ are obtained by cyclic permutation of both the rows and columns of $\mathbf{R}_1(\alpha)$, compare Mueller [1969], p. 43, the following formulas describe the relationships between the various systems, where the position unit vectors are given by equation (1.1).

Hour angle t and declination δ on the one hand, and azimuth a and zenith distance z on the other, are related by

$$\hat{\mathfrak{x}}(t, 90° - \delta) = \mathbf{R}_2(\phi - 90°)\mathbf{R}_3(180°)\hat{\mathfrak{x}}(a, z), \tag{1.2}$$

where ϕ is the previously defined geographic latitude of the observer.

The relationship between the first equatorial system and the second one

Table I.1 **Coordinate Systems of Spherical Astronomy**

SYSTEM	HORIZON	FIRST EQUATORIAL	EQUATOR	ECLIPTICAL
longitude angle	azimuth a	hour angle t	right ascension α	(ecliptic) longitude λ
latitude angle	altitude h	declination δ	declination δ	(ecliptic) latitude β
colatitude angle	zenith distance z	north polar distance* p	north polar distance* p	ecliptic polar distance* ♈
fundamental circle	horizon	equator	equator	ecliptic
zero point of longitude angle	north point	highest point of equator (no widely used special name)	vernal equinox ♈	vernal equinox ♈
left- or right-handed	l	l	r	r
line of constant longitude angle	vertical	hour circle	hour circle	longitude circle
line of constant latitude angle	almucantar*	parallel of declination	parallel of declination	parallel of latitude
special lines	$a=90°$ and $a=270°$: prime vertical; $a=0°$ and $a=180°$: celestial meridian	$t=0°$ and $t=12^{\mathrm{h}}$: celestial meridian	$\alpha=0$ and $\alpha=12^{\mathrm{h}}$: equinoctial colure*	
latitude angle $+90°$	zenith	north celestial pole (NCP)	north celestial pole (NCP)	north pole of ecliptic (NPE)
latitude angle $-90°$	nadir	south celestial pole (SCP)	south celestial pole (SCP)	south pole of ecliptic (SPE)

* used only infrequently.

("the" equatorial system) is defined by

$$\hat{x}(\alpha, 90^\circ - \delta) = \mathbf{R}_3(-\theta)\begin{pmatrix} 1 & 0 & 0 \\ 0 & -1 & 0 \\ 0 & 0 & 1 \end{pmatrix}\hat{x}(t, 90^\circ - \delta). \tag{1.3}$$

It can easily be seen that this relationship is fully defined also by the alternate

$$\alpha = \theta - t \tag{1.3a}$$

Finally the positions in the ecliptic and the equator systems are connected by

$$\hat{x}(\lambda, 90^\circ - \beta) = \mathbf{R}_1(\epsilon)\hat{x}(\alpha, 90^\circ - \delta), \tag{1.4}$$

where ϵ, the obliquity of the ecliptic, which is the angle between the mean north celestial pole and that of the ecliptic, is given by

$$\epsilon = 23°27'08.''26 - 0.''4684T - 0.''000\ 0006T^2. \tag{1.5}$$

T is expressed in units of tropical years since 1900.0.

From the relationships (1.2) through (1.5), the formulas which effect the transformations between the position of an object in any of the above discussed coordinate systems can easily be derived.

1.5 Inertial Systems

1.5.1 DEFINITION AND SIGNIFICANCE. We have stated before that *one of the principal functions of star catalogues is the definition of, and the providing of means for the realization of, an inertial system.* The way in which a star catalogue defines a coordinate system will be discussed in Section 1.6. We must be able to realize an inertial system in order to meaningfully compare theory and observation in mechanics.

At this time, the theory of mechanics is based on Newton's three postulates of motion: inertia, definition of force, and action-reaction. The first two postulates are concerned with motion of bodies and, therefore, imply the existence of a frame of reference and an independent variable called time. This is because "motion" is defined as a change of coordinates during an interval which is measured by a flow of time. It is easy to see that Newton's first postulate of motion is not valid in all coordinate systems. If we agree that the motion of a bullet coming from a muzzle represents (for a very short time) a fair approximation to the motion under the influence of no force, it is obvious that its path is curved if it is considered with respect to a rapidly rotating coordinate system. Forces which are postulated to keep Newton's second postulate of motion valid in any coordinate system are called *Inertial Forces,* such as Coriolis Forces and Centrifugal Forces.

When inertial forces are excluded, Newton's first and second postulate of motion are valid only if the motion is reckoned in an inertial system. As a matter of fact an inertial system may be defined as any reference system in which Newton's first and second postulate of motion are valid in the absence of inertial forces. More precisely, let

$$\boldsymbol{x} = \boldsymbol{x}(t) \tag{1.6}$$

be the solution of the differential equations $m\,\mathrm{d}^2\boldsymbol{x}/\mathrm{d}t^2 = \boldsymbol{F}$, which are the equations of motion of a particle, established following Newton's second postulate. Here it is essential that $\boldsymbol{F}$ is determined by the physics of the situation only and does not intrinsically depend on the coordinate system to which these equations are referred. If $\boldsymbol{F}$ were (as it is in celestial mechanics) the force generated by the gravitational attraction of two mass points, it would in any inertial system have to be written $\boldsymbol{F}_{2,1} = (-m_1m_2/|\,\boldsymbol{x}_1 - \boldsymbol{x}_2\,|^3)\,(\boldsymbol{x}_2 - \boldsymbol{x}_1)$ (where the meaning of the symbols needs no explanation) without any modifications due to an "acceleration" of the coordinate system. These modifications would, indeed, depend on the transformation to an inertial system of the coordinate system with respect to which the position vectors $\boldsymbol{x}_1$ and $\boldsymbol{x}_2$ are reckoned. These would then be the above-mentioned "inertial forces." Let us assume that a system can be found in which the particle's observed motion is actually described by the equation (1.6). We may then regard this system as an inertial system, as long as we find no example in which the observed motion of a particle cannot (by appropriate choice of the constants of integration, and so on) be made to agree with that predicted by the equations of motion, set up in accordance with Newton's second postulate of motion.

The complete, rigorous definition will also have to consider t, the *independent variable in the equations of motion*. It is commonly known as *time*, or more specifically, *Newtonian time*. The discussion above shows that its determination must be based on the comparison of observed motions with predictions made on the basis of the integration of certain equations of motion. This fact was apparently realized only in this century. Before that, time was defined purely kinematically in terms of the Earth's orientation with respect to a certain coordinate system. As theories became more sophisticated, and observations more accurate and numerous, they could no longer be adequately reconciled with each other. Thus, astronomers were forced to consider time as derived from the observation of the bodies in the solar system rather than from the Earth's orientation. This time is the best available approximation to Newtonian time and is called *ephemeris time*.

From this discussion we can conclude that the establishment of an inertial system can at best be achieved asymptotically. For in order to prove that a system is not inertial, one would have to find a particle whose motion in it cannot possibly be reconciled with Newton's second postulate of motion. In principle, one is never able to prove that any system is inertial because, if it is not, that motion which cannot be reconciled with the Newtonian postulates may not yet have been observed, and possibly may never become observed.

The question whether an inertial system exists at all is very important for physics. Just one observed example of a motion which cannot be reconciled with Newton's postulates of motion would force us to change our very fundamental concepts of physics. If the general validity of Newton's postulates of motion should ever be disproved, it is very probable that this will involve the careful and accurate observation of star positions which, originally, were made for the purpose of establishing an inertial system.

1.5.2 THE MOST GENERAL INERTIAL SYSTEM. We finally observe that the concept of the *inertial system does not imply the existence of absolute space.* Since the equations of motion whose solutions are essential for the definition of an inertial system are invariant with respect to all Galilei transformations, that is, transformations of the type $\boldsymbol{x}' = \boldsymbol{x} + \boldsymbol{a}t$, t being Newtonian time and $\boldsymbol{a}$ a constant vector, the system defined by the axes $\boldsymbol{x}'$ will be inertial if the system to which $\boldsymbol{x}$ refers is inertial. Furthermore, since the equation of the space curve described by (1.6) in natural coordinates[6] is invariant to orthogonal transformations, the most general inertial system $\boldsymbol{x}'$ is obtained from any inertial system $\boldsymbol{x}$ by the transformation

$$\boldsymbol{x}' = \mathbf{R}\boldsymbol{x} + \boldsymbol{a}t, \tag{1.7}$$

where $\mathbf{R}$ is an orthogonal matrix whose terms do, of course, not depend on the time. This may be interpreted as stating that *there are absolute accelerations* in space, even though there are no absolute positions in space.

1.5.3 CONNECTION WITH THE EQUATOR SYSTEM. The actual realization of an inertial system must proceed in two steps: the realization of any well-defined system, and the establishment of a transformation between this system and an inertial system. Since the definition and realization of the inertial system involves dynamical considerations, that is, the integration of differential equations, the easiest realization of an inertial system, in practice, would be to choose as the first system one that is only slowly accelerated (rotated) with respect to an inertial system, and which can be realized without, or almost without, recourse to dynamics.

The equator system (discussed above in 1.3) is such a system. Dynamics enters into its definition only in a not very sensitive manner since the dynamical element in its definition, namely, the orientation of the plane of the ecliptic with respect to an inertial system, changes only very slowly. In accordance with the present practice of spherical astronomy, the origin of the equator system is very close to a point with respect to which the hodograph of the Earth's orbital motion is a central ellipse. The position and orientation of this point with respect to the barycenter of the solar system changes very slowly, and in a manner which is known with sufficient accuracy. Since the barycenter of the solar system describes an orbit around the galactic center in about 2×10^8 years, the direction in which the Sun moves, that is, the direction to the solar apex, changes by one second of arc in about 150 years. The accumulated observations are at the present time neither

---

6. This implies that the curve is defined by giving its curvature and torsion in terms of the arc length as parameter, cf. Hlavatý (1939), p. 71.

sufficiently accurate nor numerous enough to verify this effect. Since nothing is known concerning a change of speed of the Sun with respect to the galactic center, we may consider the origin of the equator system to move with constant velocity in an inertial system, and thus itself a suitable origin for an inertial system. This assumption, needless to say, cannot be proved and involves some faith.

The transformation of coordinates between any two rectangular Cartesian systems with the same origin is described by the equation

$$\boldsymbol{x}_1 = \mathbf{R}\boldsymbol{x}_2,$$

where $\mathbf{R}$ is an orthogonal matrix and involves at most three independent parameters, such as Eulerian angles. Generally, these will be functions of time.

This statement is, therefore, also true about the transformation between the equator system and an inertial system. This transformation is accomplished by a continuous rotation of the equator system on all three of its axes. The angular rates of rotation are a correction to the secular rate of change of the obliquity of the ecliptic (x-axis), a correction to the presently accepted value of the precessional quantity n (y-axis), and the so-called motion of the equinox (z-axis).

It is, therefore, the accuracy with which these quantities can be computed and measured that determines the accuracy with which the equator system can be related to an inertial system. The general principles available for the determination of these angles will be discussed in Sections 2.1.6, 2.2.6.1, and 2.4.3.1.

1.6 Catalogue Systems

1.6.1 STAR POSITIONS AND THEIR ERRORS. Star positions (α, δ) listed in any catalogue have always been calculated by some statistical adjustment from measurements $(x_1, x_2, \ldots, x_m)$ in a process which usually also involved the determination of reduction parameters $(a_1, a_2, \ldots, a_n) = \boldsymbol{a}^{\mathrm{T}}$. The adjustment will also give $\mathbf{Q}$, the covariance matrix of the parameters. We shall now apply the well-known formulas of error propagation (compare Brown [1955]) to analyze the errors of the coordinates α and δ. Let the positions be calculated from the measurements by

$$\alpha = \alpha(x_1, \ldots, x_m; a_1, \ldots, a_n)$$

$$\delta = \delta(x_1, \ldots, x_m; a_1, \ldots, a_n).$$

We may usually assume that the measurements are independent of each other (that is, that they are uncorrelated) and from the parameters. Therefore, the covariance matrix $\bar{\mathbf{Q}}$ of the arguments of the functions above which determine α and δ will be

$$\bar{\mathbf{Q}} = \begin{pmatrix} \mathbf{D} & \mathbf{O} \\ \mathbf{O} & \mathbf{Q} \end{pmatrix},$$

where $\mathbf{D}$ is a square diagonal matrix of order m, containing in the main diagonal the variances of the measurements. The covariance matrix $\mathbf{P}$ of α and δ will thus be given by

$$\mathbf{P} = \begin{pmatrix} \left(\frac{\partial\alpha}{\partial\boldsymbol{x}}\right)^{\mathrm{T}} \left(\frac{\partial\alpha}{\partial\boldsymbol{a}}\right)^{\mathrm{T}} \\ \left(\frac{\partial\delta}{\partial\boldsymbol{x}}\right)^{\mathrm{T}} \left(\frac{\partial\delta}{\partial\boldsymbol{a}}\right)^{\mathrm{T}} \end{pmatrix} \bar{\mathbf{Q}} \begin{pmatrix} \left(\frac{\partial\alpha}{\partial\boldsymbol{x}}\right) \left(\frac{\partial\delta}{\partial\boldsymbol{x}}\right) \\ \left(\frac{\partial\alpha}{\partial\boldsymbol{a}}\right) \left(\frac{\partial\delta}{\partial\boldsymbol{a}}\right) \end{pmatrix} \tag{1.8}$$

where

$$\left(\frac{\partial\alpha}{\partial\boldsymbol{x}}\right)^{\mathrm{T}} = \left(\frac{\partial\alpha}{\partial x_1}, \frac{\partial\alpha}{\partial x_2}, \ldots, \frac{\partial\alpha}{\partial x_m}\right), \tag{1.8a}$$

and the vectors $(\partial\alpha/\partial\boldsymbol{a})$, $(\partial\delta/\partial\boldsymbol{x})$ and $(\partial\delta/\partial\boldsymbol{a})$ are defined entirely analogously. The covariance matrix $\mathbf{P}$ of α and δ can be written, considering the structure of $\bar{\mathbf{Q}}$, as

$$\mathbf{P} = \begin{pmatrix} \sigma_{\alpha\alpha}{}^2 & \sigma_{\alpha\delta}{}^2 \\ \sigma_{\alpha\delta}{}^2 & \sigma_{\delta\delta}{}^2 \end{pmatrix} =$$

$$\begin{pmatrix} \left(\frac{\partial\alpha}{\partial\boldsymbol{x}}\right)^{\mathrm{T}} \mathbf{D} \left(\frac{\partial\alpha}{\partial\boldsymbol{x}}\right) + \left(\frac{\partial\alpha}{\partial\boldsymbol{a}}\right)^{\mathrm{T}} \mathbf{Q} \left(\frac{\partial\alpha}{\partial\boldsymbol{a}}\right) & \left(\frac{\partial\alpha}{\partial\boldsymbol{x}}\right)^{\mathrm{T}} \mathbf{D} \left(\frac{\partial\delta}{\partial\boldsymbol{x}}\right) + \left(\frac{\partial\alpha}{\partial\boldsymbol{a}}\right)^{\mathrm{T}} \mathbf{Q} \left(\frac{\partial\delta}{\partial\boldsymbol{a}}\right) \\ \left(\frac{\partial\delta}{\partial\boldsymbol{x}}\right)^{\mathrm{T}} \mathbf{D} \left(\frac{\partial\alpha}{\partial\boldsymbol{x}}\right) + \left(\frac{\partial\delta}{\partial\boldsymbol{a}}\right)^{\mathrm{T}} \mathbf{Q} \left(\frac{\partial\alpha}{\partial\boldsymbol{a}}\right) & \left(\frac{\partial\delta}{\partial\boldsymbol{x}}\right)^{\mathrm{T}} \mathbf{D} \left(\frac{\partial\delta}{\partial\boldsymbol{x}}\right) + \left(\frac{\partial\delta}{\partial\boldsymbol{a}}\right)^{\mathrm{T}} \mathbf{Q} \left(\frac{\partial\delta}{\partial\boldsymbol{a}}\right) \end{pmatrix}. \tag{1.8b}$$

In particular, study the expressions for $\sigma_{\alpha\alpha}{}^2$ and $\sigma_{\delta\delta}{}^2$, the variances of the coordinates. If we denote by $\sigma_{ii}{}^2$ the variance of the i-th observation, we see that we may write

$$\sigma_{\alpha\alpha}{}^2 = \sum_{i=1}^{m} \left(\frac{\partial\alpha}{\partial x_i}\right)^2 \sigma_{ii}{}^2 + \left(\frac{\partial\alpha}{\partial\boldsymbol{a}}\right)^{\mathrm{T}} \mathbf{Q} \left(\frac{\partial\alpha}{\partial\boldsymbol{a}}\right). \tag{1.8c}$$

For $\sigma_{\delta\delta}{}^2$ we get an expression which is completely analogous. The contribution to the variance of any coordinate comes thus from two sources: the first term on the right hand side of (1.8c) originates from the random errors of the measurements involved and estimates the (square of the) random error of the coordinates. This is so because every position is, as a rule, calculated from a distinct set of measurements, that is, the various x do not vary continuously if one goes from one position to the next. On the other hand, every one of the parameters a_i is typically involved in the computation of every position, and the derivatives which are the components of the vectors $(\partial\alpha/\partial\boldsymbol{a})$ and $(\partial\delta/\partial\boldsymbol{a})$ are typically functions of α and δ themselves. The "accidental" errors in the a_k thus create systematic errors in the α and δ, and the *parameter variances* $(\partial\alpha/\partial\boldsymbol{a})^{\mathrm{T}}\mathbf{Q}(\partial\alpha/\partial\boldsymbol{a})$ and $(\partial\delta/\partial\boldsymbol{a})^{\mathrm{T}}\mathbf{Q}(\partial\delta/\partial\boldsymbol{a})$ *are an estimate for the expectation of this systematic error*, and depend therefore also, as a rule, on the coordinates α and δ.

Star positions that were obtained by the reduction of original observations will thus have errors of the following types:

(1) random errors caused by the propagation of the random errors in the observations themselves; first term in equation (1.8c),

(2) systematic errors caused by systematic errors in the original observations, and

(3) systematic errors caused by the unavoidable random errors of the reduction parameters obtained in the adjustment; second term in equation (1.8c).

The expectations of the errors in category (3) are, in given situations, predictable (compare Eichhorn and Williams [1963]), and will systematically falsify every position calculated from observations regardless of how free these may otherwise be of systematic errors.

1.6.2 THE CHANGES OF STAR POSITIONS. In principle, star positions may change for two reasons:

(1) The *system* to which they are referred *moves with respect to an inertial system*. This will change the objects' positions with respect to the "moving" system, even if the objects' coordinates with respect to this inertial system are constant, and

(2) the coordinates of the objects change with respect to an inertial system.[7]

Ordinarily, star positions change as a consequence of a combination of these two causes. *The change of the system's orientation with respect to an inertial system is ideally described in terms of precession and nutation.* We shall see later, however, that the precession formulas which are now used still leave a residual change of the system's orientation with respect to an inertial system. (See Section 2.4.3.)

Those changes in the stars' positions which are caused by their motions with respect to a certain inertial system are the main contributors to the *proper motions*. These proper motions, as we commonly know them, also contain that part of the position changes which are caused by the failure of the precession and nutation formulas to completely relate the equator system to an inertial system. It will be discussed later (Section 2.4.3.4) how an analysis of proper motions can contribute toward improving the accuracy of the numerical parameters in the formulas for describing the effects of precession.

One arrives at a precise definition of the proper motion in the following way: let K be a coordinate system which is (almost) inertial, for example, the equator

---

7. If these changes of the objects' positions can be described for all objects by the transformation $\boldsymbol{x}' = r\mathbf{R}\boldsymbol{x} + \boldsymbol{a}t$ where $\mathbf{R}$ is orthogonal, but neither r, $\boldsymbol{a}$ nor the elements of $\mathbf{R}$ change with time, in other words, if all objects move uniformly with respect to the old system, a new system can be found so that the positions in the transformed system would not change. However, this is a singular situation and only mentioned for the sake of completeness.

system at a certain epoch. Since at this time the period during which systematic observations of star positions have been made is very short in comparison with the stars' galactic period, the best we can do is to assume that the motion of stars in K is uniform and write

$$\boldsymbol{x}(t) = \boldsymbol{x}_0 + \dot{\boldsymbol{x}}t \tag{1.9}$$

for $\boldsymbol{x}(t)$, the position vector of the star in K at the time t. The velocity, $\dot{\boldsymbol{x}}$, of the star in K, is assumed to be constant. $\boldsymbol{x}_0$ is the star's location in K at the arbitrary initial epoch, $t = 0$. If r is the distance to the star, we may rewrite (1.9) in accordance with (1.1) as

$$\begin{pmatrix} r\cos\delta\cos\alpha \\ r\cos\delta\sin\alpha \\ r\sin\delta \end{pmatrix} = \begin{pmatrix} r_0\cos\delta_0\cos\alpha_0 \\ r_0\cos\delta_0\sin\alpha_0 \\ r_0\sin\delta_0 \end{pmatrix} + \begin{pmatrix} \dot{x} \\ \dot{y} \\ \dot{z} \end{pmatrix} t. \tag{1.9a}$$

From differentiating (1.1) with respect to the time, t, we obtain

$$\begin{pmatrix} \dot{x} \\ \dot{y} \\ \dot{z} \end{pmatrix} = \dot{r}\begin{pmatrix} \cos\delta\cos\alpha \\ \cos\delta\sin\alpha \\ \sin\delta \end{pmatrix} + \dot{\delta}r\begin{pmatrix} -\sin\delta\cos\alpha \\ -\sin\delta\sin\alpha \\ \cos\delta \end{pmatrix} + \dot{\alpha}r\begin{pmatrix} -\cos\delta\sin\alpha \\ \cos\delta\cos\alpha \\ 0 \end{pmatrix}. \tag{1.10}$$

The rate of change, $\dot{r}$, of the distance to the object is called its *radial velocity*.[8] It is usually given in kilometers per second. The time derivatives, $\dot{\alpha}$ and $\dot{\delta}$, are the object's *proper motion in right ascension and declination* respectively. Frequently, $\dot{\alpha}$ is denoted by μ or μ_α and is given in seconds of time per year or per century. Likewise, $\dot{\delta}$ is frequently denoted by μ' or μ_δ and is given in seconds of arc per year or per century. Sometimes the quantity $\mu_\alpha \cos\delta$ is called "*proper motion in right ascension on the great circle*;" it is then usually quoted in the same units as μ_δ.

The quantity $\mu = (\mu_\alpha^2 \cos^2\delta + \mu_\delta^2)^{1/2}$, usually quoted in seconds of arc per year or per century, is known as *the proper motion*. The position angle P of the proper motion is defined by $\mu_\alpha \cos\delta = \mu\cos P$ and $\mu_\delta = \mu\sin P$. Obviously the pairs $\{\mu_\alpha, \mu_\delta\}$ and $\{\mu, P\}$ have the same information value.

The formulas (1.10) through (1.12) are, however, valid only if α and δ are in radians, and if the time unit in $\dot{\alpha}$ and $\dot{\delta}$ is the same as in $\dot{r}$. Therefore one will have to introduce appropriate conversion factors unless proper motions are reckoned in the units just described.

---

8. Recently, some authors who are unfamiliar with the long established astronomical terminology have been calling this the *range rate*.

1.6.3 DISCUSSION OF PROPER MOTION. From equation (1.10) it is quite obvious that $\dot{r}$, μ_α and μ_δ are functions of the time. Since

$$\begin{pmatrix} \dot{x} \\ \dot{y} \\ \dot{z} \end{pmatrix} = \begin{pmatrix} \dot{x}_0 \\ \dot{y}_0 \\ \dot{z}_0 \end{pmatrix},$$

we have

$$\dot{r}\begin{pmatrix} \cos\delta\cos\alpha \\ \cos\delta\sin\alpha \\ \sin\delta \end{pmatrix} + \dot{\delta}r\begin{pmatrix} -\sin\delta\cos\alpha \\ -\sin\delta\sin\alpha \\ \cos\delta \end{pmatrix} + \dot{\alpha}r\begin{pmatrix} -\cos\delta\sin\alpha \\ \cos\delta\cos\alpha \\ 0 \end{pmatrix}$$

$$= \dot{r}_0\begin{pmatrix} \cos\delta_0\cos\alpha_0 \\ \cos\delta_0\sin\alpha_0 \\ \sin\delta_0 \end{pmatrix} + \dot{\delta}_0 r_0\begin{pmatrix} -\sin\delta_0\cos\alpha_0 \\ -\sin\delta_0\sin\alpha_0 \\ \cos\delta_0 \end{pmatrix} + \dot{\alpha}_0 r_0\begin{pmatrix} -\cos\delta_0\sin\alpha_0 \\ \cos\delta_0\cos\alpha_0 \\ 0 \end{pmatrix} \tag{1.10a}$$

For virtually all celestial objects, particularly those which are the subject of catalogue astronomy, α, δ, μ_α and μ_δ are much better known than r and $\dot{r}$. The equation systems (1.9a), in which $(\dot{x}, \dot{y}, \dot{z})^T$ is replaced by the expression (1.10) and (1.10a), may now be used to establish the relationships

$$\begin{aligned} \alpha(t) &= \alpha(\alpha_0, \delta_0, r_0, \dot{\alpha}_0, \dot{\delta}_0, \dot{r}_0\ ; t) \\ \delta(t) &= \delta(\alpha_0, \delta_0, \ldots; t) \\ r(t) &= r(\alpha_0, \delta_0, \ldots; t) \\ \dot{\alpha}(t) &= \dot{\alpha}(\alpha_0, \delta_0, \ldots; t) \\ \dot{\delta}(t) &= \dot{\delta}(\alpha_0, \delta_0, \ldots; t) \\ \dot{r}(t) &= \dot{r}(\alpha_0, \delta_0, \ldots; t) \end{aligned} \tag{1.11}$$

which allow one to calculate position and proper motion components (as well as distance and radial velocity) for all times when they are known at one epoch, based, of course, on the hypothesis of uniform motion.

The explicit establishment of the functions in (1.11) was carried out by Eichhorn and Rust (1970). It has thus become clear that the rigorous prediction of an object's position with respect to a specific coordinate system requires not only the knowledge of its position and proper motion components in this system at a zero epoch, but also that of the ratio of its radial velocity to its distance at this same zero epoch. The radial velocities are known only for few stars, and the distances are always known with much lower relative accuracy than the other quantities.

Due to their motions in space, the stars trace great circles in the sky (with respect to a nonrotating system). The equations of these circles (with t being the parameter) in such a system are usually given by expressing $\alpha(t)$ and $\delta(t)$ as Taylor series in time, such that

$$\alpha(t) = \alpha_0 + \left.\frac{\partial \alpha}{\partial t}\right|_{t=t_0} (t - t_0) + \frac{1}{2}\left.\frac{\partial^2 \alpha}{\partial t^2}\right|_{t=t_0} (t - t_0)^2 + \cdots, \qquad (1.11a)$$

and similar expressions for $\delta(t)$ and $r(t)$. If we remember the definitions of μ_α, μ_δ and $\dot{r}$ we see that

$$\alpha(t) = \alpha_0 + \mu_\alpha(t - t_0) + \frac{1}{2}\frac{\partial^2 \alpha}{\partial t^2} (t - t_0)^2 + \cdots \qquad (1.11b)$$

and analogously for $\delta(t)$ and $r(t)$ where the derivatives are, of course, evaluated for $t = t_0$.

The rigorous formulas of type (1.11) have in the past never been used, but the influence of the stars' inertial motions on their position change is calculated by formulas of the type (1.11b). The first derivatives of α, δ and r with respect to time do not involve distance. As first approximations we may put

$$\alpha(t) = \alpha_0 + \mu_\alpha(t - t_0),$$

and

$$\delta(t) = \delta_0 + \mu\delta_\delta(t - t_0). \qquad (1.11c)$$

These equations are usually sufficient to calculate the influence of proper motion on the positions for short time intervals $t - t_0$. The expressions for $d^2\alpha/dt^2$ and $d^2\delta/dt^2$ which one finds in most texts on spherical astronomy have been calculated under the assumption that the motion of the star is not uniform in inertial space, but uniform on a great circle on the sphere. They do not involve r and $\dot{r}$.

The complete expressions for the second derivatives in (1.11a) are

$$\frac{\partial^2 \alpha}{\partial t^2} = 2\left(\mu_\alpha \mu_\delta \tan\delta - \mu_\alpha \frac{\dot{r}}{r}\right)$$

$$\frac{\partial^2 \delta}{\partial t^2} = 2\left(\mu_\alpha{}^2 \sin 2\delta - \mu_\delta \frac{\dot{r}}{r}\right) \qquad (1.11d)$$

and one can see that the error in these second derivatives which is produced by neglecting $\dot{r}/r$, that is, by assuming that the object moves with uniform speed on a great circle, is of the same order of magnitude as that produced by neglecting the influence of μ_α or μ_δ. For critical work, for instance the construction of a fundamental catalogue (see Sections 3.1.2 and 4.1.1), the influence of $\dot{r}/r$ on a star's position must be taken into account. This is usually done by applying the *foreshortening* or *acceleration terms*

$$-\mu_\alpha \frac{\dot{r}}{r} (t - t)^2, \qquad -\mu_\delta \frac{\dot{r}}{r} (t - t_0)^2$$

to α and δ, respectively. As one can see, this is particularly important when the star is at a relatively small distance and moves fast.

Considering the accuracy which is presently attainable, the formulas (1.11b) in which the second derivatives are expressed by (1.11d) give the same results as the rigorous ones, since the influence of the higher derivatives is unnoticeable. For completeness sake, the rigorous formulas derived by Eichhorn and Rust (1970) which were mentioned earlier in this section, are given below:

$$\left.\begin{aligned}
\tan(\alpha-\alpha_0) &= \frac{\mu_\alpha(t-t_0)}{1+\left(\dfrac{\dot r_0}{r_0}-\mu_\delta\tan\delta_0\right)(t-t_0)},\\[2ex]
\tan\tfrac{1}{2}(\delta-\delta_0) &= \frac{\mu_\delta(t-t_0)+\left[1-\dfrac{r}{r_0}+\dfrac{\dot r_0}{r_0}(t-t_0)\right]\tan\delta_0}{\left[1+\left(\dfrac{\dot r_0}{r_0}-\mu_\delta\tan\delta_0\right)(t-t_0)\right]\sec(\alpha-\alpha_0)+\dfrac{r}{r_0}},\\[2ex]
&\text{where } r/r_0 \text{ was computed from}\\
\left(\frac{r}{r_0}\right)^2-1 &= \left\{2\frac{\dot r_0}{r_0}+\left[\left(\frac{\dot r_0}{r_0}\right)^2+\mu_\alpha{}^2\cos^2\delta_0+\mu_\delta{}^2\right](t-t_0)\right\}(t-t_0).
\end{aligned}\right\} \tag{1.12}$$

It should be pointed out that, eventually, *the observation of stellar positions at regular intervals over sufficiently long periods of time will allow one to calculate the stars' distances* from (1.12), provided that the radial velocities are also kept under observation. At this time, this has been done only for Barnard's star (van de Kamp [1968], Gatewood [1972]). Note, however, that the equations (1.12) are valid only for a uniformly moving star, which is a hypothesis that may not hold for time intervals that are long enough to solve the equations (1.12) for the distance r.

The computation of the proper motion components from the positions (and, strictly, also the distances) of an object at two epochs could be carried out rigorously by computing $(\dot x, \dot y, \dot z)^T$ from (1.9a), in which $\{r, \alpha, \delta\}$ and $\{r_0, \alpha_0, \delta_0\}$ as well as t would be known, from which, by (1.10), $\mu_\alpha = \partial\alpha/\partial t$, $\mu_\delta = \partial\delta/\partial t$ and r could be calculated. This procedure would, however, in most cases be unnecessarily and even unjustifiably complicated. It is discussed further in Section 2.4.2. Also, r is seldom available with the necessary accuracy.

Equation (1.10) illustrates, however, the frequently overlooked fact that the proper motion components μ_α and μ_δ are functions of time, since α and δ change with time.

Within the framework of presently achievable accuracy, the following procedure gives precise results:

Calculate preliminary values of μ_α and μ_δ from (1.11c) and use them to compute the second derivatives of the position components which are, of course, the first derivatives of the proper motion components from (1.11d). If $\dot{r}/r$ is not known, the term which contains it will have to be left out. Then, use the thus calculated $\partial^2\alpha/\partial t^2$ in formula (11.b) to calculate a new μ_α, and analogously get a new μ_δ. Repeat this process, if necessary. Needless to say, the μ_α and μ_δ, thus obtained, refer to the epoch t_0. The application of this process will occasionally be required near the pole where the tan δ, which occurs in μ_α, assumes large values. For a rigorous iteration procedure, see Section 2.4.2 below.

Also useful are the *formulas by Fabritius* (see Schaub [1950], p. 166f).

The position changes due to the acceleration, that is, rotation of the coordinate system with respect to an inertial system are known as the phenomena of *precession and nutation.* Assume that the orientation of K at T_0 coincides with that of an inertial system. At the time T, the system has rotated on its origin so that the coordinates of the same point $\boldsymbol{x}$ in the system at T_0 and at T are related by

$$\boldsymbol{x}(T) = \mathbf{P}(T, T_0)\boldsymbol{x}_0(T_0) \tag{1.13}$$

where $\mathbf{P}(T, T_0)$ is the matrix which performs the operations of precession and nutation, that is, which transfers to coordinate system from the orientation at $\mathbf{T}_0$ to the orientation at $\mathbf{T}$. By applying this transformation to equation (1.9) we obtain

$$\mathbf{P}(T, T_0)\boldsymbol{x} = \mathbf{P}(T, T_0)\boldsymbol{x}_0 + \mathbf{P}(T, T_0)\dot{\boldsymbol{x}}t$$

which shows that the object's velocity is influenced by precession and nutation in exactly the same manner as are its Cartesian coordinates.

The rates at which the *proper motion components change due to precession* are given by

$$\frac{\partial^2\alpha}{\partial t\partial T} = \frac{\partial\mu_\alpha}{\partial T} = n(\mu_\alpha \cos\alpha \tan\delta + \mu_\delta \sin\alpha \sec^2\delta)$$

$$\frac{\partial^2\delta}{\partial t\partial T} = \frac{\partial\mu_\delta}{\partial T} = n\mu_\alpha \sin\alpha \tag{1.14}$$

where n is the well-known *precession in declination,* given by

$$n = 2004.''685 - 0.''8533\tau - 0.''00037\tau^2,$$

where τ is expressed in tropical centuries after 1900.

1.6.4 EFFECTS PRODUCED BY THE FINITE SPEED OF LIGHT (Aberration). The finite speed of light in combination with the velocity of the observer and that of the object under observation produces the phenomenon of *aberration.*

The light from a celestial object which the observer sees at the moment of observation is light which left the object at some instant in the past and which traveled in such direction that the observer is just now in a position to intercept it.

The point where the object was when the light now being observed left it will be called the *emission point.*

According to the principle of relativity, the observed phenomena must be the same regardless to which inertial system the coordinates of the objects involved (that is, emitter and observer) are referred. Consider the situation where both emitter and observer move with different constant velocities with respect to an inertial system. There must not be any intrinsic difference between the observed phenomena, in particular the directions in which the object is seen, that is caused by first fixing the origin of an inertial system in the (unaccelerated) observer and then in the (also unaccelerated) object, as long as the axes of these two inertial systems remain parallel.

There is, however, a difference in the nomenclature depending on where the origin of the coordinate system is and how it moves. With respect to a certain inertial system, aberration effects caused by *motion of the emitter* are called *planetary aberration,* those produced by the *motion of the observer, stellar aberration.* If the motion of either emitter or observer is accelerated with respect to the inertial system, the aberration effect produced thereby can unambiguously be classified as planetary or stellar aberration, respectively. When, however, emitter and receiver move uniformly with respect to an inertial system, this separation is no longer possible and a certain effect can then be rendered either planetary or stellar aberration, depending on the location of the origin of the inertial system with respect to which the motions of observer and of emitter are considered.

When, for example, emitter and observer move uniformly and have zero relative velocity, the direction in which the emitter appears with respect to the observer is, according to what was said above, not affected by aberration effects. Yet, by considering emitter and receiver as moving uniformly with respect to an inertial system, there will be both planetary and stellar aberration affecting the direction of observation. These effects, however, will be equal and in opposite direction and thus cancel each other.

According to the definition above, *planetary aberration is the displacement from the direction to the actual location of the star* (*toward the geometric position*) *toward the direction of the emission point.* Note that the location of the emission point with respect to an inertial system depends on the motion of this system. In order to compute the effects of planetary aberration (sometimes also referred to as *light time correction*), we would have to know for how long the light from the star has been under way, and this in turn would require that we know the star's distance. As we said before, this distance is for objects outside the solar system never known with the same degree of accuracy as are the coordinates α and δ which determine the position. The effects of planetary aberration for bodies outside the solar system could therefore never be calculated very accurately. The positions which we find in star catalogues therefore do not describe the *geometric positions,* that is, the directions in which the star is now, but rather the *true positions* of their emission points at their respective epochs of observation.

Stellar aberration produces, generally speaking, the *difference between the true position* (as just referred to) and the *apparent position* which is generally the direction from which the light from the star seems to come.[9]

Since we have seen that the location of, and therefore the direction to, the emission point depends on the motion of the inertial system which is the basis of our considerations, we may choose the origin of this system such that its origin moves with the same velocity as the barycenter of the solar system so that the only motions of the observer, who is on the Earth, would be a combination of the Earth's annual revolution and diurnal rotation, giving rise to *annual* and *diurnal aberration* respectively. Since, in this system, the entire motion of the Earth would be composed of very nearly periodic motions, there will be no uniform component of the Earth's velocity in space, and thus no *secular aberration.*

In this system, the *true position* would *define* the *coordinates of the emission point* which occupies a somehow defined average of the observed positions of the star. By exactly spelling out how this average is taken, we shall now precisely define the concepts "true position" and "apparent position."

Consider the triangle which has one apex at the emission point, another one at the observer and the third at the vertex of a vector $\boldsymbol{v}$ which is defined as follows. It originates at the emission point, is in the direction of the observer's velocity and referred to the inertial system in which the coordinates are reckoned. The length of this vector $\boldsymbol{v}$ equals the distance traveled by the observer while the light travels from emitter to observer.

This vector $\boldsymbol{v}$, representing the velocity of the observer, has two well-distinguishable components. They are:

1) The relative velocity $\boldsymbol{v}_1$ of the Earth's center with respect to the barycenter of the solar system. Note that it is the velocity of the Earth's actual center which enters this definition, and not the mean velocity of the Earth-Moon system's barycenter.

2) The relative velocity $\boldsymbol{v}_2$ of the observer with respect to the center of the Earth.

We can now form triangles of the type described above using $\boldsymbol{v}_1$ and $\boldsymbol{v}_2$, respectively, in lieu of $\boldsymbol{v}$. The *total stellar aberration* is the difference between the directions from observer to the star's emission point and the vertex of the vector $\boldsymbol{v} = \boldsymbol{v}_1 + \boldsymbol{v}_2$ which originates at the emission point. The difference between the directions to the emission point and the vertex of $\boldsymbol{v}_1$ is the *annual aberration*, and the difference between the directions to the vertices of $\boldsymbol{v}_1$ and $\boldsymbol{v}_2$ is the *daily* or *diurnal aberration.*

The geocentric direction to the apex of $\boldsymbol{v}_1$, which originates in the emission point, defines the *apparent position* of the star.

The apparent position of a star is therefore not the position in which the star is

9. In this discussion it is to be understood that the light is moving in a vacuum and not in the Earth's atmosphere, which would produce in addition the effects of *atmospheric refraction.*

seen, but rather the direction in which the star would be seen if the observer moved parallel with the center of the Earth.

As we shall see below in Section 2.2.8, the accuracy of star positions obtained in routine astrometric procedures has improved steadily in the course of time. Accordingly, the methods for the calculation of the effects of annual aberration, that is the computation of $\boldsymbol{v}_1$, became more accurate. Initially, annual aberration was computed (following F. W. Bessel, the father of modern spherical astronomy) from the theory of the Earth's unperturbed heliocentric motion. Development in series of powers of small quantities was then and, unfortunately in many instances, still is, the standard procedure in spherical astronomy. As a consequence, the effects of annual aberration on the coordinates of the stars were developed in a series in the small quantity $|\boldsymbol{v}_1|$, of which only the first power was retained. If we consider the hodograph of the Earth's heliocentric motion, that is, the locus of the vertices of $\boldsymbol{v}_1$ during a revolution of the Earth, we notice that it is almost a closed curve, but not symmetrical with regard to the origin because of the eccentricity of the Earth's orbit. The place in which we see the star follows a path in the sky which is a projection of this hodograph.

If we compute the effect of $\boldsymbol{v}_1$ on the star positions by a first order approximation on the basis of unperturbed heliocentric two body motion, the equation of the hodograph is essentially

$$v_{1\mathrm{y}} = A + B \cos t$$

$$v_{1y} = A' + B' \sin t$$

where t is the time in units of $1/2\pi$ years. In this approximation the hodograph is an ellipse, whose center has the coordinates $\{A, A'\}$. These quantities are functions of e and Π, the eccentricity and the longitude of the perigee, respectively, of the Earth's orbit. If we put in the system of the ecliptic

$$\boldsymbol{v}_1' = \begin{pmatrix} A \\ A' \\ 0 \end{pmatrix} \quad \text{and} \quad \boldsymbol{v}_1'' = \begin{pmatrix} B \cos t \\ B' \sin t \\ 0 \end{pmatrix}, \tag{1.15}$$

so that $\boldsymbol{v}_1 = \boldsymbol{v}_1' + \boldsymbol{v}_1''$, we see that $\boldsymbol{v}_1''$ changes periodically with time and that $\boldsymbol{v}_1'$ undergoes a slow secular change because e and Π change only due to the perturbations of the Earth's orbit and in the case of Π, due to precession. The aberrational effect produced by $\boldsymbol{v}_1'$ is called the *elliptic aberration*, or *E-terms*, or *perigee terms* of the aberration. Because of its very slow rate of change it has traditionally been regarded as constant, and was not removed from the apparent positions of the stars when calculating the true positions. *True positions*, which we shall later define yet more precisely, still contain therefore the effects of the stars' planetary aberration, and the elliptic aberration. The true position is thus strictly not in the direction of the emission point, but points to the apex of the vector $\boldsymbol{v}_1'$ originating at the emission point.

Eventually, the accuracy standards of certain star positions had become so high that it was internationally agreed upon that from 1960 onwards, the effects of annual aberration should be computed with an error of no more than 0.″001. Up to then, the just described first order theory based on the heliocentric unperturbed motion of the barycenter of the Earth-Moon system had been used as a model for the computations of the *Besselian day numbers* and star positions published in the various national ephemerides. The new requirement made it necessary to put the computation of these quantities on the basis of the perturbed motion of the Earth's center with respect to the barycenter of the solar system. This required a new model for $\mathbf{v}_1$ and consideration of the second power of $|\mathbf{v}_1|$. The development of the details falls beyond the framework of this book. They can be found, for instance, in Woolard and Clemence (1966), pp. 120–127. In the new practice, the vector $\mathbf{v}_1$ is calculated on the basis of the perturbed motion of the Earth's center around the barycenter of the solar system and split as follows

$$\mathbf{v}_1 = \mathbf{v}_1' + \mathbf{v}_1'''$$

where $\mathbf{v}_1'$ is defined as before and $\mathbf{v}_1'''$ the remainder necessary to make up the improved $\mathbf{v}_1$. The theory is now carried to the second power of $|\mathbf{v}_1|$.

The true position of a star, then, is the apparent position, as defined above, *freed from the effects of annual aberration, and calculated before* 1960 *on the basis of* $\mathbf{v}_1''$, *and after* 1960, *on the basis of* $\mathbf{v}_1'''$ *only.*

Note, however, that even after the change in computational practice, the effect of $\mathbf{v}_1'$, which is calculated from the same formulas as in the old, that is, to first order heliocentric theory for the Earth-Moon system's barycenter, is still not removed from the observed positions in the reduction from apparent to true positions. On the other hand, the geometric center of the hodograph described by $\mathbf{v}_1'''$ unlike that of $\mathbf{v}_1''$ is not at the origin of $\mathbf{v}_1'''$. This means that the sophistication of the theory has uncovered another constant aberration effect. In contrast to the other constant aberration effects, this one is removed from the apparent position in the reduction to true positions.

This is a completely arbitrary and somewhat incongruous procedure, especially since $\mathbf{v}_1'$, as we have pointed out, is not rigorously a time constant. The introduction of the more accurate theory in 1960 would have been an ideal opportunity to define the true positions rigorously as in the direction of the emission point by removing also the elliptic terms of the aberration from the apparent positions. These terms were originally left in the apparent positions only for reasons of computational convenience in a time when the most efficient calculation aids were tables of logarithms and the slide rule, and when it saved a lot of calculators' time to develop every small effect as a series in powers of small quantities that was broken off as soon as higher order terms became insignificant. By carrying this practice into the present, incongruities result, such as the one discussed above. One might paraphrase a dictum about the theory of sets,[10] ascribed to the mathematician Kronecker, by

10. Die Mengenlehre ist eine Krankheit, von der sich die Mathematik hoffentlich noch einmal erholen wird.

observing that series developments are a disease from which spherical astronomy will hopefully one day recover.

This author feels that it should have been decided to calculate the aberration effects from a rigorous theory using a $\boldsymbol{v}_1$ which is the entire velocity vector of the Earth's center with respect to the solar system's barycenter.

The vector $\boldsymbol{v}_2$ which produces the diurnal component of aberration can easily be calculated as a function of the observer's distance from the Earth's axis and the star's hour angle. Modifications of the formulas used for the calculation of diurnal aberration might become necessary when accurate position observations should some time be made from stations that have an appreciable motion with respect to the Earth's surface.

1.6.5 TRUE POSITION AND MEAN POSITION. Apparent positions and true positions are referred to the same coordinate system, namely the *true*, that is, actual *equator system* as oriented at the epoch of the observation. The matrix $\mathbf{P}(T, T_0)$ from (1.13) can be written:

$$\mathbf{P} = \mathbf{N}(T)\mathbf{S}(T, T_0)\mathbf{N}^{-1}(T_0) \tag{1.13a}$$

where all terms that are periodic in time are contained in **N**. **N** thus performs that part of the transformation which is due to *nutation* whose effects are strictly *periodic* in time. The matrix **S** performs the transformation caused by the *precession of the equinoxes* whose effects are strictly *secular*. If $\hat{\boldsymbol{x}}_t(T_0)$ represents the *true position* at T_0, then $\mathbf{N}^{-1}(T_0)\hat{\boldsymbol{x}}_t(T_0)$ represents the *mean position* at T_0. This is the position with respect to the so-called *mean coordinate system at* T_0. This system shares the secular changes of orientation with the true system, but not the periodic changes. In order to calculate the position $\hat{\boldsymbol{x}}_m(T)$ with respect to the mean coordinate system as oriented at T (the mean position, equator and equinox T) we apply to $\hat{\boldsymbol{x}}_t(T_0)$, the true position at T_0, the transformation

$$\hat{\boldsymbol{x}}_m(T) = \mathbf{S}(T, T_0)\mathbf{N}^{-1}(T_0)\hat{\boldsymbol{x}}_t(T_0),$$

while the transformation between the true positions at T_0 and T, respectively, is of course performed by

$$\hat{\boldsymbol{x}}_t(T) = \mathbf{N}(T)\mathbf{S}(T, T_0)\mathbf{N}^{-1}(T_0)\hat{\boldsymbol{x}}_t(T_0). \tag{1.13b}$$

Star catalogues list mean positions exclusively. They are always referred to a *mean heliocentric*[11] *coordinate system* as oriented at some epoch T that used to be chosen close to the average of the epochs of the observations, but in this century,

---

11. Since stars are observed from the Earth, their directly determined positions refer to a geocentric system or one which has its center in the observer. For stars sufficiently close to the Earth, their observed coordinates undergo noticeable periodic changes during a year which are the effects of *annual parallax*, and which must, of course, be removed from the observations by transforming them to a system which is located in the solar system's barycenter, before they are suitable for inclusion in a catalogue.

standard epochs 1900, 1925, 1950 and 1975 are used. Since the term *epoch* is reserved for the instant which indicates the *star's location with respect to an inertial system*, the *orientation* of the coordinate system is indicated by *equator and equinox* or sometimes only by *equinox*, so that, for example, one speaks of a mean position for equator and equinox 1950. Since this is the generally used term, it will occasionally be used in this book also, although the author feels that "orientation" instead of "equator and equinox" is shorter and more descriptive.

A star position $\hat{x}(t, T)$ is thus characterized by two times: T, the time of orientation, and the epoch t, the time to which, in an inertial system, the position refers; thus $\partial\hat{x}/\partial t$ is related to the proper motion and $\partial\hat{x}/\partial T$ to precession and nutation.

Epoch, and equator and equinox, are usually not the same. Transformation between epochs is made by proper motion, while mean positions are transformed from one orientation (equator and equinox) to another by precession.

1.6.6 SYSTEMS DEFINED BY A CATALOGUE. We have pointed out in Section 1.6.1 that catalogued star positions are subjected to accidental as well as systematic errors. This fact is very important when one considers in which way, and how accurately, star catalogues define a *frame of reference* (coordinate system). It is clear, and we shall discuss this further in more detail in Section 2.2, that the determination of an object's position involves not only the determination of the position with respect to the frame of reference in question but also, at least implicitly, means for the realization of this system. One means of determining star positions in a system is by measuring their positions not directly with respect to the system, but with respect to other stars (reference stars) whose positions in the system were determined previously. This implies that the system is then defined by the positions of the reference stars. If the system is realized in this way, it is clear that the systematic errors of the reference star positions will also affect the newly derived positions, superimposed, of course, upon the errors produced by the process of obtaining the new star positions.

A coordinate system is completely defined by a great circle (plane) and a point on (direction in) it. Suppose we knew from a catalogue at least the position of two stars not diametrically opposite one another. By observing directly or indirectly, for instance by means of a photographic plate, the directions to (positions of) these stars, we can find the directions of the axes of the reference system to which the catalogued positions of the stars refer. Inaccuracies in the catalogued star positions will, of course, propagate themselves and cause the derived directions to the coordinate system's axes to be inaccurate. In what follows we shall speak of the *fitting of a coordinate system to star positions in a catalogue.* This is to be understood in the sense of the just described process of observing stars whose positions are listed in the catalogue in question. If the stars were observed directly at, say, a transit circle, the fitting process implies that we can determine the instrument's errors of orientation and adjust its setup in such a way that the axes on which the telescope of the transit circle turns are parallel to the axes with respect to which longitude and latitude angle of the system in question are reckoned. These are, of course, the Earth's axis and one perpendicular to it, in the east-west direction.

Also implied is that we are able to make the zero points of the instrument's latitude and longitude angles equal to the zero points of the angles of the system which the meridian (or other) instrument intends to emulate.

This means that the circle measuring the latitude angle (altitude circle), that is, the declination, can be set to indicate "zero" at the fundamental circle of the system in space, that is, when the instrument's optical axis points toward the equator, and that we can indicate the time when, say, the x-axis of the system (which points in the direction to the vernal equinox) would coincide with the telescope's optical axis.

If the observations were made photographically, we will (after the fitting), at least theoretically, be able to indicate the image points of the system's axes on the photographic plate.

The unavoidable errors associated with the measuring process cause difficulties, of course. We shall, therefore, at least for the moment, assume that the instrument is an ideal one and that the errors of the measurements are negligibly small as compared with the errors in the catalogued star positions. In order to avoid inessential complications, we shall also assume that all catalogued positions refer to the same epoch, namely, that at which the measurements are made.

Since three coordinates (of two stars) are sufficient to unambiguously determine a coordinate system, two or more complete positions overdetermine a system. Since we assume the catalogued positions to be affected by accidental errors, the system fitting has become an adjustment problem.

Consider S, the set of star positions in a catalogue. Let $\{S_i\}$ be a set of subsets S_i of S, and let K_i be the coordinate system fitted to the stars in S_i. Due to the nature of any adjustment process, the various K_i will all be different. If the subsets S_i are all samples whose elements were taken randomly from all regions of the sky that are covered by the catalogue, the K_i should differ only slightly due to the accidental errors of the positions of the stars in the various samples.

One could regard *the system K represented by the catalogue as that which results when every catalogued position is involved in the adjustment*. Apart from the fact that in practical situations this is virtually never possible, since, in general, only the stars in a restricted region are involved, the *adjustment residuals*, that is, the catalogued position components minus the reconstructed ones, can be expected to *show systematic trends* in any sufficiently large region. The basic question is: If we use positions from this catalogue to realize the equator system, where will its axes point? Since in a process of this type position of stars in only a limited region will be involved, the result will vary depending on the region. The system represented by the positions listed in a catalogue C thus has no obvious definition.

To illustrate the problem, consider a simply connected region R_1 that contains the subset S_1 of the set of stars whose positions are in the catalogue C. S_1 must consist of a least two stars so that the problem of fitting a system to these positions is a genuine adjustment problem. Let R_1 be of simple shape and sufficiently small so that the adjustment residuals, after the fitting of the system K_1, do not show any significant nonrandom pattern. Now change the region S_1 to S_1' by including (or excluding) one star close to its border. The system K_1' fitted to S_1' will be different from K_1, although it represents the system defined by the catalogue C in the same region as K_1, since it is easy to change S_1 to S_1' in such a way that their geometric centers are identical. We can, moreover, obtain a system K_1'' which is different from K_1 without changing the sample of the stars in S_1 of their positions, by using a *different adjustment model* for the fitting of the system.

If, for example, R_1 is the area covered by a photographic plate and S_1 is the set of the reference stars in it, K_1 is determined by using the measured coordinates of the images of the reference stars and relating them to their catalogued coordinates through some projection model. If this model is changed, then the system K_1'' defined by the same stars on the basis of the new model will also differ from K_1.

Thus we realize: *The coordinate system defined by the positions of stars listed in a catalogue depends (a) on the region from which the sample stars are chosen, (b) the selection of stars within this region, and (c) the mathematical model chosen for the adjustment. There is consequently no unique coordinate system defined at a certain point by the positions of the stars in a catalogue.* Estimates for this system can of course be found and, although they should not be too different from each other, they will differ according to the approach taken.

We could, at least in theory, uniquely define a coordinate system by the positions catalogued in C by acknowledging as completely valid only such adjustments that involve all stars in the catalogue C, and which were made on the basis of a suitably chosen standard model. In practice, however, such adjustments are nearly impossible so that in every concrete example only approximations to the "rigorous" system will be found.

1.6.7 THE SYSTEMATIC ERRORS OF A CATALOGUE. Suppose, by considering coordinate systems in regions around points, that we have obtained estimates for the systems defined by a certain catalogue at every point (point of reference) in the region which the catalogue covers. As the point of reference is changed, the directions of the axes of the associated system may or may not change. If one obtains the same system, regardless of what region it is fitted to (that is, regardless of the location of the point of reference), the catalogue is called *self-consistent.* This does not mean that the positions in the catalogue are free from systematic errors. These can still appear due to the fact that the catalogue system's orientation can (and, in general, will) be different from that of the physical system which it is intended to represent. Self-consistent catalogues can, in practice, exist only to a certain degree of approximation.

One can now perform the following mental experiment: Fit a system to the entirety of the catalogued star positions, by an adjustment model which does not allow for deviations of the catalogue from self-consistency, and consider the adjustment residuals, that is, catalogued minus reconstructed star positions. Due to the deviations of the catalogue from self-consistency, which was deliberately not allowed for in the adjustment model, the average residuals in a region around a point of reference will show correlations with the coordinates of this point of reference and may also vary systematically with the magnitudes and the spectral characteristics of the stars. If a model is found which represents these systematic deviations of the individual positions with respect to a mean coordinate system so that they no longer show any significant nonrandom distributions, we may regard this model as a representation of the catalogue's *systematic errors.* They will be discussed in detail in Section 3.2.3 below.

Note that in practice this experiment is not rigorously feasible, because in order to compare the catalogued positions with the "real"[12] positions, we first have to establish these real positions. This is, of course, impossible since all we can do is to compare the catalogued positions with those observed in a system that is realized by an instrument, such as a transit circle, or a photographic plate. It is *only possible to establish the systematic differences* between the positions of the same stars as given in two catalogues, which is the same as to describe the differences between the coordinate systems fitted to them as functions of position in the sky, and, when warranted, magnitude and spectral characteristics.

As the instruments for the realization of the astronomical coordinate systems become more accurate and define the actual system with smaller and smaller errors, the star catalogues established with their aid become more self-consistent and more representative of the actual system which they are intended to represent.

Catalogue astronomers, therefore, are somewhat reluctant to use the term "systematic errors" of such and such a catalogue, because this would imply knowledge of the actual system. Since the systems defined by two different catalogues will invariably differ, we prefer to speak instead of *systematic differences between the positions in two catalogues* which is the same as referring to the different systems defined by them, or the *systematic corrections which reduce the positions in one catalogue to the system of another.*

Catalogues that were computed with great pains from a large source of material with special emphasis on removing systematic errors and establishing the accidentally and systematically most accurate star positions which can be established at that time are called *fundamental catalogues.* The *fundamental systems* which they define are of course not the actual, ideal equatorial system but constitute the best obtainable approximation to it then available. The fundamental systems and their mutual relationships will be discussed in more detail below in Sections 4.1 and 4.2.

---

12. A statistician would have called these the "true" positions. Since this technical term has been preempted in Spherical Astronomy for a different meaning, we cannot use it here.

Chapter 2

THE ACQUISITION OF ASTRONOMICAL DATA

2.1 The Observations

2.1.1 THE PRINCIPLE OF ASTRONOMICAL MEASUREMENTS. In the previous chapter we have seen that the coordinates defining star positions are essentially polar coordinates on a unit sphere. Therefore a point is uniquely determined by its direction cosines within any coordinate system that can be established by observations. All astronomical measurements are directly or indirectly measurements of angles and, in some instances, directions. The only, but very notable, exception to this is the direct determination of distances within the solar system by radar measurements (compare Muhleman, Holdridge and Block [1962]) which led to an increase in the accuracy of the solar parallax by several orders of magnitude. The difference between angles and directions is as follows: Consider two points on the sky (or anywhere) at different positions; the angle between the radius vectors from the observer to them is an "angle"; their direction cosines in a certain coordinate system define the directions. In general, *directions are not invariant with respect to an orthogonal transformation* (that is, rotation) of the underlying coordinate system, but *angles are.* The reason for this is that directions are derived from the angles which the position vector makes with the coordinate axes. Directions are, therefore, derived from angle measurements. We will be concerned here with the accurate measurement of angles.

The concept of the angle is very intimately connected with that of the rotation, that is, it is the measure of rotation. (While it is impossible to establish an absolute unit for length, it is possible to establish one for angular measure.) Analyzing the measuring process, we note the following: we bring the "pointer," that is, an apparatus capable of defining a direction (for instance, a sight, the optical axis of a telescope or such) in coincidence with one of the position vectors. We then rotate it on a well-defined axis; frequently in the plane defined by the first and the second position vector until it coincides with the second position vector, and measure the angle between.

The measurement of an angle therefore involves two elements: the *settings* (that is, getting the pointer to coincide with the sides of the angle) and *determination proper of the angle* between the directions defined by the settings. The total error of the angle measurement is consequently produced by a combination of the errors of these elements.

2.1.2 SETTING. The setting, in its traditional form, consists of the establishment of coincidence between a point on the image of a celestial object (in the case of a star this point is ideally the center of its diffraction pattern) and a mark, such as an illuminated line, or the middle between two lines or similar devices. The establishment of coincidence is therefore subjected to a variety of errors, the sources of which are partly in the atmosphere, partly physiological and partly due to mechanical or other imperfections of the measuring device.

Without thereby seriously restricting the validity of the following discussion, we shall consider the celestial object whose direction is to be determined as a point source outside the atmosphere, a "star."

If there were no atmosphere, we could consider all the light rays from the star as strictly parallel and, except for the effects of diurnal aberration, in the direction of the star's apparent position. The index of refraction of the atmosphere, however, is different from unity and increases in general towards the Earth's surface. The statistical average of the broad features of this increase can be fairly well computed in terms of several parameters at the observing station, such as air temperature, and so forth, see Section 2.2.5.2 below, but relatively little is presently known about the rate, size and average duration of the fluctuations from this statistical mean which produce *scintillation* and the phenomenon of "*seeing*" (recently called "*shimmer*" by some authors) in the starlight. The passage through the atmosphere therefore deflects the starlight from the direction to its source's apparent position in a manner which can almost, but not quite, be exactly calculated.

The effects of this "bad seeing," that is, the fluctuations of refraction in time and space, combined with the diffraction at the cell of the objective, cause the star image in the telescope to appear as a small, pulsating, dancing disk. The quieter the atmosphere, the smaller and the more regular, uniform and quiet is the appearance of the image of a star in the eyepiece.

Errors in pointing may be introduced by a tendency of the observer to actually "set" off center while judging that he set on the center of the image. Modern devices have rather successfully combated this "*personal error*" (see Section 2.2.1 below). Finally, the accuracy of the setting can be impaired if the marks on which the setting is made, say illuminated spider webs, do not represent the theoretically intended direction; for example, if the spider webs forming crosshairs sag because of excessive atmospheric moisture and thus fail to define the viewing instrument's optical axis.

2.1.3 ROTATION. Measuring the angle between two directions always involves a rotation. The viewing sight may be rotated on an *axis within the instrument* and the angle may be read on a divided circle, or the instrument may retain its position with respect to the Earth and rotate, carried by the Earth, on the *Earth's axis*. The angle on which the Earth rotated can be measured by the time interval between the first and the second setting, since the angular velocity of the Earth may be regarded as constant and known, particularly over intervals of only a few hours. The *measurement of an angle* is thus *reduced to the measurement of a time interval.*

In most cases, a single measuring process consists of a combination of rotations, partially performed by the measurer on an axis within the instrument and partially by the Earth.

As was pointed out in Section 1.1.2, astronomical coordinates, that is, directions on the sphere, are completely defined by two angles, for instance, hour angle and declination, or some other combination of two angles in different planes. However, the astronomer is ultimately interested in the coordinates, or at least directions, of the astronomical objects in an inertial system. Therefore, the establishment of coordinates of astronomical objects consists of two parts: (1) the measurement of a pair of angles which define the position in a coordinate system that is relatively easy to realize, for instance, the system of the horizon at the observing station, and (2) the establishment of the transformation between this coordinate system and an inertial system, for instance, the equatorial system as oriented at a certain epoch. These tasks are often performed simultaneously, for example when the geographic latitude of the observing station is the by-product of declination observations.

2.1.4 TRANSIT OBSERVATIONS. Consider two (not necessarily great) circles on the sphere, and their poles. Let one of the circles be defined by a sight whose axis passes through the center of the sphere and which makes a constant angle with the axis of the circle, so that it is always directed to some point on the circle. Let the other circle be defined by the path described by a celestial object in its daily motion. One can immediately see that the two circles may intersect in no more than two points, and that the position of the intersection points (that is, their spherical coordinates in a certain system) and the angle (great circle arc) between them, or the arc segment between them as measured on one of the circles, will yield relationships from which the star's spherical coordinates in an arbitrary coordinate system can be determined, if a sufficient number of other data are known. A general theory covering this situation is given below in Section 2.2.2.3.

Examples are as follows: let the circle defined by the sight be the meridian. During one sidereal day, every star passes twice through the meridian. Since the meridian through a certain place is carried by the Earth, one might say that the meridian sweeps over every star twice; once at upper and once at lower culmination. Nothing can be gained in this case by measuring the time difference between two successive meridian passages of an object, because this is always twelve hours due to the geometry of the situation. However, by measuring the objects' altitude at both upper and lower culmination, its declination and the geographic latitude of the observing station can be computed. Likewise, the right ascension of the object can be obtained by recording the sidereal time of its meridian passage at upper or lower culmination. If the sidereal time is unknown, but a timepiece is available, the right ascension difference between two objects can be measured; it is equal to the sidereal time interval between their respective upper (or lower) culminations.

The angle of rotation of the instrument between the measurements of the altitudes at upper and at lower culmination must be measured by use of a graduated circle mounted on an axis within the instrument. The other rotation, the one which carries the meridian past the star, is performed by the Earth, and therefore its magnitude must be determined by the measurement of a time interval because the angular velocity of the Earth is, for these purposes, assumed to be constant. After all, sidereal time is nothing but the time interval between the upper culmination of the vernal equinox and the epoch in question. The *geographic latitude* of the place of observation is one of the parameters in the transformation from the system in which the measuring is done (horizon system) to the system with respect to which the coordinates are reckoned (equator system). It is a by-product of the measurements. Another of these parameters, namely, the zero point of the sidereal time scale, must be derived or known from independent sources. Meridian observations will be more fully discussed below (Sections 2.2.1, 2.2.2).

Another example put to practical use in the *Danjon prism astrolabe* discussed below in Section 2.2.4 is as follows: the circle defined by the measuring instrument (the sight) is the almucantar of 30° zenith distance.[1] The hour angle at which the star passes through the almucantar is a function of the declination and the geographic latitude. From the measured difference between the hour angles at successive eastern and western passes through the almucantar, the declination of the object in question can be computed when the observing station's geographic latitude is known. This hour angle difference is, of course, equal to the sidereal time interval between the two passages. If the sidereal times of the passages through the almucantar themselves are recorded, not only their difference but also their average can be taken. The latter is, of course, equal to the sidereal time of the object's meridian passage, that is, its right ascension. Note, however, that this is not the way in which the prism astrolabe is used in practice.

Some other interesting cases were discussed in a work by Niethammer (1947). Particular attention is paid in Niethammer's work to the creation of observing conditions (selection of stars, distribution of their coordinates, and so forth) which will yield the most accurate results, which may be either the spherical coordinates of the star, or the geographic coordinates of the observing station, or a combination of both.

The coordinate systems used in astronomy are systems of spherical polar coordinates with two angles that measure rotations whose axes are perpendicular to each other.

In the *transit circle* mode of astronomical observations (see Section 2.2.1), this

---

1. If, generally, z is the zenith distance of an almucantar, and ϕ the geographic latitude of the observer, only stars whose declinations satisfy the inequalities $180° - \phi - z \geqq \delta \geqq \phi - z$ when $\phi + z \geqq 90°$, and $\phi + z \geqq \delta \geqq \phi - z$ when $\phi + z \leqq 90°$, will pass through this almucantar.

coordinate system is realized by the Earth whose axis defines the axis of the longitude angle. *Errors* will be introduced in the measurements by deviations of this setup from the ideal conditions just outlined. Rotation on the longitude angle is provided by the rotation of the Earth which cannot be controlled by the observer, and is measured in terms of one of the angles describing the Earth's orientation in space, namely, time. The rotation in latitude is on the axis of the instrument, at the observer's discretion, and is measured by a graduated circle. All the imperfections of the latter will appear as errors in the measurement of the latitude angle.

Quite generally, errors in any transit observations will thus have two principal sources. For one, the *circle* on which the positions in the sky lie to which the instrument can be pointed, and through which we observe the transit may (or rather, will) *not be oriented exactly* to correspond to the ideal model (this will be discussed below in Section 2.2.2); as a matter of fact, this closed curve on which transits may be observed may not even rigorously be a plane curve. Secondly, the *timepiece* may (or rather, will) *not give the correct and accurate time.* Besides, there are many other sources of errors which influence observations of this type, such as imperfections of the measuring device in general, including the human operator whenever he is a part of the measuring process. Examples for these are the uncertainty of the definition of the direction to the object by the measuring device due to its unavoidable mechanical imperfections and uncontrolled outside (for example, thermal) influences, and the failure of the recording device (objective or human) to perceive and record the event (transit) at the exact instant it occurs. Some of these effects produce systematic errors of a type that can be instantly eliminated by an appropriate arrangement of the measurements, or later through a careful reduction of the material.

2.1.5 OBSERVATIONS WITH CIRCLES. Beside the rotation of the sight through the Earth's movement in a transit observation, there are the measurements of rotation angles where the sight is moved at the discretion of the observer. Usually, the sight is rigidly connected to a graduated circle (or circle section), ideally mounted in such a way that its center passes through the axis of rotation of the sight and its plane is perpendicular to this axis. Readings are then taken before and after the rotation by means of a mark which does not participate in the rotation, and, under ideal circumstances, the difference between these readings yields the angle.

The description of the measuring process immediately leads to the *error sources*: the *axis* on which the sight rotates may not always remain parallel to itself, or may generally not be perfectly defined. The *center of the circle* may not coincide with the axis of the pointer: this means that the circle is *eccentrically* mounted, and the axis of the pointer may not be perpendicular to the plane of the circle. The graduation marks of the circle may be inaccurately placed (*"division" errors*) and the *indexes* at which the circle is read, may *not* be *immovable.*

Except for uncertainties in the definition of the pointer's axis—the errors in setting and reading, and the like—all the sources enumerated above give rise to systematic errors, which can be determined and removed from the measurements.

A good general discussion with instructions for the determination of instrument errors was given by Podobed (1965). An improved method for the determination of errors of circle divisions was recently given by Høg (1961).

2.1.6 THE RECORDING OF TIME. Angles swept over by using the Earth as the vehicle which carries the pointer are measured by the time interval between two events, usually the transit of objects through circles on the celestial sphere. This implies that quantities which are measured in terms of time intervals can be directly measured only as accurately as the time intervals themselves.

Anticipating information from Section 3.3.3, we see, for instance, that the accidental accuracy of relative right ascensions from single observations corresponds at best to a rms. error in the order of $0.''15$. In order to reach this accuracy, the times of the observed meridian transits must be recorded with a corresponding accuracy, which is about $0.^s01$. This requires not only timepieces from which at any epoch the time can be realized with this accuracy, but also a method of ascertaining and recording the exact instant of an astronomical event (as, for instance, a transit).

Modern *quartz clocks* and *atomic clocks* record time which deviates from uniform time (that is, Newtonian time, approximated by ephemeris time) by less than one part in 10^{10} which is much more accurate than required for transit observations. At present, the limitations of the accuracy of absolute star positions (in particular of right ascensions) are not imposed by the accuracy limitations of uniform time.

Before quartz clocks (and atomic clocks) were generally available for astronomical observations, *pendulum clocks* were the most accurate timepieces. They keep time based on the well-known relationship between the period of a free swinging pendulum in a vacuum and its length (plus the gravitation), and on the fact that for small amplitudes the period of the oscillation of a pendulum is almost independent of its amplitude. The difficulties of keeping accurate time by pendula are enormous: their amplitudes are never zero. Even if they could be operated in a complete vacuum, some energy would always be dissipated so that their amplitudes would diminish to an eventual standstill. Therefore periodic impulses must keep the pendulum swinging. These must be extremely gentle and, above all, uniform. Variations in air pressure change period, and amplitude, and variations in air temperature change the pendulum's length and thus its period. Ingenious devices have been employed to counteract these effects; in particular, the change of pendulum length with temperature. Pendulum rods were manufactured of materials with extremely small temperature expansion coefficients, namely *Invar metal* and *quartz*. (Even the influence of the changing position of the Sun and the Moon, which produces as impressive a phenomenon as the tides, is sufficient to cause a noticeable perturbation in the period of a free swinging pendulum.) With all the compensating devices and other precautions, the best pendulum clocks still keep time which is proportional to absolute time (that is, the independent variable in the equations of motion) only within a few hundredths of a second per day. Nowadays, pendulum clocks are completely obsolete, and astronomers use only quartz and atomic clocks as timepieces.

In any case, *the angular speed of the Earth is not constant,* and neither is the axis on which the Earth turns firmly anchored within the Earth (*polar motion*). *Uniform (ephemeris) time* is, therefore, *not directly usable* as a measure which is proportional to an angle by which the Earth has rotated on its (movable) axis. In order to properly evaluate transit observations which involve time recordings, the relationship between uniform time, as represented for example by ephemeris time, and the angular position of the earth as represented by, say, *universal time* (UT0), must be established empirically because, at present, there is no theory available which would accurately predict their relationship. This is done by observing stars with known right ascensions; recently, the *zenith telescope* has emerged as the most accurate instrument for this purpose.

This situation is somewhat of a vicious circle. If the right ascensions are determined assuming that the Earth rotates uniformly, while, in fact, the rotation of the Earth is nonuniform, the right ascensions thus determined will contain systematic errors which depend on the right ascension itself. Determining, in turn, the time from these systematically falsified right ascensions will introduce an error in the time which will be propagated into the newly-determined right ascensions, and so the cycle is perpetuated. The discovery of the nonuniformity of the Earth's rotation led to the recognition of this difficulty and to the development of methods that were unaffected by this source of errors. (See also Section 2.2.6.2.) Since quartz and atomic clocks only measure uniform time, their emergence could not have led to a breakthrough in the accuracy of star positions determined from transit observations.

As mentioned above, one also must not forget that the instantaneous axis on which the Earth rotates moves within the Earth's body. This causes the so-called "polar motion," that is, the motion of the geographic poles on the surface of the Earth, which was discovered in 1884 by F. Küstner from the analysis of meridian observations. Its most obvious effect is that the geographic latitude of every place on the Earth changes. The topocentric orientation of the actual meridian changes also. Furthermore, it becomes necessary to designate some point on the Earth through which the zero meridian goes, for instance, that point which defines the locus of zero geographic longitude on the Earth's surface. This point, which is fixed with respect to the Earth's body, was chosen more or less at the average location of the intersection of the instantaneous zero geographic meridians (through the Airy transit circle at the old Greenwich Observatory and the Earth's instant north pole) with the equator. Obviously, polar motion will also change a point's geographic latitude, and, in particular, the longitude difference between two locations. All this must be considered when the relationship between universal time and uniform (ephemeris) time is defined and established.

Advances in technology have tremendously increased the accuracy with which the time (indicated by a certain timepiece) of an event can be recorded. The oldest procedure is the *eye-ear method*: while the observer was looking through the telescope, he listened to and counted the ticks of a pendulum clock (or a chronometer which had been carefully compared with a pendulum). He then estimated the time at which the event occurred to a tenth of a second and recorded it manually.

The invention of the *chronographs* relieved the observer from counting seconds: the swinging pendulum (or some other device governed by it) activated a contact which caused a visible mark to be made at regular time intervals (seconds) on a strip of paper moving (ideally) with uniform speed, or an equivalent medium of recording. The beginning of a full minute was marked in some special way, say, by leaving out the mark for second 60. The observer would activate a contact at any time which also produced an "event mark" on the paper strip. Measuring the position of the event marks between the second marks, later resulted in a recording accuracy in the order of $0.^s01$. Subsequent devices, such as Repsold's *self-registering micrometer*, relieved the observer even from the task of triggering the event marks. Thus they contributed substantially to the reduction of accidental errors and the elimination of certain systematic ones, notably the *personal equation* (bias) and the *magnitude equation* in right ascension. However, they did not increase the accuracy of recording itself.

Progress in this respect was made only when *printing chronographs* were introduced. These, instead of generating marks, actually print the time of the event (to $0.^s01$, usually) directly. Even more modern devices, regulated by quartz clocks, make it easy to record the instant of an event with an accuracy of $0.^s001$ and better.

2.2 The Acquisition of Star Positions by Visual Observations

2.2.1 THE PRINCIPLE OF MERIDIAN OBSERVATIONS. The sidereal time of upper culmination of a celestial object equals its right ascension, that of lower culmination its right ascension $+12^h$, and the zenith distance z is related to the declination δ and the geographic latitude ϕ as follows: On both hemispheres ($\phi \gtreqless 0$),

$$\text{for upper culmination south of the zenith, } \delta = \phi - z,$$

and (2.1)

$$\text{for upper culmination north of the zenith, } \delta = \phi + z,$$

On the northern hemisphere, ($\phi > 0$) for lower culmination (north of the zenith) $\delta = 180° - \phi - z$, and for lower culmination (south of the zenith) on the southern hemisphere ($\phi < 0$), $\delta = z - \phi - 180°$.

Therefore, an instrument which registers the time of meridian passage of celestial objects, and measures their zenith distances during their meridian passages (as far as they occur above the horizon) would be ideally suited for the determination of star positions. This fact was appreciated very early by the astronomers and one may indeed regard Tycho Brahe's *mural quadrant* as the *predecessor of the transit circle*. Although other instruments have made heavy inroads into the determination of star positions by direct observation (that is, not by measuring a record obtained by photographic and other means), the transit circle has been the most frequently-used instrument for the determination of right ascensions and declinations for almost 200 years.

Essentially, a *transit circle* (in Europe called a *meridian circle*) is a telescope which is rigidly mounted at an angle of 90° to an axis (instrument axis) which ideally always remains in the east-west direction and, therefore, perpendicular to the Earth's axis. Thus, the "sight axis" of the telescope can ideally be directed towards any point at the meridian, but nowhere else.

The transit circle therefore *emulates the equator system*. The rotation of the sight line of the telescope on the longitude angle, that is the hour angle, is performed by the rotation of the Earth to which the instrument is rigidly connected, and measured in terms of the Earth's orientation on its axis, that is sidereal time. The rotation of the sight line on the latitude angle, that is declination, is performed by rotation on the instrument axis, and measured by a graduated circle.

Rigidly connected to the instrument axis is at least one *graduated circle*. (Normally there are two, one on either side of the telescope, but only one is customarily used for the observations.) The angular position of this circle is recorded at four (and sometimes even six) microscopes each 90° (or 60°) apart which are rigidly connected to the instrument support and are thus, ideally, immobile. At least two microscopes would be necessary to eliminate the effect of circle eccentricity, but the extra ones serve to introduce redundancy and may help to avoid parts of the circle which are known to be "bad." This is desirable since the errors of the circle division marks can only be determined with an accuracy corresponding to rms. errors which range between about 0.″05 and 0.″10, so that the redundant readings serve to decrease the effect of the inaccuracy of the determination of the circle division errors.

If the circle is intended to indicate polar distances (or declinations), the circle reading which corresponds to the pole must be known. This reading can be established in principle by taking the mean of the circle readings obtained during the upper and lower culmination of the same circumpolar star, although refraction and flexure complicate this procedure.

In order to also calculate the geographic latitude of the observer from these measurements, the circle reading corresponding to the zenith (*zenith point*) must be known. This is determined by *nadir observations into a mercury basin*, which actually defines (if it can be regarded as undisturbed) the plane of the horizon and thus one axis in the system in which the observations are made.

2.2.2 THE SOURCES OF ERRORS IN MERIDIAN OBSERVATIONS.

The following factors cause meridian instruments to actually yield star coordinates in a system different from the one which they are intended to emulate.

2.2.2.1 Failure of the instrument axis to be perpendicular to the axis of the Earth.

Even if the axis of the instrument were not in the observing point's east-west direction, the instrument system could still emulate the equator system if the instrument axis were in the plane of the equator. In this case, the sidereal time of the object's transit through the sight line of the telescope would have to be corrected by the angle between the instrument axis and the east-west line to give the right ascension.

A nonperpendicularity of the instrument axis to the Earth's axis renders the coordinate system, which the instrument realizes, a nonrectangular one. Therefore, it cannot be transformed into the equator system by an orthogonal transformation.

The angle between the instrument axis and the Earth's axis will have two components: one which depends on the basic mounting of the instrument, and another one which depends on the degree of perfection with which the instrument itself is made to conform to its ideal.

The first-mentioned *mounting errors* will be considered under the basic assumption that a rotation of the instrument on its instrument axis does not change the direction of the latter. We may thus imagine that two points on the instrument axis define its position with respect to the Earth. The direction of the instrument axis may then be described by giving the azimuth and altitude of one of its ends (usually the western one) in the horizon system, or it's hour angle and declination in the equator system. In actual instruments, the direction of the instrument axis is defined by its *bearings*, which are anchored to the Earth by (usually a pair of) *piers*. Any motion of these piers with respect to each other and/or with respect to the crust of the Earth will change the direction of the instrument axis. The aim is, therefore, to construct piers for meridian instruments in such a fashion that these changes are eliminated or at least minimized. Effects of thermal expansions on the direction of the instrument axis should be avoided. (Needless to say, this is generally impossible.) Even if we maintain the direction of the instrument axis with respect to that portion of the Earth's crust on which it is mounted, its orientation with respect to the equator system will be changed (1) when the local section of the Earth's crust changes its orientation with respect to the average local horizon system, and (2) when the average local system of the horizon changes its orientation with respect to the first equator system. Effects of the first type are, for instance, produced by the *continental tides* and frequently described as "*change of the vertical*"; effects of the second instance are produced by changes of the geographic coordinates of the observing station, such as changes in the shape of the geoid (say, through earthquakes) but mostly by the wandering of the Earth's axis with respect to the Earth's body which is the phenomenon well known as *polar motion* or *wandering of the pole* described in Section 2.1.6 above.

Practical astrometrists apply various methods for the determination of azimuth and altitudes (the latter is, in this connection, usually called *level*) of one end of the instrument axis. We have discussed in Section 2.2.1 the determination of the observer's geographic latitude from meridian observations. The level can be measured by a *spirit level*, or more accurately by nadir observations into a mercury basin. The azimuth of the axis can be computed from appropriately arranged observations of stars (usually a combination of transits of stars close to the celestial pole and of stars near the equator). Measurements of this type are made almost nightly on meridian instruments in use. Because of their minute effects, it is very difficult by any analytical process to isolate the causes of changes in the direction of the instrument axis with respect to the local horizon system, be they secular or periodic. For further details, the reader should consult the original publications of the various investigators. Detailed modern descriptions are found in the books by Zverev (1950, 1954) and by Podobed (1965). It will suffice here to say that the situation is very complicated, and many effects are probably produced by causes whose presence we do not at this time even suspect.

Take as an example the *variations in the geographic latitude* of the observing stations which had been observed for quite some time before they were accorded any significance, and led Küstner, in 1884, to the discovery of the wandering of the pole. Remember, however, the exact definition of the "local" direction of the Earth's axis as given in Section 1.2.2. It is quite feasible that future generations of observers will be able to determine geographic latitude much more accurately than the present one. Their observations may then, perhaps, lead them to confirm the nonrigidity of the Earth from latitude observations.

The second type of mounting errors concerns the *deviations of the instantaneous orientation of the instrument axis from the "average" instrument axis*, whose orientation was just discussed above.

These deviations occur because the instrument axis is mechanically defined by the *pivots*, which are usually circular cylinders of hardened steel with a diameter of 5–8 cm, shaped as perfectly as possible. Each of these rests on two flat pieces of softer material (the "*wyes*") which are made of agate, bronze or brass and are carried on the same supports on which are mounted the circle microscopes. (Most of the weight of the moving parts of the instrument is counterbalanced by special devices, the pressure of the pivots on the wyes is maintained just strong enough to keep pivots and wyes always in firm contact so that the position of the axis is well defined.) Deviations of the form of the pivots from the shape of spherical cylinders in the region where they are in contact with the wyes will, of course, alter azimuth and altitude of the axis when the telescope is pointed to different altitudes. It is apparently not yet possible to make the pivots so perfectly cylindrical that the axis remains parallel to itself when the instrument is rotated. Various methods have therefore been employed to measure very exactly the amount of the deviations of the pivot's shape from that of an ideal cylinder. Once these are known, their influence on azimuth and level of the axis can be computed (or tabulated for easy application). Discussion of the methods used for the determination of these pivot errors are found, for instance, in the introductions to various catalogues, and in papers dealing especially with these subjects. The most recent discussion was given by Scott (1966), Section II.

2.2.2.2 Failure of the instrument's sight line to be perpendicular to the instrument axis. This is called *collimation error*, usually denoted by c, and it may be measured with special *collimator telescopes* placed north and south of the main telescope. In a transit circle, the line of sight used by the observer is established by the position of the middle thread or some arbitrarily chosen reading of the right assension micrometer screw near the center of the field. It is to this line that the observer initially refers the "*time of transit*". The collimation error is the angular distance between the middle thread and a line which passes through the optical center of the objective, and which is perpendicular to the instrument axis.

2.2.2.3 THE GENERAL THEORY OF TRANSITS. A relationship between the coordinates in the first equator system $\hat{\boldsymbol{x}}(H, 90° - D)$ of the point toward which the sight line is directed (near upper culmination; H and D being its hour angle and declination respectively) and the characteristics of the instrument, as discussed so far, is found as follows: In the first equator system, let $\hat{\boldsymbol{x}}(6^{\mathrm{h}} - h, 90° - d)$ be the direction indicated by the west end of the instrument axis. This may be regarded as the y axis in the instrument system in which coordinates $\boldsymbol{x}'$ of points are found from their coordinates $\boldsymbol{x}$ in the first equator system by the transformation $\boldsymbol{x}' = \mathbf{R}_1(d)\mathbf{R}_3(-h)\boldsymbol{x}$.

If $90° + c$ is the angle between the sight line and the instrument axis, it is clear that the coordinates of the point toward which the optical axis is directed when the (correctly zeroed) declination circle reads δ, are in the instrument system $\hat{\boldsymbol{x}}_p' = (\cos c \cos \delta, -\sin c, \cos c \sin \delta)^{\mathrm{T}}$. Note that δ reads "zero" when the optical axis is in the y'-z' plane. The coordinates of this point in the first equator system are, of course, given by $\hat{\boldsymbol{x}}(H, 90° - D)$, so that

$$\hat{\boldsymbol{x}}(H, 90° - D) = \mathbf{R}_3(h)\mathbf{R}_1(-d)\hat{\boldsymbol{x}}_p' \tag{2.2}$$

from which

$$\begin{aligned} \cos D \cos H &= \cos \delta \cos h \cos c - \sin h(\cos d \sin c + \sin d \cos c \sin \delta) \\ \cos D \sin H &= -\cos \delta \sin h \cos c - \cos h(\cos d \sin c + \sin d \cos c \sin \delta) \\ \sin D &= -\sin d \sin c + \cos d \cos c \sin \delta\ . \end{aligned} \tag{2.2a}$$

These formulas are rigorous no matter how large c, d, and h are and can, therefore, be used as the basis of the theory of any transit instrument. If the instrument is mounted conforming closely to an ideal meridian instrument, one obtains as an approximation, considering that in this case H, h, d, and c will be small, the well-known *formula of Bessel*:

$$H = -h - c \sec \delta - d \tan \delta \tag{2.2b}$$

and also

$$D = \delta.$$

If the orientation of the west end of the instrument axis is given in terms of its azimuth and inclination rather than h and d, the formula analogous to equation (2.2b) is known as *Tobias Mayer's formula*.

2.2.2.4 DEFORMATIONS OF THE COORDINATE SYSTEM REPRESENTED BY THE INSTRUMENT. In the previous subdivisions of Section 2.2.2 we have assumed that the instrument was essentially rigid. Actually, *instruments are not rigid*. The effects which the deformations of the instrument have on the coordinate system can be categorized according to different principles. On the one hand, the effects of *non-elastic* and *elastic deformations* can be *juxtaposed.* Only that component of the non-elastic deformations which results in a permanent change of the instrument's characteristic is accessible to systematic investigation; the other irregular, non-elastic deformations will have to be regarded as a source of accidental errors of the data (coordinates) being acquired on the instrument. On the other hand, we may divide the deformations into those which change the orientation of the instrument axis and the collimation, and those which change the relationship between the circle reading and the direction of the sight line. The former can be (and sometimes are) handled by including their effects in the determination of the effective values of the orientation parameters of the instrument axis and the collimation. The latter are at this time considered to be principally due to the *sagging* (*flexure*) of the telescope tube and of the circle under their own weights. Undoubtedly, however, there are other sources, too, such as the sagging of other parts of the instrument, effects of *thermal expansion,* and so on. The best way to make these innoxious is not to eliminate them in the reduction, but to design the instrument in such a way that these phenomena will not occur, or at least not have any significant effect on the measurements, but this is of course never possible in full rigor.

Unfortunately, astronomers and instrument designers had thought, up to the beginning of the twentieth century, that the effect of *flexure of the telescope tube* would be nil if the objective lens end and the eyepiece end of the instrument were constructed very similarly; in particular, if they were both equally heavy. It was assumed that, under these circumstances, objective end and eyepiece (or micrometer) end would bend by an equal angle, so that the axis of the telescope, in effect, would remain parallel to itself and thus produce the same circle reading that it would have produced if there were no flexure. After this assumption was proved to have been incorrect, the standard procedure for the elimination of flexure effects was to assume that they were proportional to the sine of the zenith distance. However, during the last two decades it was recognized that the dependence of the effect of flexure on zenith distance is a function which cannot readily be described by a simple (trigonometric) function. A good account of the history of this phenomenon was given by Schmeidler (1952).

In order to minimize *flexure of circles,* they should be made not too large, light, but still with maximum structural strength. Anisotropisms in the material of which they are made must be carefully avoided. Smaller size, of course, also means—ceteris paribus—less angular accuracy of the marks. Designers, at present, regard diameters of 65–75 cm as optimium for metal circles, while glass circles are considerably smaller, with no resulting loss of accuracy.

2.2.2.5 Modern observations, general discussions. Modern meridian circles are *automated* in many ways. The *reading* of the *circle microscopes* is often recorded *photographically*; at the U. S. Naval Observatory, for instance, the films taken by the cameras attached to the circle microscopes are evaluated on a fully automatic measuring machine which punches the results directly onto cards. A direct hookup of a scanner to a computer is planned. During the visual observation of a star's transit, its image in the self-registering Repsold-type eyepiece micrometer is kept between two vertical parallel threads which are driven with a speed equal to that of the star which is a function of declination and is preset. The observer only makes small manual guiding corrections to adjust for the unavoidable inaccuracy of the motion of the threads and to keep the star during the transit between them. As the threads travel, the instant of their passage through certain hour angles is automatically recorded, and from these data, the accurate time of meridian passage is obtained.

Various descriptions of a transit circle have been published. The most recent is that by Watts (1960). Podobed (1965), Zverev (1950, 1954), and Scott (1966) give detailed instructions for the determination of the instrumental errors (these should more properly be called instrument parameters), and so does Dick (1963). Cohn (1907) presented a fairly comprehensive exposé of the instruments and methods used for the determination of stellar positions at that time, and their accuracies and errors. His article is still very well worth reading.

2.2.3 VERTICAL CIRCLE AND TRANSIT INSTRUMENT. When star positions in any system are determined from systematic observations on an instrument, various *instrument errors* (or parameters) will be introduced. The latter are mostly due to and describe the unavoidable imperfections in design, construction and orientation of the measuring instruments. The effects of these errors can in principle be compensated in three ways: (1) The effect of the error is *neutralized by an appropriate design* of the measuring process. (2) The parameter describing the error is *directly measured* and its effect calculated and removed from the measurements. (3) The error is *calculated* from the effect it produces in a judiciously, and preferably purposely, designed series of measurements. The astrometric profession has held almost as an article of faith (first apparently formulated by F. W. Bessel, the father of modern astrometry on visual instruments) that the order of preference for the procedures for removing the effects of instrument errors from the measurements is that given above. Thus, if at all possible, the measurement procedure should be arranged in such a way that the effects of certain errors in the instruments on the measurements cancel. If this is impossible, the errors should be directly measured, and only as a last resort should they be computed from their effects in the measurements.

Examples for each of these procedures occur in the reduction of measurements at a transit circle. The declination circles, for instance, are read on at least one pair of diametrically opposite measuring microscopes, so that the effect of the eccentricity of the axis of the circle is almost completely neutralized. The inclination of the axis of the telescope to the plane of the horizon is directly measured with a spirit level, and the deviation of this axis' azimuth from the east-west line is computed from a combination of transit observations of stars at the different declinations (ideally at the equator and at the pole).

One of the *principal sources of errors in the observed right ascensions* is the above-discussed *collimation error*, that is, the failure of the line of sight of the telescope to be perfectly perpendicular to the instrument axis. The influence of this error on the observed transit times could easily be eliminated by recording the transit time slightly east of the meridian (at a vertical mark that is set off the optical axis in the telescope), and then, while the object approaches the meridian, quickly turning the instrument on an axis that points accurately to the zenith, and pointing the telescope to the appropriate zenith distance so that the passage of the same object can now be recorded again at the same mark which, this time, will point somewhat west of the meridian. It is fairly obvious that the average of the two recorded transit times, the first east and the second west of the meridian (if the object is observed in upper culmination) will be the time of transit through the meridian and free from the effect of the collimation error.

According to the above-mentioned norm established by Bessel, this would be the preferred procedure of dealing with the collimation error. A typical transit circle, however, is too heavy and unwieldy and even too sensitive an instrument to be reversed during the course of every transit observation. The collimation error of a meridian circle is therefore determined with special collimator telescopes (compare Section 2.2.2.2) and its effects are removed; and furthermore, in the course of an observing program, which will usually take several years to complete, the instrument is reversed every few months (or weeks) so that, if possible, every star in the program is observed in both positions of the transit circle's axis.

It is usually the size of the declination (or zenith distance) circle which necessitates the bulkiness of the transit circle. Furthermore, the few seconds available between east and west transit are too short to get the instrument back into observing position after it has been reversed. On an instrument on which transits only are observed, one may dispense with the accurate declination circle and the associated bulk, and one has thus an instrument which is light enough to be reversed in the course of every transit. Instruments of this type are called "*transit instruments.*" They are generally smaller than transit circles but can, of course, be used for the determination of right ascensions by observing meridian transits; however, they are predominately used by astronomical geodesists in the field for purposes directly connected with the determination of the geographic position of the observing station. Obviously, by observing the meridian transit times of stars with known right ascensions, the local sidereal time becomes known, and thus also the longitude of the observing station, provided the Greenwich sidereal time at the instant of the observed transit is known from reading (directly or indirectly) a sufficiently accurate timepiece.

Another nowadays obsolete use of the transit is for the determination of declinations and latitudes, as follows: Equations (1.1), (1.2) and (1.3a) yield for observations in the prime vertical

$$\begin{aligned} \cos\delta\cos t &= \cos\phi\cos z \\ \cos\delta\sin t &= \pm\sin z \\ \sin\delta &= \sin\phi\cos z, \end{aligned} \tag{2.3}$$

where the upper sign in the second equation holds for $a = 90°$ (observations in the east), and the lower for $a = 270°$ (observations in the west). From equations (2.3) we get immediately

$$\cos t = \tan \delta \cot \phi, \tag{2.3a}$$

a relationship from which either δ, ϕ, or t can be computed if the other two are known. As t, the hour angle, is half the time interval between the same object's passages through both the eastern and the western half of the prime vertical (and the exact purpose of the transit observations is to establish this interval), ϕ can be computed from observing the time interval between both prime vertical transits of stars with known declinations. Conversely, declinations can be computed from observations at a station with accurately known latitude. Several series of declination observations by this method have been carried out and have proved the accuracy of the method. Its principal advantage is that regular astronomical *refraction has no influence* on the time of transit of any object through any vertical. Note that z, the only quantity subject to refraction, does not occur in equation (2.3a). The main disadvantages are: (1) the instrument must be very carefully leveled, and (2) the long time interval between the east and west observations of the same star. Another disadvantage is that the number of stars that can be observed is limited.

Transit instruments used by geodesists are usually fairly small (with telescope apertures in the order of three inches and focal lengths of about three feet) so that they can fairly easily be transported from one site to the next and set up and adjusted without too much trouble. Instruments used for purely astronomical purposes are almost always larger otherwise only the brightest stars could be observed at them, and they also are not usually moved around from site to site.

Most transit instruments have "broken" telescopes, that is, the objective is carried by a tube which is mounted at the center of the horizontal axis of the instrument at a right angle to it. Inside the hollow axis is a mirror, or a prism, which reflects the light passing the objective towards one end of this axis right through the pivot, at the end of which there is the usual eyepiece micrometer or, in more modern instruments, some photographic or electronic device for the recording of the transit.

As in all other cases, the instrument errors must be carefully taken into account. A good description of the procedures employed for this purpose was given by Dick (1963).

The *vertical circle* is that modification of the transit circle on which *altitudes* (zenith distances) but no transit times *are observed.* Such an instrument lacks, therefore, a carefully positioned and adjusted east-west axis and can be used at all azimuths, but is not provided (as a theodolite or universal instrument would be) with a very accurate azimuth circle. In common with the transit circle, it has a very accurate circle for measuring zenith distances. Counterbalances and other devices keep most of the weight off the guiding bearings on the vertical axis. A vertical circle is more movable than a transit circle. It can thus be, and usually is, reversed during the measurements on a particular object. Therefore, knowledge of the zenith point (or nadir point) of the declination circle for the purpose of measuring zenith distances is unnecessary. The vertical axis must, however, rigorously be directed to the zenith, or otherwise the point toward which the vertical axis is directed must be accurately known. This condition is assured by constantly keeping the position of the axis under surveillance by means of spirit levels. Vertical circle observations in the meridian are frequently employed for the determination of absolute declinations. The current internationally-adopted fundamental system of astronomical positions, namely, the FK4, relies heavily on declination observations made over many years at the Ertel vertical circle at the Pulkovo Observatory. Podobed (1965) gives a somewhat more detailed description of the vertical circle as well as a discussion of the observations made at it.

2.2.4 THE PRISM ASTROLABE. Instruments that are designed to *register the transits of stars through a particular almucantar* are also capable of determining right ascensions and declinations. Of all instruments of this type that have been proposed, the *impersonal astrolabe* developed by Danjon (1955) is the only one which is still used and producing first-rate results.

Danjon's prism astrolabe is used to register the times when a certain star passes the 30° zenith distance almucantar, on the western as well as on the eastern hemisphere of the sky. It can, theoretically, be used for the determination of right ascensions and declinations in the following way. According to the third equations of the system which one gets by solving equations (1.2) for $\hat{x}(a, z)$,

$$\cos z = \sin \phi \sin \delta + \cos \phi \cos \delta \cos t. \tag{1.2a}$$

If θ_1 is the sidereal time when a star passes the almucantar in the east and θ_2 the time when the same star passes in the almucantar in the west, the hour angle t at which this happens is clearly $\frac{1}{2}(\theta_2 - \theta_1)$. This is inserted in equation (1.2a), which, solved for δ, gives

$$\delta = -\arctan(\cos \phi \cos t) + \arcsin[\cos z(1 - \cos^2 \phi \cos^2 t)^{1/2}]. \tag{1.2b}$$

From this, δ can be found if ϕ is known. Conversely, ϕ can be found with this instrument for observing stars of known declination. (Of course, $z = 30°$ always.) The right ascension of the star is $\frac{1}{2}(\theta_2 + \theta_1)$, since this equals the sidereal time of the star's meridian passage.

The special properties of Danjon's astrolabe have rendered it, within the last ten years, an important instrument for the improvement of the systematic accuracy of stellar coordinates.

In practice, prism astrolabes are not used according to equation (1.2b) but rather under the assumption that z and ϕ are also unknown. Under these circumstances, the contribution of this type of instrument is the establishment of systematic errors $\Delta\alpha$ and $\Delta\delta$ of the coordinates of the observed stars rather than their coordinates themselves. Even so, it is easier to establish the $\Delta\alpha$ than the $\Delta\delta$. Débarbat and Guinot (1970, Chapter IV) give a detailed discussion of how this is done. The chief disadvantage of this instrument is that, due to technological limitations, the size of its objective lens is rather small so that only relatively bright stars (brighter than 6^m, say) can be observed with it.

Another difficulty is that most of the time only one almucantar transit per star can be observed, so that, even with known ϕ and z, only a relationship of the form $f(\alpha, \delta) = 0$ can be established, in effect giving a celestial line of position.

Only in the years to follow will it be possible to definitively assess the value of the prism astrolabe (or, more generally, the method of equal altitudes) as a tool in fundamental astrometry.

2.2.5 THE DETERMINATION OF ABSOLUTE DECLINATIONS.

A position determination is called *absolute* when it in no way makes any use of already known positions of celestial objects.

2.2.5.1 Instrument errors.

As described in the preceding sections declinations can, in principle, be determined by observing the stars' transits through the prime (or any other) vertical, or through a certain almucantar, but chiefly by measuring their altitudes at culmination. In virtually all cases, the geographic latitude of the observing station (or the circle reading for $\delta = 90°$) enters as an unknown parameter into the reduction, and must, therefore, be either known from independent sources, or the observations must be arranged in such a way that the geographic latitude is either eliminated or determined in the course of the observations. The latter is the most prevalent practice. On transit circles, this is done by observing circumpolar stars at upper and lower culminations. The sources of systematic errors of the instrument were discussed before (Section 2.2.2). The uncertainties of instrument independent parameters, for instant refraction, also contribute systematic errors to the positions finally obtained (see Section 1.6.1) from the reduction of the instrument readings.

For measurements on meridian instruments (transit circle, vertical circle), *flexure* and *circle division errors* are those imperfections chiefly responsible for errors in the declinations. As stated above, this can be experimentally determined and, to a considerable extent, measurements can be freed from their influence.

The exact determination of circle division errors is extremely laborious and time consuming—see, for instance, Levy (1955), Høg (1961)—and sometimes carried out under conditions that are at least slightly different from those under which the stars are observed. In some instances they have even been made in daytime, with the observer constantly standing near the circle, and radiating body heat from a different direction than during the observations on the stars. Similar considerations hold for the laboratory-style determination of flexure (compare Podobed [1965], pp. 68 ff.). It helps to appreciate the difficulty of the problem if one considers that, fundamentally, we are concerned with the measurement of large angles to an accuracy of better than $0.''1$. On a circle with a diameter in the order of half a meter, this corresponds to a few tenths of a micron at the periphery. Within such small quantities one simply cannot rely on the stability of an instrument as heavy and complicated as a modern transit circle.

2.2.5.2 REFRACTION. Sources outside the instrument which contribute to errors in the declinations are chiefly the *refraction* and the *uncertainties in the determination of the geographic latitude* of the instrument.

Theoretically, the index of refraction of the atmosphere along the path of the light from a celestial object to the observer must be known to calculate the exact amount of the *refraction*, that *is*, the *difference between the directions of the light ray before it enters the atmosphere*[2] *and when it reaches the observer*. The index of refraction depends on the density of the air which, in turn, depends on the pressure, the temperature, and their gradients, that is the manner in which they change with elevation, namely, height above the geoid. It turns out that one relationship between the physical characteristics of the atmosphere, say that of the temperature as a function of elevation, must be established before the density or the index of refraction as a function of elevation can be deduced (compare Bemporad [1907], Garfinkel [1944, 1967]). The manner in which atmospheric temperature varies as a function of elevation is, however, not stationary. There are diurnal variations (nights are colder than days) and even considerable variations within an hour and fractions thereof. It follows that, ideally, the temperature gradient of the atmosphere should be kept under constant surveillance during the observations. This obviously would be quite expensive, and would require a great deal of effort. The assumption of a model atmosphere is a suitable substitute, since it can be shown that local density irregularities, as they occur in practice, seldom have any noticeable influence on refraction.

It can be shown that the total refraction would depend only on the index of refraction of the atmosphere at the observing station, if the equidensity surfaces of the atmosphere were planes parallel to that of the horizon. In that case, all that is necessary to calculate its amount is the accurate knowledge of the index of refraction of the atmosphere at the observing station, either by direct measurement or by calculation. The actual atmosphere, is, of course, not so structured. This idealization, however, is useful for calculating a good approximation to the refraction at small zenith distances. For small zenith distances, the formula $z - z_o = R \tan z$, which is rigorous for a plane parallel atmosphere, gives fairly accurate values for the refraction $\rho = z - z_o$. The quantities z and z_o are the topocentric apparent and the observed refracted zenith distances of the object, respectively. R, the *constant of refraction*, depends on the air temperature and barometric pressure at the observing station and is about one minute of arc.

If we assume a more realistic model of the atmosphere in which the equidensity surfaces are concentric spherical shells, the formula for the computation of refraction

2. Unfortunately, no commonly used name is available for the position defined by this direction. It differs from the apparent position by diurnal aberration and might thus be called *topocentric apparent position* in contrast to the "geocentric apparent position" which would be the same as the apparent position. In this author's opinion, there is little danger that this juxtaposition of "topocentric" and "geocentric" would, in this context, be construed as relating to the diurnal parallax.

is sometimes written in the form[3]

$$\rho = A \tan z + B \tan^3 z + C \tan^5 z + \cdots.$$

Even in this more sophisticated formula, A and B depend only on the index of refraction—and, therefore, on the meteorological conditions—at the observing station; the constitution of the atmosphere (that is, the density gradient) becomes essential only for the computation of the coefficients from $\tan^5 z$ on. This is called the *theorem of Oriani and Laplace*, and means that even the second order approximation to the exact formula for computing the effect of refraction depends only on the conditions at the observing station.

Thus it is clear why the theoretical values of the effect of refraction are not very sensitive to changes in the model for the constitution of the atmosphere, especially if one does not come too close to the horizon. In general, one will avoid making precise astrometric observations at zenith distances of over 70°. For this reason the refraction tables of the Pulkovo Observatory (Orlov 1956) are still useful and widely used, although they are based on Gyldén's theory from the year 1860 when the constitution of the atmosphere could be observed directly, only up to elevations that were accessible through climbing mountains, or through balloon flights. Garfinkel's (1944) theory yields remarkably accurate values, while the more sophisticated 1967 version, still largely untested, should be even more accurate.

2.2.5.3 SIMULTANEOUS DETERMINATION OF REFRACTION CORRECTION AND INSTRUMENT LATITUDE. In practice, the latitude of the instrument and the constant of refraction are computed from observations of lower and upper culminations of circumpolar stars. The refraction theory thus serves only to establish a mathematical model for the way in which ρ depends on z (or z_o).

This means that the refraction at the observed zenith distances is computed (by means of a refraction theory) except for a factor, so that the ratio of the refractions at upper and lower culmination of the same circumpolar star, but not their individual values, are assumed to be known.

Consider (without restricting generalities) culminations north of the zenith only; z_u and z_l be the topocentric apparent zenith distances of a star of declination δ at upper and lower culmination respectively, and let the refraction corrections to these zenith distances be $\rho_u \Delta R/R$ and $\rho_l \Delta R/R$. ρ_u and ρ_l represent the refractions calculated with the assumed (preliminary) value of the refraction constant R, and ΔR its unknown correction.

---

3. This necessarily leads to a value of $\rho = \infty$ at the horizon. Garfinkel (1944) avoids this by developing ρ as a series in the tangent of an auxiliary angle, which reaches about 1.3 at the horizon. In his most modern development (Garfinkel 1967), which allows one the calculation of the refraction for zenith distances to 180°, series developments are avoided altogether.

From equations (2.1) we then have for every star observed in both culminations

$$z_u = \delta - \phi - \rho_u \frac{\Delta R}{R}$$

$$180° - z_l = \delta + \phi + \rho_l \frac{\Delta R}{R}. \tag{2.1a}$$

In (2.1a), δ will, of course, be different for different stars, while ϕ and ΔR may (with appropriate precautions) be taken as constant. The minimum situation for the solution of (2.1a) is the observation of two stars in both culminations. Then there will be four equations in the four unknowns δ_1, δ_2, ϕ, and ΔR. If there are more than two stars available, this will be an adjustment situation.

The necessity of observing the upper and lower culminations of circumpolar stars for the purpose of absolute declination determinations accounts for the fact that most of these were made at observatories at fairly high latitudes where a large section of the sky is circumpolar. The latitude ϕ is by no means constant, but varies due to the motion of the instantaneous axis of rotation with respect to the Earth's body (that is, polar motion), and possibly movements of the instrument with respect to the body of the Earth, such as continental tides. Experience has shown that reductions in which frequently and individually determined instrument latitudes—the values of ϕ which result from the formal solution of (2.1a)—are used, instead of geographic latitudes computed from the known positions of the Earth's instantaneous poles, yield superior absolute declinations.

In order to measure zenith distances, the circle reading corresponding to $z = 0$ must be known. This is determined several times during the night by *nadir observations into a mercury basin*. Note, however, that z is not needed except rather approximately for the calculation of ρ, and ϕ is not needed at all as long as the circle reading corresponding to $\delta = 0$ is known, which could be established from equations analogous to (2.1a). Since it is easier, however, to observe into a mercury basin than make a complete determination of ϕ from (2.1a), the zenith distance observations are used as an intermediary.

In spite of all precautions, absolute declination determinations are still never completely free from systematic errors. Absolute declinations of the same stars determined during the same period on different instruments at different places sometimes show systematic differences of up to $2''$. In older series of observations, errors were often introduced due to inadequate accounting for flexure, and even in modern times, flexure is the most difficult effect to allow for. Except for observations at the prism astrolabe, observations for the determination of absolute declinations outside of the meridian never acquired much importance and will, therefore, not be discussed here any further.

2.2.5.4 DYNAMICAL IMPROVEMENT. Fortunately, the information yielded by the measuring instruments about the absolute declinations may be supplemented by conditions imposed by the dynamics of the solar system. The bodies of the solar system (Sun, Planets, Minor Planets) move, except for perturbations, in planes through the barycenter of the solar system. If the positions of a certain object are observed with respect to a system of positions, the equation of the plane in which the object moves can be determined in this system by analyzing the observations. If the declinations in this system are affected by declination-dependent systematic errors, the observations will show systematic residuals if the planes of motion are properly forced to pass through the barycenter of the equatorial system. The analysis of the declination observations of many bodies in the planetary system, carried out over a long period of time will then establish those systematic errors in the declination system that still remain after careful consideration of all known sources of systematic errors.

This procedure is, of course, restricted to the range of declinations which are assumed by bodies in the planetary system. Even including the Minor Planets they only reach a belt along the ecliptic, and can essentially be used only to determine the *equator point*, that is the average amount by which the declinations are in error at the equator.

It would be interesting to speculate on the state of absolute positional astronomy if bodies of the solar system or suitable artificial bodies would cover all declinations uniformly. Theoretically at least, absolute positions could then be established as follows: Determine the star positions with high accidental accuracy on appropriate instruments but do not take great pains to eliminate the effects of any possible sources of systematic errors; however, see to it that the structure of the systematic errors is not too complicated. The *rectification of the system*, that is, the elimination of the systematic errors, could then be achieved entirely by analyzing a sufficiently numerous set of observations of the solar system (and artificial) bodies.

Under the present conditions, this is a utopic scheme, since the series of observations available would be neither numerous enough nor would they reach into sufficiently high declinations to serve the purpose intended, and, finally, the model for the adequate description of the systematic errors would require too many parameters (that is, be too complicated) to allow their determination in this way only.

However, the method could be applied with advantage for the removal of systematic errors of simple structure which will still remain even in positions that were obtained after all possible known error sources had been carefully removed.

In particular, Dneprovsky (1932), Numerov (1933) and Brouwer (1934, 1935) have proposed observations of Minor Planets for the purpose of determining the systematic errors of the declination system. An analysis of the accuracy of this method was given by Clemence (1948a). Of all bodies of the solar system (with the exception of comets and meteoroids), the Minor Planets cover the widest declination range and are the most numerous group of bodies. They look like stars, so that there would be no errors introduced by the differences in brightness and appearance as between the major bodies of the solar system and the stars.

First results of this method have been published by Pierce (1970); see also Section 5.2.1.

2.2.6 THE DETERMINATION OF ABSOLUTE RIGHT ASCENSIONS.

2.2.6.1 THE PRINCIPLE OF FUNDAMENTAL DETERMINATION OF RIGHT ASCENSIONS AND THE OBLIQUITY OF THE ECLIPTIC. *Right Ascension* is defined as the *eastward angle between the hour circle through the vernal equinox and the hour circle through the object in question.* The right ascension of an object is therefore equal to the sidereal time at its upper culmination. This is easily determined by a transit circle, and the problem is thus reduced to the absolute determination of sidereal time. The vernal equinox, in turn, is defined as one of the intersection points of the ecliptic and the equator. Because the ecliptic is defined in terms of the apparent path of the Sun in the sky, we can expect that the *determination of absolute right ascensions will involve observations of either the Sun itself or of objects in the solar system whose position with respect to the Sun are accurately known.*

In order to determine the position of the vernal equinox (that is, in essence, sidereal time) we note that we obtain from equation (1.4), for $\beta = 0$ after eliminating λ, the relationship

$$\sin \alpha_\odot \tan \epsilon = \tan \delta_\odot \tag{2.4}$$

between right ascension and the declination of any point on the ecliptic and the obliquity of the ecliptic, ϵ. This relationship would rigorously apply to the Sun if its ecliptic latitude $\beta_\odot$ were always exactly equal to zero. However, this is not so. $\alpha_\odot$ and $\beta_\odot$ are geocentric and therefore slightly different from the values they would have if referred to the barycenter of the Earth-Moon system, the point through which the ecliptic passes. In addition, perturbations of the Earth's motion (due mainly to Venus and Jupiter) will appear as perturbations of the Sun's motion and thus make it appear to stray from the ecliptic by a small amount (see Section 1.3.3). The latitude of the Sun, however, remains so small (usually under 2″) that it can be very accurately computed, even if the parameters on which it depends (that is, mainly the orbital elements and the masses of the orbits of the Moon, Venus, and Jupiter) were only approximately known. Thus the actual situation is, in principle, not different from that in which equation (2.4) would be rigorously satisfied.

Equations (2.4) can, in principle, be used for *finding the obliquity of the ecliptic* and for *setting a clock to sidereal time.* Once such a *clock* is available, the problem of determining absolute right ascensions is essentially solved. This is done as follows:

Suppose we have a clock that shows uniform time, whose *correction* k (that is, that amount that must be added to the reading a to obtain sidereal time) is given by $k = k_0 + k'a$. k' is called the *clock rate.* Assume that on a transit circle on at least three occasions, the declinations δ_1, δ_2, and δ_3 of the Sun were measured absolutely.

If a_1, a_2, a_3 are the clock readings at the meridian transits on the occasions when the declinations were measured, equations (2.4) will read

$$\sin[a_i(1 + k') + k_0] \tan \epsilon = \tan \delta_i. \qquad (i = 1, 2, 3) \tag{2.4a}$$

These three equations may be solved for the three unknowns k_0, k' and ϵ. With k_0 and k' available, the clock readings can immediately be converted to sidereal time, which is the basic quantity necessary for the determination of right ascension from transit observations. The determination of the obliquity of the ecliptic is a by-product of this process.

The rate k' could alternatively be found from observing the transits of a fixed object in the ecliptic, such as a star with no proper motion, so that one may assume that the equivalent of a sidereal clock with zero rate is available. In this case, only two declination observations and clock readings at meridian transit are necessary to obtain ϵ and k since the equations (2.4a) then become

$$\sin(a_i + k) \tan \epsilon = \tan \delta_i. \qquad (i = 1, 2) \tag{2.4b}$$

From (2.4b), k and ϵ can be computed.

It is obvious from equation (2.4) that ϵ can best be determined from observations of $\delta_\odot$ at two successive solstices. At such times, $\sin \alpha_\odot = \pm 1$ and therefore give us the widest range, namely 2, in the coefficients of $\tan \epsilon$. Also, at these times $\sin \alpha_\odot$ changes only very slowly. On the other hand, $\alpha_\odot$, which is determined by its sine will be determined with the highest accuracy when its sine changes most rapidly. This occurs when $\alpha_\odot$ is close to an integer multiple of 12^h, that is, near the equinoxes.

Unfortunately, due to the perturbations which the various planets (mainly Venus and Jupiter) exert on the Earth's orbit, ϵ does not remain constant but changes slowly, at a rate which can be calculated from the theories of celestial mechanics. The predicted and observed rate of change of ϵ differ by about $0.''3$ per century, see for instance Aoki (1967), also Fricke (1970) and Lieske (1970). As mentioned above, it is a requirement for the determination of absolute right ascensions to have a clock which runs not proportionally to Newtonian (ephemeris) time—see Section 2.1.6—but proportionally to the angular speed of (that is, "in phase" with) the Earth. *Daily and seasonal variations of the clock rate* will thus appear as systematic errors in the right ascensions.

Even so, sidereal time (namely, the hour angle of the vernal equinox) itself is not proportional to the orientation angle of the Earth. The reason for this is that due to precession and nutation (see Section 1.6.5) the planes of neither ecliptic nor equator retain their orientations with respect to an inertial system (see Section 1.1.3). (For a more thorough—and causal—discussion of these phenomena see Danjon [1959]). The most recent, and complete, discussion of *precession* and *nutation* was given by Woolard (1953). One of their consequences is a retrograde (that is, eastward) motion (*regression*) *of the vernal equinox* on the equator with a period of approximately 25,900 years (sometimes called a *platonic year*). Due mainly to nutation, the rate of this regression is not constant and, therefore, sidereal time is not strictly in phase with the Earth's rotation. The calculation of the deviation of sidereal time from uniformity thus requires, besides knowledge of the non-uniformity of the Earth's rotation, the knowledge of the parameters of precession and, most important, of nutation.

As pointed out above, the clock correction k, determined as just described, will not change as long as we have a perfect timepiece. *Quartz* and *atomic clocks* are nearly perfect for the purposes of fundamental astronomy, although, because of irregularities in the Earth's motion, their relationship to its rotation must be constantly monitored.

As long as pendulums were the best available clocks, there was always the danger of fluctuations in rate. This is least damaging if the right ascensions of the objects (stars) are close to that of the Sun, so that all observations occur within a relatively short interval of time. In that case, daytime observations are unavoidable. Once the absolute right ascensions of stars have been determined by relating them directly to the Sun, they could in turn be used for determination of the clock correction and thus the determination of the right ascensions of other objects.

Obviously, the involvement of the Sun itself in the observations leads to all sorts of difficulties. Its position is difficult to observe. The intensity of its light must be weakened by means of screens or filters and, even then, one can accurately observe only the positions of the limbs, but not that of the center of the Sun. Besides, it is almost inconceivable that the direct radiation of the Sun should not have any influence upon the performance and the characteristics of the instrument. Likewise, the instrument and the observer may not perform in the same manner during daytime and nighttime observations.

The method of determining the vernal equinox directly from observations of the Sun has really only historical importance. Very few right ascensions were actually determined by direct comparison with the Sun, principally those of Altair and Procyon which lie near the equator at right ascensions about 12^h apart, by Bessel and Piazzi.

Once the right ascension of any object is established, it may itself be used for finding the correction to a sidereal clock and thus for establishing the means necessary for finding right ascensions from meridian transit observations.

N. Maskelyne, one of Bradley's successors as Astronomer Royal and Director of the Greenwich Observatory, observed the absolute right ascension of 36 bright stars which he called "*fundamental stars*."

The daily and annual fluctuations of the "effective" clock rate (composed of influences of daily fluctuations of the observer, the instrument and the actual clock rate fluctuations) will, of course, exercise their full influence upon the right ascensions observed in this manner.

This is the reason why the right ascensions of Altair and of Procyon (which are about 12^h different) were determined as often as possible through a year. In this way, the stars will, in the course of a year, be observed in all right ascension differences from the Sun and at all times of the day, and their observed right ascension differences can be averaged over the whole year.

2.2.6.2 THE DETERMINATION OF ABSOLUTE RIGHT ASCENSION DIFFERENCES. Much easier than the determination of absolute right ascensions, which necessitates the setting of a sidereal clock from observations of transits of the Sun, is the *determination of absolute right ascension differences.* For this, one needs only a clock that is in phase with the rotation of the Earth and a perfectly set-up, error-free, that is, ideal, transit instrument (or transit circle).

Neither are available. The observations will therefore have to be subjected to an adjustment in order to compensate for (or rather eliminate) the deviations of the clock from being in phase with the Earth's rotation.

As mentioned before (Section 2.1.6), the rotation of the Earth is not uniform. The deviations from nonuniformity are only partially regular. It is known that *the Earth's rotation is generally slowing down* (at least, in part, because of *tidal friction*) and that the angular speed during the summer in the northern hemisphere is slower than in winter, due to the seasonal increase of the Earth's moment of inertia which possibly occurs because of the melting of the ice around the north pole. Even after removal of these trends, the Earth's rotation is irregular (that is, its angular speed is not constant) and the relationship between universal time UT0, which is in phase with the rotation of the Earth, and uniform time must be kept under constant surveillance. In practice, this is done in part by relying for uniform time in the short range on quartz and atomic clocks, and in the long range on the motion of the Moon, which is used to establish the relationship between ephemeris time and UT2, a measure of universal time from which the annual fluctuations have been removed. For details see Mueller (1969) Section 5.4. The concept of ephemeris time has recently been severely criticized, see, for example, Mulholland (1972).

In order to remove from the observations any seasonal or diurnal irregularities of the clock, the program stars are observed at all times during all seasons. The raw observations are then subjected to an adjustment to make them self-consistent. Several methods have been devised for this; they are described in detail, for instance, in the books by Newcomb (1906), Zverev (1963), Podobed (1965), and in the introductory remarks of star catalogues in which the observations were made in a fundamental manner. For example, see various volumes of the publications of the U.S. Naval Observatory. The final positioning of the vernal equinox is then accomplished with the use of observations of various bodies of the solar system and will be discussed in Section 2.2.6.4 below.

Before the observations can be adjusted to make them self-consistent and to remove from them the influence of clock irregularities, they must be reduced to an ideal instrument. This means that the effects of flexure, division errors, and the noncircularity of the pivots must have been removed from the observations, and that the set-up errors of the instrument as well as the collimation error must be known.

The effects of this latter group of errors are removed from the observations by equations (2.2a) or equivalent ones. The collimation c is determined by means of special collimators although, in principle, observations of stars could also be used for this purpose. The orientation angles h and d are equivalent to the azimuth a and level i of the instrument axis. The azimuth a can be absolutely determined from observations of stars and i can be found by means of a spirit level or mercury basin.

Since the absolute determination of the azimuth a requires some effort, transit circles (and transit instruments) are often used in conjunction with a *mire*, that is a meridian mark on which the instrument can be easily trained several times a night in order to check on daily variations of the instrument's azimuth. The fundamental azimuth of the mire must be determined at regular intervals from observations of stars. This can be done in several ways.

In principle, the vertical through the pole, that is, the northern half of the meridian, can be found from the observation of the azimuth of circumpolar stars in (*greatest*) *eastern and western elongations*.[4] This is quite analogous to finding the altitude of the celestial pole from altitude observations of circumpolar stars in upper and lower culmination Fortunately, neither refraction nor flexure, the two main obstacles to accurate declination observations, have any influence on the azimuth. Since on a transit circle stars can be observed only very close to the meridian, it is clear that only stars very close to the pole (such a star is called a *polarissima*) can be used for the determination of azimuths, since only these will be in the instrument's field at their greatest elongations. In this method, no times need be registered.

One can also easily see that an analysis of the observed transit times of a circumpolar star, the pole (Polaris itself is mostly used for this purpose) will yield the azimuth of the instrument axis. In the absence of collimation error, and if the azimuth of the instrument's west axis were exactly 270°, the time intervals between successive upper and lower culminations of a star would be exactly 12^h. Any deviation from this value indicates an azimuth error, a collimation error, or a combination of both. (The collimation error, however, was supposedly determined independently.) The effect is obviously largest for stars closest to the pole. These methods are just two examples. There are many more for finding the orientation parameters of the instrument. In practice, most azimuths are determined as follows:

Polarissimae, which can be observed at elongation in the field of view of a transit circle, are usually too faint to be seen. The use of mires is better. In a program one would normally have 15 or 20 circumpolars, $\delta \geq 80°$, which can be observed at both culminations. Each circumpolar observation may be used with an equatorial star to give a value of a, the azimuth of the instrument. If, at about that time, a reading on the mire was made, its azimuth could be deduced. If one is careful, in the course of 3 or 4 weeks, to observe each circumpolar star an equal number of times at both culminations, the resulting mean of all azimuths of the mire is quite well determined. The interval (3 or 4 weeks) is short enough so that polar motion has, at most, a linear variation during the period. With the azimuth of the mire so determined, the azimuth of the instrument may be deduced for each time a reading was made on the mire. (Note: Balancing of upper and lower transits

---

4. Refers to these positions of the star in the horizon system at which the azimuth of the object differs most from that of the North Celestial Pole.

of each circumpolar star insures that the azimuth of the mire is free of errors in the provisional α's adopted for the circumpolar.) In this manner, absolute azimuths may be determined. The method was first used at Pulkova and later adopted at most other observatories doing fundamental work. It has been used quite regularly at Washington, Greenwich, and the Cape.

2.2.6.3 PARTIALLY ABSOLUTE RIGHT ASCENSION OBSERVATIONS. It is clear that irregularities of the clock (or, generally, an undiscovered phase difference between the Earth's rotation and the clock) will introduce errors in the right ascensions that depend in essence on the right ascensions themselves; they are, in the notation familiar to catalogue astronomers, errors of the form $\Delta\alpha_\alpha$. On the other hand, the first equation of (2.2b) shows that the instrument set-up errors produce effects which vary mostly with declination (or zenith distance) and thus will, if uncorrected, produce errors of the form $\Delta\alpha_\delta$ in the measured right ascensions. Uncorrected deviations of the pivots from circularity will also produce errors of the type $\Delta\alpha_\delta$.

Some investigators will produce right ascensions that have been reduced with careful attention to the adjustment and correction of the clock rate; these right ascensions will then be free of errors of the form $\Delta\alpha_\alpha$. If, however, the instrumental set-up errors are not just as carefully investigated, in particular if the effects of the deviations of the shape of the pivots from strictly circular cylinders are not carefully removed, the observations will be affected by errors of the form $\Delta\alpha_\delta$. Positions observed in this way are called partially absolute. The right ascensions discussed above would, for instance, be absolute with respect to right ascension, but not with respect to declination. It is, of course, also possible to produce right ascension observations free from errors of the form $\Delta\alpha_\delta$, but not of $\Delta\alpha_\alpha$.

In most cases, when not specifically stated otherwise, the term "absolute" when applied to right ascensions actually means only absolute right ascension *differences*. It is clear that declination observations can also be partially absolute, that is, made to exclude only errors of the form $\Delta\delta_\alpha$ or $\Delta\delta_\delta$ but not both.

2.2.6.4 THE CORRECTION OF THE EQUINOX. Although the first catalogues of absolutely determined star positions[5] were observed by getting the clock correction directly from observations of the Sun, it later became the established practice to observe only absolute right ascension differences and find the position of the equinox among the stars from the observation of the planets or the Sun.

While in principle—see Section 2.2.6.1, especially equation (2.4)—the right ascension of the Sun can be calculated from its measured declination under the assumption that the Sun is rigorously in the ecliptic, it can also be calculated by the methods of celestial mechanics. If we thus obtain α and δ from mechanical theory, we may use this information in equation (2.4), or an analogous one, to derive the

---

5. These will be called "absolute" catalogues to distinguish them from fundamental catalogues which are something completely different; see Sections 4.1 and 4.2.

orientation of the equatorial coordinate system. In the equations below we assume that the Sun's (or planet's) latitude is exactly zero. (This is, of course, never rigorously true, and we must therefore assume that appropriate corrections have been applied to reduce the observations to this idealized case.) From equation (2.4) we obtain by differentiation,

$$\sec^2\delta \, d\delta = \cos\alpha \tan\epsilon \, d\alpha + \sin\alpha \sec^2\epsilon \, d\epsilon. \tag{2.5}$$

The differentials $d\delta$, $d\alpha$, and $d\epsilon$ may be regarded as the sums of components originating from various sources. One could, for instance, write

$$d\alpha = d_0\alpha + \sum \frac{\partial\alpha}{\partial E_k} dE_k + v_\alpha$$

and

$$d\delta = d_0\delta + \sum \frac{\partial\delta}{\partial E_k} dE_k + v_\delta. \tag{2.6}$$

In (2.6), $d\alpha$ and $d\delta$ are the differences between the observed and predicted right ascensions and declinations of the body (Sun or planet). $-d_0\alpha$ and $-d_0\delta$ may be regarded as constant biases that enter the observations somehow. In the traditional language of meridian astronomy, they are called *correction to the equinox* and *equator point correction*, respectively. $d_0\alpha$ and $d_0\delta$ could, however, also be regarded as nonconstant systematic errors in the measured right ascensions and declinations of the stars that provide the reference system. These errors can be represented by mathematical models with the appropriate parameters, such as, for instance,

$$d_0\alpha = A + B\sin\alpha + D\cos\alpha$$

and so forth. For the form assumed for $d_0\delta$ by Nyrén, see below in this section. The E_k are the parameters used for the calculation of the predicted α and δ. There are, besides the osculating elements of the Earth's orbit, also other quantities, as for example, the elements and masses of certain other planets, since these also have an influence on the predicted position of the body in question. The v_α and v_δ are the accidental errors in the Sun's (or planet's) measured right ascension and declination. Needless to say, with a series of transit circle observations taken during a single program it would be impossible to solve for any masses. In these considerations, we must, of course, assume that the Sun's (or body's) right ascensions (and possibly even declinations) were measured with respect to an already existing system of reference stars whose systematic errors are to be found.

The equations (2.5) with $d\alpha$ and $d\delta$ given by (2.6), as they originate from every set of observations, are then used (with $v_\alpha = 0 = v_\delta$) as the set of condition (observation) equations for the determination of $d_0\alpha$, $d_0\delta$, $d\epsilon$ and the dE_k.

While the discussion above describes in full generality the principle used in the orientation of the equator system, the practical applications were always based on some specialized case. In particular, some or all of the dE_k were mostly neglected.

Examples are as follows: Newcomb (1906, p. 325 ff.) shows in a very elegant discussion how one may arrive at a practically rigorous $d_0\alpha$ (which he calls E) even if all dE_k and the $d_0\delta$ are neglected. Nyrén's (1876) study is also interesting since he assumed

$$d_0\delta_k = d\phi - d\rho \tan z$$

where $d\phi$ is a correction to the latitude of the observing station, and $d\rho$ one to the refraction constant. z is, of course, the observed zenith distance. Beside $d\phi$ and $d\rho$, $d_0\alpha$ and $d\epsilon$ are the only unknowns in Nyrén's investigation. At observatories at geographic latitudes lower than, say, 40°, the separation of $d\phi$ and $d\rho$ will be practically almost impossible. Brouwer's (1935), Clemence's (1948a), and other similar proposals for the use of Minor Planets for orienting the equator system and finding the systematic errors of the fundamental system with respect to which the Minor Planets' positions are observed, are also special cases of the general case as described in this section.

At this time absolute right ascension differences can be determined with a rms. error in the order of $0.^s01$, and the position of the equinox is probably still uncertain by about $0.^s01$.

No matter how carefully the observations are made, small systematic errors will still remain. The safest way to arrive eventually at an accurate system of absolute positions and, incidentally, proper motions (*fundamental system*), is to make absolute observations on as large a variety of instruments as possible. This requirement is very hard to satisfy around the south celestial pole, since in the Earth's southern hemisphere, the Cape Observatory is practically the only one at which long series of positions have been observed. Fortunately, the situation has recently become better with the establishment of southern meridian stations by several observatories.

2.2.7 THE VISUAL DETERMINATION OF RELATIVE POSITIONS. In this section we shall be concerned only with those positions which were observed within a regular program for the determination of relative positions and their eventual publication in the form of a catalogue. Only if the observations were carried out systematically in large numbers and by uniform methods is there a chance of investigating their systematic errors and reducing them to a well-defined fundamental system. We shall, therefore, not deal here with occasional determinations of the positions of a few scattered objects, either on transit circles or other instruments.

Virtually all visual relative star positions that were obtained during systematic and numerous series of observations were observed on transit circles, as described in 2.2.1. A *relative position*, in contrast to an absolutely fundamentally determined position, is one that *was* essentially *determined with the use of, and in reference to, other already known positions.* These "*reference positions*" could be either absolute, or relative themselves.

Nowadays most relative star positions are found photographically (see Section

2.4.8). Before 1925, however, the photographic techniques had not been sufficiently developed for this purpose, and the vast bulk of star positions were observed by relative observations on transit circles. Although the photographic method is now applied to most tasks which were formerly handled by visual relative determinations of positions on transit circles, a majority of the observations on transit circles, even today, consists of the determination of relative star positions.

The reason why most visually determined positions are relative is that they require much less effort and labor than absolute positions. The principle is very simple: Zenith point, clock correction and clock rate, as well as the set-up errors (azimuth) of the transit circle, are determined from the observations of "fundamental" stars whose positions are regarded as known and usually extracted from a fundamental catalogue. The level of the instrument axis may still be determined by means of a spirit level or mercury basin. The reduction parameters thus determined are then used for the calculation of the relative star positions.

Soon, thereafter, there emerged a procedure for this which was widely followed, namely, the observation of relative positions in *zones*. This means the observation of the program stars (usually all stars brighter than a certain magnitude,) within a narrow strip, bordered by parallels of declination one or a few degrees apart, covering a right ascension range of several hours. Occasionally and regularly at a typical ratio of about 20:1 for programs to fundamental stars, a number of the latter will be observed within a declination interval of, say, 50° or so within the zone, and some close to the pole. In more recent decades, however, rather drastic changes were introduced in this type of work; also, notably in the programs that were carried out for the determination of the positions of the reference stars for the photographic zone catalogues (AGK2A, AGK3R, SRS; see Sections 4.3.2, 4.3.4 and 4.4.1, respectively). The zones vary from 10° to 50° in width, with the fundamental stars extending 5°, at least, beyond the zone limits. Stars at declinations widely outside the zone limits were observed for the determination of Bessel's n, or the azimuth error. The ratio of "program" stars to fundamental stars nowadays is usually about 3:1. The collimation error is always determined by laboratory methods, while the level error may or may not be determined by laboratory methods. The parameters necessary for the reduction of the observations of the program stars or zone stars are then obtained from an analysis of the observations of the fundamental stars whose positions, as pointed out above, are regarded as known and usually extracted from a fundamental catalogue. As the instrument is moved only very little during the observations of the zone stars, one assumes that the variations in the systematic errors which affect the instrument, or which are in the Fundamental Star Catalogue, are distributed in an essentially linear manner. This makes it possible to use a rather simple analysis of the fundamental stars each night for finding the necessary reduction parameters. A simple model often used with narrow zones for the computation of the position coordinates from the readings on the fundamental stars is *Bessel's formula* (2.2b). For the right ascensions, this may be rewritten in the form

$$\alpha_i = T_i + \Delta T + c \sec \delta_i + d \tan \delta_i, \tag{2.2c}$$

where T_i is the clock reading at the meridian transit of the i-th star, c and d are collimation and declination of the instrument axis, respectively. ΔT consists of the clock correction, and h, the excess of the instrument axis' hour angle over 6^h. In this method, clock correction and h cannot be separated, nor is it necessary to separate them. They could only be separated if, in addition to the observations on stars, the level were measured. Detailed discussions are found in special works on meridian observations.

Equation (2.2c), applied to the fundamental stars, is used as the model condition equation for a least squares solution (or some other method of adjustment) in which the α_i and δ_i are assumed known, the T_i are observed and ΔT, c and d are the unknown parameters (c, however, is usually measured directly). Applied to the program stars after ΔT, c and d have been found, it is used to calculate the α_i from the observations T_i and δ_i.

For the calculation of the declinations δ_i one assumes, in the case of narrow zones, that

$$\delta_i = \delta_{i0} + M, \tag{2.7}$$

where δ_{i0} is the circle reading during the meridian passage of the i-th star, and M the so called *equator point correction.* (2.7) is applied—mutatis mutandis—to the declinations completely analogously as (2.2c) is to the right ascensions.

Essential improvements in the technique of relative observations were introduced by F. Küstner around 1900. He almost eliminated magnitude equation (that is, the dependence of ΔT on magnitude) by using a *coarse diffraction screen* in front of the objective to reduce the brightness of bright stars close to that of faint stars.

Another essential improvement was *Repsold's self-registering eyepiece micrometer*, already mentioned above (Section 2.1.6 and 2.2.2.5). During the last decade, the observer's judgment was completely excluded from the observing process by the use of photoelectric and electronic devices (compare Klock [1970], and Laustsen [1967]).

Experience has, nevertheless, shown that in spite of this differential procedure (and further sophistications in the reduction techniques designed to eliminate systematic differences between overlapping zones), systematic differences between the derived field stars positions and those of the fundamental stars still remain. This can only be partly due to physiological differences in the perception of the fundamental (on the average two to three magnitudes brighter) and the field stars during the observations. In part, these systematic differences must be expected as a consequence of the existence of the "parameters variance" that is the second term in equation (1.8c) as discussed in Section 1.6.1. This means that the accidental errors of the statistically determined ΔT, and so forth, are propagated as systematic errors into the derived α_i and δ_i. They will also be due to inadequacies of the equations (2.2c) and (2.7) as error models, and maybe to some other still unsuspected errors. If the sources of these errors could be determined, the errors themselves could probably be eliminated.

For the same reasons, one can also expect systematic differences between the positions of the reference stars that were recomputed with the adjusted ΔT, and so forth, from the observed T_i and δ_{i0} by means of formulas (2.2c) and (2.7), respectively. According to what was said above, error models more adequate than (2.2c) and (2.7) would be the obvious way to eliminate these systematic differences. These, however, almost invariably involve an increase in the number of adjustment parameters and, accordingly, an increase in the computing labor. Today, with computers available, this is a minor consideration. However, it was a major one for the investigators whose only computing aids consisted in logarithms and other tables, and slide rules. (Even desk calculators were not very common before about 1930.)

Among the various ingenious methods devised to eliminate the systematic differences between the coordinates calculated from the observations and the coordinates of the reference stars from the fundamental catalogue, we mention Küstner's *series* (Küstner 1900).

During a so-called series, only fundamental stars are observed, and the systematic differences between calculated and catalogued positions are carefully established so that appropriate corrections can later be applied to the positions of all stars obtained in the conventional way on the instrument in question. Experience has shown that this procedure yields also a more reliable determination of the basic instrument errors than the analysis of fundamental star observations during one zone only. During the execution of a program for the determination of relative positions, Küstner series are observed as often as necessary to keep a check on the errors of the instrument and the behavior of the *instrument system,* in general. Needless to say, the observation of series is not a cure-all for all the problems one encounters in a differential program.

2.2.8 THE ACCIDENTAL ACCURACY OF RELATIVE VISUAL POSITIONS. Theoretical considerations with the purpose of finding and analyzing the accidental—and systematic—errors of relative star positions have little meaning so long as the contributing factors to the total error are not clearly determined. One error source is the parameter variance of the reduction constants.

The technical advances in the manufacture of instruments and the introduction of advanced observing and evaluating methods have brought a decrease of the accidental errors to about one-tenth of their value 200 years ago. Table II-1 illustrates this.

The increase of the mean error of an observed declination with the zenith distance is quite drastic. At the Pulkova vertical circle in 1885, the dependence of the mean error of a typical declination observation on zenith distance is illustrated by Table II-2. This is caused by the worsening of seeing and by increasing irregular, short-term refraction anomalies. This shows that there is little value in making astrometric observation at altitudes lower than 20°.

As will be discussed again in Section 3.2.4, the accidental errors of relative star positions cannot be indefinitely decreased by increasing the number of times a star is observed at a certain instrument. Table II-3 gives the mean errors of relative star positions obtained by certain observers at certain observatories, depending on epoch and on the number of individual observations.

Table II-1

Accidental Errors of Relative Star Positions[6]

		ϵ_α	ϵ_δ	Rem.
Greenwich	1755	0.160	1.92	1
Dorpat	1830	0.10	1.4	2
Berlin	1885	0.033	0.40	3
Washington	1950	0.012	0.20	4

REMARKS:

(1) On Bird's mural quadrant, observed by Bradley, reduced by Auwers. The values in this Table are valid only for $\delta = 0$ (in R.A.) and $z = 0$ (in Declination). For $\delta \neq 0 \neq z$, we have $\epsilon_\alpha^2 = 0.160^2 + (0.083 \sec \delta)$,^{2} and $\epsilon_\delta^2 = 1.92^2 + (0.085 \tan z)$.^{2} δ: declination, z: zenith distance.

(2) Reichenbach meridian circle, observations at equator.

(3) 8-inch transit circle by Pistor and Martin.

(4) 6-inch transit circle.

Table II-2

Mean Errors of Pulkova Declination Observations in Dependence on the Zenith Distance

z	15°	25	35	45	55	65	72.5	77.5
$\mu(\epsilon_\delta)$	0.″26	0.″28	0.″29	0.″33	0.″39	0.″49	0.″59	0.″83

6. The Tables in this section were adapted from Cohn's (1970) article, which contains a rather detailed discussion of the errors of meridian observations and their sources. The R.A. errors are in units of seconds of time; the declination errors in units of seconds of arc.

Table II-3[7]

Comparison of Mean Errors $\epsilon_\alpha \cos\delta$ and ϵ_δ of Positions Which are the Mean of n Observations (*after Cohn*)

	Bradley	*T. Mayer*	*Dorpat*	*Åbo*	*Greenwich*				*Pulkowa*			*Berlin*				*Lick*	*Bonn*	*München*	*Königs-berg*
n	1755	1755	1830	1830	1830	1840/45	1860/64	1880/90	1845	1865	1885	1875	1889 a	1889 b	1905	1895	1900	1900	1900
1	$0^s.16$	$0^s.22$	$0^s.086$	$0^s.060$	$0^s.14$	$0^s.12$	$0^s.070$	$0^s.054$	$0^s.060$	$0^s.041$	$0^s.037$	$0^s.031$	$0^s.026$	$0^s.033$	$0^s.031$	$0^s.049$	$0^s.034$	$0^s.049$	$0^s.019$
2	.12	.19	.070	.046	.12	.09	.049	.037	.043	.029	—	.024	.019	.026	.023	.037	.025	.036	.014
5	.07	.14	.046	.032	.07	.05	.032	.025	.029	.021	—	.016	.014	.019	.016	1024	.017	.023	.010
10	.05	.13	.034	.026	.06	.04	.024	.019	.022	.016	—	—	.012	.016	.012	.017	.013	.017	—
20	.04	.13	.028	.022	.05	.03	.019	.015	.017	.013	.011	—	.011	—	.010	.013	.011	.013	—
1	$1''.3$	$2''.0$	$0''.89$	$0''.89$	$1''.5$	$1''.25$	$0''.89$	$0''.72$	$0''.41$	$0''.45$	$0''.41$	$0''.48$	$0''.30$	$0''.39$	$0''.39$	$0''.48$	$0''.39$	$0''.56$	—
2	0.9	1.5	.72	.72	—	0.89	.63	.56	.32	.33	.32	.35	.23	.31	.30	.35	.31	.39	—
5	0.7	1.3	.51	.48	—	.51	.39	.36	.23	.23	.22	.24	.19	.23	.21	.23	.23	.31	—
10	0.6	0.9	.39	.36	0.48	.39	.31	.28	.20	.19	.18		.17	.19	.17	.19	.19	.26	—
20	0.5	—	.34	.30	0.36	.31	.26	.23	.18	.16	.15		.16		.15	.16	.17	.23	—

7. This table covers only the catalogues of the 18th and 19th centuries but is representative of modern catalogues, also, as far as the increase of the accuracy with the number of individual observations is concerned.

2.3 Photographic Positions

2.3.1 HISTORICAL REMARKS. The first time a photograph of a star was made was almost exactly in the middle of the last century. In 1850, Vega (α Lyrae) was photographed on the 10-inch refractor of the Harvard College Observatory. (This telescope is still in use.)

The photographic technique was applied to astrometric tasks only very hesitantly. Except for investigations of limited regions, notably star clusters (compare, for instance, Rees [1906]), no positions were determined photographically before about 1880.

Around 1880, astrophotography was rather well developed. In particular, the experiments of the Henry brothers at Paris were so encouraging that the Astronomical Community planned the establishment of a catalogue of the positions of all stars down to at least 11-th magnitude with an rms. error of 0.″3 which was, at that time, approximately equal to that of a good meridian observation. This gigantic enterprise, known as the *Astrographic Catalogue* (see Section 5.5) in which almost twenty observatories had participated, was finished in 1963. Voluminous series of photographic position catalogues were and continue to be published by the Royal Observatory, at the Cape, the Yale University Observatory and the Observatories at Hamburg-Bergedorf and at Bonn. These will be described below in Sections 5.1, 5.2, 5.3.

2.3.2 THE PRINCIPLE OF PHOTOGRAPHIC ASTROMETRY. Except under special circumstances, positions obtained from the evaluation of information on photographic plates will allow one *only* the derivation of *relative star positions.* This is so because positions can be derived from measurements on photographic plates only with the use of a set of known (estimates of) positions of stars whose images are among those recorded on the plate.

We assume that we have a plate on which the images of the stars are recorded as reasonably small and approximately circular blackened areas. The rectangular, or other well-defined, coordinates (x, y) of the center of each image are measured. These are called the *plate coordinates* or *measured coordinates.* For every plate there also exist relationships

$$f(x_i, y_i; m_i, c_i; \alpha_i, \delta_i; a_1, \ldots, a_n) = 0$$

$$g(x_i, y_i; m_i, c_i; \alpha_i, \delta_i; a_1, \ldots, a_n) = 0 \tag{2.8}$$

between the measured coordinates x_i, y_i of the image of the ith star, its magnitude m_i and color index c_i, and its coordinates α_i, δ_i, on the one hand and a set $\{a_k\}$ of parameters, on the other hand. If one considers all plates obtained on the same camera (telescope), $\{a_k\}$ will generally consist of some elements that vary from plate to plate, others which vary from plate to plate but are in part constrained, and finally some that are constant for all plates obtained on the same telescope.

Photographic relative positions are now obtained in the following way: Estimates of the spherical coordinates α_i and δ_i of a subset of the stars (*reference stars*) imaged on the plate in question are available from some source, usually from another already-existing catalogue. These can be used in the equations (2.8), which thus become equations of condition for a least squares (or other) adjustment in which the x_i, y_i, α_i and δ_i are regarded as observations, and the $a_1, \ldots, a_n$ as the unknown parameters. Needless to say, there must be enough reference stars to make the system generated in this way at least determinate or, for a true adjustment problem, overdetermined. The m_i and c_i are assumed to be known for every star. Once estimators of the a_k have become known through the adjustment, they may be used in the equations (2.8), which may be solved to give, as functions of the x_i, y_i, m_i, c_i and the a_k, the α_i and δ_i of every star whose image was measured.

Note that this procedure is mathematically analogous to the determination of relative positions on a meridian instrument as described above in Section 2.2.7.

The various methods of photographic astrometry differ mainly in the model assumed for the equations (2.8) and in the method used for the adjustment.

2.3.3 GNOMONIC PROJECTION. In all applications of photographic astrometry that had as their aim the construction of a star catalogue, the geometry of the *gnomonic projection* has been used as the basis of the model for the relationships (2.8). This means that it is assumed that the camera, at which the plate was exposed, has the same projection properties as a *pinhole camera*. The derivation of the formulae (2.8), under these circumstances, is given below. In this case it is expedient to define intermediary variables (ξ, η), so-called *standard coordinates*, which are functions of α and δ and two of the parameters a_k, namely α_0 and δ_0. They are the coordinates at which a plane representing the plate is tangential to the sphere. The point (α_0, δ_0) is called the *tangential point*, and is the point in which a line normal to the plate will intersect the optical center of the objective lens. If the lens-tube-plate assembly is well adjusted, this line will coincide with the camera's optical axis and usually penetrate the plate close to its geometrical center. The stars on the sphere and their images on the plate lie then on straight lines through the center of the sphere, which coincides with the optical center of the objective system.

The *standard coordinates* (or *normal coordinates*) (ξ, η) are defined as follows: The ξ, η system is a rectangular Cartesian system in the plane of the plate.

The positive η axis is the tangent on the image of the meridian through (α_0, δ_0) at (α_0, δ_0) and points toward the image of the north celestial pole.[8] The positive ξ axis which intersects the η axis in (α_0, δ_0) is a tangent to the image of the parallel through the tangential point, and points east. ξ, η with respect to a specified (α_0, δ_0) are usually given in seconds of arc, minutes of arc or radians. The standard coordinates in radians of an object at (α, δ) with respect to the tangential point

8. This definition breaks down if one of the celestial poles is the tangential point. However, standard coordinates can be defined also when one of the celestial poles is chosen as the tangential point, by, for example, requiring the ξ axis to point toward the vernal equinox.

(α_0, δ_0) are given by

$$\xi = \frac{\cos\delta \sin(\alpha - \alpha_0)}{\sin\delta_0 \sin\delta + \cos\delta_0 \cos\delta \cos(\alpha - \alpha_0)};$$

$$\eta = \frac{\cos\delta_0 \sin\delta - \sin\delta_0 \cos\delta \cos(\alpha - \alpha_0)}{\sin\delta_0 \sin\delta + \cos\delta_0 \cos\delta \cos(\alpha - \alpha_0)}. \tag{2.9}$$

The spherical coordinates (α, δ) of a star whose standard coordinates are (ξ, η) with respect to (α_0, δ_0) are obtained by solving (2.9) for $\alpha - \alpha_0$ and δ. One obtains

$$\cot\delta \sin(\alpha - \alpha_0) = \frac{\xi}{\sin\delta_0 + \eta \cos\delta_0}$$

$$\cot\delta \cos(\alpha - \alpha_0) = \frac{\cos\delta_0 - \eta \sin\delta_0}{\sin\delta_0 + \eta \cos\delta_0}. \tag{2.10}$$

Needless to say, the formulas (2.9) and (2.10) are completely rigorous.

2.3.4 RELATIONSHIP BETWEEN STANDARD AND MEASURED COORDINATES. The fundamental problem in connection with setting up equations (2.8) is now the establishment of a relationship between the measured plate coordinates (x, y) of the stars' images and their standard coordinates (ξ, η). If x and y are strictly rectangular coordinates of the stars' images on the plate, the measured rectangular coordinates (x, y) and the standard coordinates (ξ, η) will be connected by a *similarity transformation*:

$$\xi = Ax + By + C$$
$$\eta = -Bx + Ay + D, \tag{2.11}$$

where A, B, C, D are a set of "*plate constants*". In this case, frequently termed the "four constant model," the (ξ, η) must be computed from the observed, that is refracted topocentric apparent positions which describe the directions in which the stars are seen. One cannot use (2.11) if (ξ, η) are computed from mean positions.

Usually, the plates are measured in such a way that the x-axis is almost parallel to the ξ-axis which will (if x and y are strictly rectangular) render the y-axis almost parallel to the η-axis. This will make B very small compared to A. Putting $A = \rho \cos\sigma$, $B = \rho \sin\sigma$, we can rewrite equations (2.11)

$$\xi = \rho(x \cos\sigma + y \sin\sigma) + C$$
$$\eta = \rho(-x \sin\sigma + y \cos\sigma) + D. \tag{2.11a}$$

ρ is called the "*scale factor,*" and σ the "*orientation angle.*" C and D are called the *zero point corrections.*

The parameters (a_k) in this case are A, B, C, D, α_0 and δ_0 (or $\rho, \sigma, C, D, \alpha_0$ and δ_0). Very often, especially when the format of the plate is less than $2° \times 2°$, one may either assume α_0, δ_0 to be known, or determinable by extraneous methods and thus not subjected to an adjustment. For wide fields, however, (say, over $5° \times 5°$) the extraneous methods for the determination of α_0 and δ_0 yield approximate values of these quantities and, thus, they must be corrected differentially. It can be shown (Eichhorn 1970) that the following equations are rigorous to the first order

$$\xi = Ax + By + C + \cos\delta_0\, d\alpha_0 \cdot x^2 + d\delta_0 xy$$
$$\eta = Ay - Bx + D + \cos\delta_0\, d\alpha_0 \cdot xy + d\delta_0 y^2, \tag{2.13}$$

where $d\alpha_0$ and $d\delta_0$ are corrections to the assumed (α_0, δ_0), provided that the plate was so oriented during measurement that the x- and the ξ-axes as well as the y- and the η-axes were nearly parallel, and that (x, ξ) as well as (ξ, η) are approximately expressed in radian units. ξ, η are, of course, referred to the approximate tangential point (α_0, δ_0). If $d\alpha_0$ and $d\delta_0$ turn out to be large, it is recommended that the calculation be repeated with the newly calculated ξ, η, now referred to the tangential point with the coordinates $\alpha_0 + d\alpha_0$, $\delta_0 + d\delta_0$. This may (if necessary) be repeated until further repetitions result in no change in the assumed α_0 and δ_0.

The adjustment itself is usually carried out by minimizing $\sum[(\xi - \xi_c)^2 + (\eta - \eta_c)^2]$, where ξ and η are the standard coordinates computed from (2.11), (2.13) or (2.14), and ξ_c and η_c those computed from the available α and δ of the reference stars. This is not exact from the standpoint of error theory, but the results obtained in this way are practically indistinguishable from those in which the α and δ of the reference stars and the measured x and y are regarded as observations. However, computers are now generally available and one probably ought to abandon approximate methods in favor of more rigorous ones whenever this will result even in an only small increase in accuracy.

A detailed systematic discussion of various aspects of photographic astrometry on the basis of gnomonic projection was given in Zurhellen's Dissertation (1904)—now quite obsolete—and in two articles by König (1933, 1962). There one also finds references to further literature on the subject.

2.3.5 CONSIDERATION OF REFRACTION AND ABERRATION. The requirements that the (ξ, η) must be computed from observed, that is refracted topocentric apparent positions (see Section 2.2.5.2), necessitates the application of corrections for precession, nutation, annual, and, for extreme accuracy, also diurnal aberration, as well as refraction to the mean α and δ of every star before they are used for the calculation of the ξ and η. This is a simple but laborious procedure which can easily be performed with the aid of an electronic computer, provided the data necessary for the calculation of refraction are available, but is prohibitive without one. Various authors (compare König [1933]) have thus constructed formulas to express the influence of the nonorthogonal components of the reduction from mean to observed position (that is, aberration and refraction) in terms of the standard coordinates themselves. (It is evident that the effects which lead to orthogonal transformation, namely, nutation and precession, do not change the form of the basic relationship (2.11)).

The reductions from (ξ_m, η_m), the standard coordinates computed from the mean positions to (ξ_o, η_o), those computed from the observed position, are given in the form

$$\xi_o - \xi_m = \Delta\xi(\xi_o, \eta_o)$$

and

$$\eta_o - \eta_m = \Delta\eta(\xi_o, \eta_o); \tag{2.12}$$

(see, for example, König [1933], pp. 522–533 and p. 540). In practice, the $\Delta\xi$ and $\Delta\eta$ are, at least for reasonably small field not larger than $5° \times 5°$ and not too large zenith distances, given as polynomials of the third order in ξ_o and η_o. Since the equations (2.11) are easily inverted, $\Delta\xi$ and $\Delta\eta$ can be transformed into Δx and Δy (which will also be polynomials of the third order in x and y). These are the corrections to the measured x and y which reduce them to what they would have been if the starlight had come from the mean instead of the observed positions.

Provided the appropriate parameters temperature and air pressure, and so forth, during the exposure are available, the Δx and Δy can be precomputed and applied to the measured x and y, thus rendering equations (2.11) valid between the (corrected) x and y and ξ_m and η_m.

2.3.6 DEVIATIONS FROM GNOMONIC PROJECTION. Even under the most favorable conditions, the assumption of a perfect gnomonic projection can only approximate the actual circumstances. In some rare instances, this approximation is true for all practical purposes; much more often, however, things are not so simple and clear cut.

First, there is no optical system in practical use which unites in a single point all light rays which enter the front component of the objective from the same direction. The cylinder of those rays which enter the front component of the objective from a certain direction is transformed not into a cone, but rather into a ruled surface which narrows to almost a point. We now consider the cylinders of rays representing all directions (within the solid angle of the expected optical performance of the lens), and the points of "minimum diameter" of the associated ruled surfaces. The latter (focal points) will lie on a surface called the *caustic* or the *focal surface*. The projection would be gnomonic if the following three conditions were satisfied: First, the "areas of minimum diameter" must be points; second, the caustic must be a plane; and third, every focal point must be on the principal ray.

None of these conditions are strictly satisfied in any actual optical system. Deviations from the third condition are classified as *distortion*. In the broader sense, this may be applied to all deviations of the projection properties from strict spherical symmetry around the optical axis. Independently of the optical system, *film shifts* during the processing of the photographic emulsion carrier (plate) would produce effects of the same nature. These, however, can be checked by a special technique (réseau) which will be discussed later. (Section 5.5.1)

Even if the caustic were rigorously a plane (which usually it is not), the *photographic emulsion* might *not coincide with the focal surface* at all points—which would, in the ideal case, be the *focal plane*.

The deviations of the images from ideal point shape have been classified into several categories. For the purpose at hand, it is sufficient to state that inside the intersection figure of the photographic emulsion with the ruled surface formed by the rays, the intensity of the light, which is uniform and homogeneous (disregarding for the moment the disturbance it experiences in the atmosphere due to bad seeing) before it meets with the front component of the objective lens, will be not only nonhomogeneous within the intersection figure, but the spectral distribution of the light is different for every point within this intersection figure. Apart from this imperfection of the image caused by the optical system, there is another one peculiar to astronomical photography. *Bad seeing*,[9] as it is known, will "blow up," that is, enlarge the stars' images. Furthermore, the relative position of telescope and sky cannot be perfectly maintained during the sometimes-quite-long exposures due to mechanical and human limits in guiding. This will produce "smeared out" images, which are also caused by the fact that no optical system has a perfectly sharp focus. This, together with the fact that the stars photographed are of different magnitudes and spectra, along with the peculiarities of the process of photographic image formation, may render the positions of the stars' images on the plate (by this we mean the places where they are measured, for example, suitably defined centers of blackening) dependent not only on the geometry of the situation, but also on their magnitudes and spectra expressed in terms of color indices. This approach leads to the establishment of a relationship in the form (2.14) between (ξ, η) and now not only (x, y), but also (m, c).

There are, in principle, two schools of thought as to how to take into account the deviations from gnomonic projection in astrometric work. One could, for instance, carefully investigate the functional form of all possible deviations from gnomonic projection and accordingly modify equation (2.13). The parameters describing these effects would then have to be determined either in the course of the adjustment or by extraneous methods. This is the approach which was traditionally advocated by the *German School* of photographic astrometry, founded essentially by *F. Küstner*. In modern times, much of what has been known to astrometrists for decades, including matter described in König's (1933) article, was independently developed by analytical photogrammetrists (compare Brown [1965, 1968]). The other possibility is not to analyze the individual sources that contribute deviations from the ideal gnomonic projection, but to generalize formula (2.13) by writing

$$\xi = \sum a_{ikln} x^i y^k m^l c^n$$
$$\eta = \sum b_{ikln} x^i y^k m^l c^n. \tag{2.14}$$

It must then be decided empirically which of the a_{ikln} and b_{ikln} are to be included in the model (2.14), either by investigating adjustment residuals produced by less complicated models, or in some other way.

The simplest of the formulas (2.14) is the so-called six-constant model, where one assumes that the affine equations

$$\xi = Ax + By + C; \qquad \eta = Dx + Ey + F \tag{2.14a}$$

adequately describe the relationship between the standard coordinates (computed from mean places) and the measured coordinates.

9. For a discussion of *seeing*, see Protheroe (1961).

This approach was mainly advocated and applied in large scale by *F. Schlesinger*, the founder of the *American School* of photographic astrometry. In what follows we shall call it the *empirical approach*, as compared to the *model approach* which is followed by the German School and the photogrammetrists.

The advantage of the empirical approach is mainly its simplicity. In contrast to (2.12) and (2.13), one may regard (2.14) as a relationship between the measurements and standard coordinates ξ_m, η_m computed from mean positions, since formulas can then incorporate the corrections for aberration and refraction discussed in Section 2.3.4. The parameters in these corrections are in this case not computed from the available physical and geometric data, but are incorporated in the $\{a\}$ and $\{b\}$ and obtained by the adjustment. The increase in the number of adjustment parameters which unavoidably accompanies this procedure requires, however, a corresponding increase in the number of reference stars, if the expectancy of the systematic errors of the resulting ξ and η is to be held at the same level as in a model approach with fewer parameters. This increase of the expectation of the systematic error as the reduction model gets more complicated was discussed by Eichhorn and Williams (1963). If reference positions are plentiful and accurate, one may thus, with advantage, follow the empirical approach which is easier to apply than the model approach.

This is not to say that the model approach will under all circumstances give results with higher systematic accuracy. In a properly carried out empirical approach only those terms will be carried which really contribute significantly to the removal of systematic trends from the residuals. (A criterion for deciding whether or not a certain term should be carried in the model was given in the above-mentioned paper by Eichhorn and Williams [1963].) It may very well happen that a simple empirical model can remove the systematic trends from the residuals just as effectively as a more complicated "rigorous" one. In the latter case the empirical approach would yield the results with the smaller systematic errors. By and large, though, the model approach is to be preferred when the achievement of highest systematic accuracy, without regard to effort, is the ultimate goal.

Under no circumstances should one forget that it is impossible to eliminate completely all systematic errors from any data that depend on parameters, which themselves were the result of an adjustment. The parameter variance—the second term in equation (1.8c)—always provides an estimate for the expectation of the square of the systematic error, no matter whether the model function was established on the basis of a geometrical and physical model, or just as an empirical interpolation function.

2.3.7 THE MODELING OF THE RELATIONSHIP BETWEEN MEASURED AND STANDARD COORDINATES. The first order aberrations of standard optical systems are well known and can be modeled.

First, we assume that the measured coordinates are true rectangular Cartesian coordinates with equal units of length on each axis. The corrections which must be applied to the directly measured coordinates to make them conform to this requirement are *characteristics of the coordinate measuring machine* and may be found by standard methods. For an example of a thorough investigation of a measuring machine see, for instance, Eichhorn (1955).

As pointed out above, the relationship (2.13) would then (to first order accuracy, and under certain assumptions) perfectly describe the geometry of the gnomonic projection, provided the ξ and η are related to observed coordinates.

Astigmatism of the objective is not a *metric distortion* (that is, it effects the shape of the image rather than the location of its center) and thus has no influence on the model.

Radial distortion originates at the intersection point of the optical axis with the focal plane and its origin will, therefore, also be very close to the tangential point. We shall also choose this to be the origin of the x-y system. This aberration is radially symmetrical and usually modeled in the form

$$\Delta x = x \sum_{k=1}^{n} R_k (x^2 + y^2)^k$$

$$\Delta y = y \sum_{k=1}^{n} R_k (x^2 + y^2)^k. \tag{2.15}$$

Unless the radial distortion is very strong, and for the field sizes used in photographic catalogue astrometry, only the first terms of (2.15) are necessary for appropriate modeling of radial distortion, so that it suffices normally to put $n = 1$ and write

$$\Delta x = Rx(x^2 + y^2), \qquad \Delta y = Ry(x^2 + y^2). \tag{2.15a}$$

In the notation of (2.14) we would have $R = a_{3000} = a_{1200} = b_{2100} = b_{0300}$.

Decentering distortion occurs when the components of the objective are not perfectly centered, and is thus in theory avoidable. In practice, however, few objectives are so perfectly aligned that decentering distortion will be completely unnoticeable. This aberration was traditionally ignored by astrometrists; König, for instance, does not mention it in either his 1933 or in his 1962 survey article. From an investigation of the residuals it was, however, quite apparent (Kohlschütter 1957) that the objective used for the Bonn zone of the AGK2 (see Section 5.1.3.3) was decentered, although no account of this fact was taken in the reduction.

The most careful investigation of this aberration was made by Brown (1966), who finds that the appropriate corrections to the measured x and y coordinates are

$$\Delta x = \{2[P_1 x^2 + P_2 xy] + P_1(x^2 + y^2)\} \left[1 + \sum_{k=1}^{n} P_{k+2}(x^2 + y^2)^k\right]$$

$$\Delta y = \{2[P_1 xy + P_2 y^2] + P_2(x^2 + y^2)\} \left[1 + \sum_{k=1}^{n} P_{k+2}(x^2 + y^2)^k\right]. \tag{2.16}$$

Again, under the conditions imposed by catalogue photographic astrometry, the P_{k+2} may usually all be assumed to equal zero. Furthermore, one can see by comparison with equation (2.13) that the terms in the first bracket (which are multiplied by 2) are of exactly the same structure as those which model the correction to the tangential point. If, therefore, the model includes the provisions for correcting the tangential point anyway, the above decentering terms may be combined with them. Thus, one needs only to add the terms

$$\Delta x = P_1(x^2 + y^2); \qquad \Delta y = P_2(x^2 + y^2) \tag{2.16a}$$

to the model to account for the presence of the effects of decentering distortion.

It is worth noting that the coefficients which describe the distortions are very insensitive to changes in the ambient temperature, and may thus be regarded as more or less constant for any one objective.

This statement is not true for *coma*, which is, in effect, a dependence of the focal length (scale) on magnitude. It is radial and accounted for by the terms

$$\Delta x = Smx; \qquad \Delta y = Smy. \tag{2.17}$$

Experience shows that the coefficient S is rather sensitive to the rate of change of the ambient temperature during the exposure. Thus it cannot be regarded as a constant for any one objective, although the values obtained in practice will have a tendency to cluster around a central value. Eichhorn and Gatewood (1967) showed how this type of behavior can be properly taken into account.

The *standard model* for the conversion of measured to standard coordinates which considers corrections to the position of the tangential point, radial and decentering distortion, and coma, is then from (2.14), (2.15a), (2.16a) and (2.17)

$$\begin{aligned}\xi &= Ax + By + C \qquad + px^2 + qxy + P_1(x^2 + y^2) \\ &\qquad\qquad + Rx(x^2 + y^2) + Sxm + Tm \\ \eta &= Ay - Bx \qquad + D + pxy + qy^2 \qquad + P_2(x^2 + y^2) \\ &\qquad\qquad + Ry(x^2 + y^2) + Sym \qquad + Um.\end{aligned} \tag{2.18}$$

This must not be regarded as a "minimum" model, nor will the projection properties of the lens always be adequately modeled with only the terms in (2.18). We shall later, on occasion of the discussion of the photographic catalogues, discuss the models used in the various investigations in more detail.

2.3.8 VARIATIONS OF THE ADJUSTMENT TECHNIQUES. It was stated above (Section 2.3.3) that the principle according to which the adjustment of the observations is performed, is to require $\sum [(\xi - \xi_c)^2 + (\eta - \eta_c)^2]$ to be a minimum. Also, we have so far considered only the finding of the parameters $\{a_k\}$ in (2.8) by means of comparing predicted and observed positions (or standard coordinates) of reference stars. This *classical method* was developed mainly with a view on economy in the numerical calculations. The emergence of computers has, however, made this a secondary consideration.

2.3.8.1 FINDING MAGNITUDE DEPENDENT TERMS BY MEANS OF THE OBJECTIVE GRATING TECHNIQUE. Some of the *most dangerous effects* in photographic astrometry are *magnitude dependent terms* in (2.8). The reason for this is that the average magnitude of the reference stars is typically about two to three magnitudes brighter than that of the field stars. The magnitude dependent terms in (2.8) will affect the field stars as well as the reference stars. Their coefficients, however, are found by analyzing material derived from stars which cover only part of the magnitude range to which they are later going to be applied. The effects of the magnitude terms are thus "extrapolated" to the fainter field stars, which is clearly an undesirable situation.

This can be rectified by use of a *coarse objective grating*, an ingenious device first employed (for other purposes) by E. Hertzsprung. The grating consists of a series of equally spaced bars of uniform thickness which are placed in front of the objective. The telescope then generates, beside the central image of every star, a sequence of diffraction spectra, symmetrical to and in line with the central image. Through the use of appropriate combinations of plates and filters, the sensitivity of the photographic emulsion can be restricted to a few hundred Angstroms, so that at least the first order and sometimes even second and third order diffraction spectra have a starlike appearance. Since the components of a pair of diffraction spectra are always located symmetrically with respect to the central image, the average of their coordinates is very nearly equal to the coordinates of the corresponding central image. The blackening produced by a diffraction image of a star of magnitude m is that which would have been produced by the central image of a star of magnitude $m + \Delta m$. The quantity Δm is called the *grating constant*. Ideally, it is invariable over the plate and the same for stars of all magnitudes. It can be either determined empirically or computed from the dimensions of the components of the grating (Bucerius 1932).

The comparison of measurements of central image and diffraction image positions can now be used in the following way: Let the measured positions x (the y's are quite analogous) of an image of magnitude m be given by

$$x = x_0 + \Delta x(x, y, m; a_1, \ldots, a_n), \tag{2.19}$$

where x_0 is what x would be for magnitude zero, and Δx its correction to magnitude m. All the parameters $a_1, \ldots, a_n$ relate to the magnitude dependence of x. If the subscripts u and l refer to the two images (upper and lower, respectively) of a pair of diffraction spectra and c to the corresponding central image, we have, of course, apart from the accidental errors of the measured x,

$$\tfrac{1}{2}(x_{0u} + x_{0l}) = x_{0c},$$

and therefore, from (2.19),

$$x_c = \tfrac{1}{2}(x_u + x_l) = \Delta x(x, y, m) - \Delta x(x, y, m + \Delta m), \tag{2.20}$$

since the left-hand side of (2.20) is directly available from the observations. These equations will be available for every star that produces a central image and a pair of diffraction spectra, and can now be regarded as equations of condition for obtaining the parameters $\{a_k\}$. Thus, the magnitude dependence of the ξ (and, of course, η)

can be established without any recourse to reference stars, and from investigating positions of stars that extend over the entire magnitude range covered by the field stars.

A further discussion of the grating technique, especially concerning the problems connected with it, was given by Eichhorn (1963 and 1964).

2.3.8.2 THE PLATE OVERLAP TECHNIQUE. Consider the patterns in which plates are usually taken that are to serve for the contruction of a star catalogue. They will almost always *overlap* either in a way that the edge of one plate lies in (or near) the center line of the neighboring one (*edge-in-center-line pattern*), or such that the corner of one plate is at (or near) the center of a neighboring one (*corner-in-center pattern*). In both cases, every star will thus ideally be imaged on two plates.

If the reductions of the plates are performed by the classical method, the values for the positions of the same star obtained from different plates will differ from each other: first, because of the accidental errors inherent in the location and the measurement of the images and, second, because of the parameter variances of the individual plates. Traditionally, the accepted definitive position of a star is the mean of all the individual ones.

The effects of the parameter variances connected with the individual plates will produce systematic differences between the positions of the same stars as derived in an overlap region from the different plates. Since a star can have only one correct position at any one time, this means that the "best" positions derived from the adjustment on the individual plates are no longer regarded as the best ones if they are, in the end, overridden by their mean. One must, therefore, look for a different adjustment principle which will not only produce plate constants that leave almost no systematic differences between the positions of the same stars derived from overlapping plates, but which will also directly produce unique results for the positions.

These aims are achieved by the plate overlap technique, first proposed by Eichhorn (1960). Lacroute (1964) suggested a variant of Eichhorn's technique, and several papers and applications have appeared since.

The principle of the plate overlap method is this. Under the assumption that a model equation of the type (2.8) describes the relationship which exists between the spherical coordinates (α, δ) of the stars, the measurements (x, y) and the plate parameters, the adjustment is performed regarding not only the sets of parameters $\{a_k\}_j$ on all of the plates P_j as adjustment parameters, but also the coordinates $\{\alpha, \delta\}_i$ of every star S_i. The equations (2.8) serve now as before as the set of condition equations. Now, however, images of the star S_i will occur on more than one plate so that the systems of equations, pertaining to different plates, are no longer independent. In addition, condition equations of the form $\alpha_i = \alpha_{ic}$; $\delta_i = \delta_{ic}$ are considered for the reference stars, where α_{ic} and δ_{ic} from some catalogue are regarded as observations. For details see Googe, Eichhorn, and Lukac (1970).

The whole set of condition equations leads to one huge system of normal equations in all unknowns: the spherical coordinates of all stars and the parameters for all plates. Fortunately, the *matrix* of the normal equation system is of the "*banded-bordered*" type which facilitates the solution of the system. For details see, for example, the paper by Brown (1967).

2.3.9 THE PHOTOGRAPHIC CAMERAS (TELESCOPES). For making photographs of small fields (say, up to 2° × 2°), relatively simple telescopes with two component objectives are used. In the earlier days, when the art of sensitizing photographic emulsions to almost any wave-length interval was not yet perfected, telescopes which were intended to serve mainly as astrocameras were provided with *photographically corrected* objective lens systems.[10]

A telescope which is mainly used for stellar photography is called an *astrograph*. The use of photographic objectives, as they were called, became unnecessary when plates became available which could be used with standard *visual objectives* in conjunction with *yellow filters*. By this technique, almost any astronomical telescope becomes suitable as an astrocamera. It has been shown (see van de Kamp [1967]) that the use of the visual region of the spectrum for photographic purposes minimizes the bad effects of *atmospheric dispersion*, while the exposure times required to record the same stars are on the average about four times longer than when photographically corrected lenses are used. A fuller discussion of the relative merits of photographically vs. visually corrected two-component objectives is found on pp. 85–97 of van de Kamp's book.

Occasionally, astrometric photographs were secured on large *parabolic reflectors* (in particular, with the 60-inch reflector of the Mount Wilson Observatory, by the late Adriaan van Maanen). Such instruments have the advantage of the *complete absence of chromatic aberration*. This advantage, however, is more than compensated by the presence of other aberrations, particularly coma. These aberrations are very detrimental to the potential accuracy of the positions derived from the resulting photograph if their effects are not carefully considered. Unfortunately, none of the investigators who worked in the past with plates taken on large parabolic reflectors paid much attention to coma and similar effects. Potentially, the giant reflectors are still very valuable for special photographic astrometric tasks involving very

---

10. It is well known that the so-called *chromatic aberration*, that is, the dependence of the effective focal length of an objective on the wave length of the light, can be controlled to a certain extent by using for the objective lens system components ground from glasses with different dispersions. When only a single lens is used, the color curve (that is, the curve describing the dependence of the effective focal length on the wave length of the incoming light) has a fairly constant slope. When two components form the objective lens system, the color curve can be made to have a more or less flat minimum in the neighborhood of a given wave length. If the sensitivity of the photographic emulsion is restricted to the neighborhood of this wave length, chromatic aberration will have a minimum effect on the definition of the photograph.

Telescopes with the minimum focal length in the blue region of the spectrum are called *photographically corrected telescopes*. Ordinary or visual two-component lens telescopes have their minimum focal length in the yellow region of the spectrum.

faint stars because of their tremendous light-gathering power, and the fact that the entire spectrum can be used for the formation of the photographic image. A 61-inch reflector especially designed for astrometric work has been constructed for the U. S. Naval Observatory at Flagstaff (Strand 1960), and gives superb results.

Two-component objective lens systems become unsuitable for making photographs of fields larger than 2° × 2°. In the language of the photographic astrometrists, cameras designed to photograph fields from, say 4° × 4° on up, are called *wide angle cameras.* A field of 20° × 20° would be considered an extremely wide angle field. Usually, the objective lens of a wide angle camera has four components. This type of telescope is at the present time used for almost all photographic catalogue work, nearly always with a focal length of about 2000 mm which gives a scale of 100″ to 1 mm on the plate.

Typical exposure times for photographic position catalogue work are in the neighborhood of five minutes. At the present time, no sidereal drive is reliable enough to keep the telescope pointed with sufficiency accuracy. Therefore, photographic telescopes are frequently rigidly mounted together with and parallel to a visual telescope, which is set to a "*guiding star.*" Very delicate manual corrections to the pointing of the telescope, which are sometimes effected by small electric motors, compensate for inaccuracies of the sidereal drive and keep the guiding star on the crosswires during the exposure. Although the two-telescope principle (separate camera and guiding telescope) has been used in the design of giant telescopes (for instance, the 80 cm photographic–50 cm visual Steinheil double refractor of the Potsdam Astrophysical Observatory in Germany), it is seldom used on instruments with focal lengths over five meters. Because of *differential flexure*, the optical axes of the two systems do not retain their relative orientations during the exposure, especially if the photograph is made at a large hour angle. This effect must be experimentally investigated and corrected for, see, for instance, Küstner's (1920) investigation of this effect for the five-meter focal-length astrograph of the Bonn Observatory. In addition, it must be noted that a long telescope tube with a considerable moment of inertia will not readily follow the delicate manual guiding corrections to the sidereal drive. For longer focal lengths, say, over five meters, the telescope is usually moved by the sidereal drive alone, and guiding is done by small shifts of the plate holder with respect to the main tube. In this one-tube design, a guiding star is usually chosen at the edge of the field and viewed through an adjustable eyepiece which is rigidly attached to the plate holder. In this way, there cannot be any effect of differential flexure because the photograph originates from the same objective lens system which forms the image of the guiding star. As to how frequently it is necessary to manually override the sidereal drive to guide well, the author can quote the following example from his own experience. Working on a normal astrograph (focal length 3400 mm), it was possible to leave the telescope for two to three minutes without running into much danger of the guiding star "slipping" from the crosshairs; on the other hand, a 9500 mm focal length refractor could not safely be left unguided for as little as twenty seconds.

Recently, *automatic guiding devices* have been constructed and put into use at some observatories. The guiding may, for example, be accomplished by putting a four-sided light sensitive prism into the beam of the guiding star which activates minute correcting motors whenever the balance of the light on the four sides of the prism becomes disturbed by the unavoidable inaccuracies of the telescope drive. These devices are, of course, custom made and therefore their cost is still prohibitive for a small observatory. Furthermore, a skilled observer can probably achieve just as good images as one obtains by automatic guiding. The increase in positional accuracy by their use is possibly significant but has not yet been established, and, therefore, they must at this time be regarded as conveniences for the observer rather than necessities.

2.3.10 THE ACCURACY OF PHOTOGRAPHIC ASTROMETRIC POSITIONS. The accuracy of photographic positions depends on the *measuring accuracy*, the *intrinsic accuracy of the plate* (that is, the accuracy with which the measured relative positions of the images correspond to the actual relative positions in the sky), the *accuracy with which the system of the reference stars can be realized* and, finally, the *focal length* of the telescope-camera.

Measuring machines have improved considerably during the last half century. Some of the early photographic astrometric work was done with measuring machines that were rather crude. They provided for a least count of as much as 5 microns and thus produced an average measuring error of two to three microns.

Most modern measuring machines of standard type have two long lead screws, are *digitized* (that is, are provided with an apparatus for automatic reading and recording of the measurements) and give a least count of one micron. The accuracy with which a good measurer can repeat settings on the same image depends considerably on the quality of that image, and runs from a dispersion of about one micron (and, in exceptional cases, even less) for very good, well-defined circular images and excellent measurers, to about three microns for fair images, with about 1.5 microns as the average. Note that the sharp images usually obtained on short focus (below 3.5 meters) telescopes can be measured more accurately than the typically larger and less regularly-shaped ones obtained on long focus instrument (that is, those with focal lengths of 7 meters or more). *Modern semiautomatic and fully automatic measuring machines* which do not rely on the human eye and human judgment for determining what constitutes a "correct" setting, will show a dispersion of the differences made on the same image of considerably less than one micron and typically about one-half micron.

The intrinsic accuracy depends, as does the measuring accuracy, to some extent on the *quality of the images* which, in turn, is a function of the quality of the camera's optical system, the focal length and seeing. There is some evidence that film shifts which occur during processing (and possibly even during a storage period of several years or decades) are of such an order that they correspond to a dispersion of about one to two microns. With the influence of image quality taken into account, the intrinsic accuracy of the positions rendered by photographic plates corresponds to an error dispersion of from 2 to 4 microns, with the typical value being about 2.7 microns.

The accuracy with which the system of the reference stars can be realized depends on the model for the condition equations, the number of reference stars and, the accuracy of the reference star positions themselves. Strong irregularities in the distribution of the reference stars on the plate decrease the accuracy of the reference frame.

The dependence of the systematic errors on the reduction model was investigated by Eichhorn and Williams (1963). Generally, the uncertainty of the system realized on the plate, measured by the *parameter* (that is, *plate constant*) *variance*, increases toward the edges and especially toward corners of the plate and also with the complexity of the condition equation model. If the latter includes magnitude as a variable, the parameter variance will also depend on the star's magnitude.

When a simple, four-constant linear reduction model—equation (2.11)—is used, only about four reference stars are required to achieve a systematic accuracy corresponding to the dispersion of the reference star adjustment residuals (averaged over the plate). This number rises to about 12 when quadratic and distortion terms are used in the adjustment model.

The focal length determines the scale of the photograph, that is, the conversion factor from linear to angular measure. If a distance l in the focal plane (at the tangential point) of a telescope with focal length f (in the same units as l) subtends an angle λ measured in seconds of arc, we have $\lambda = 206265l/f$.

Also, as mentioned above, the focal length influences the accuracy of the plate measurements because there is, by and large, an unavoidable deterioration of image quality with increasing focal length.

2.4 Data Derived From Observed Positions

2.4.1 RIGOROUS DEFINITION OF PROPER MOTION. Proper motions were defined above in Section 1.6.2 as essentially those changes in the positions of the stars caused by their motions with respect to a certain inertial system. The origin of this inertial system is at the barycenter of the solar system. One can also consider the motions of the stars in a system with respect to which the average velocity components of the stars in the solar neighborhood are equal to zero. Chandrasekhar (1942) has pointed out (p. 5) that this so-called *local standard of rest* is by no means rigorously defined. If we use it, the stars change their heliocentric positions and thus show a proper motion, primarily for two reasons: first, the motion of the Sun, which is responsible for that proper motion component which is called *parallactic motion*, and second, the motion of the star itself, which gives rise to the so-called *peculiar motion*. Clearly, the parallactic motion is systematic and depends on the positions of the stars as well as their distance, and can, therefore, be used for the determination of the *secular parallaxes*. Details can be found in any treatise on stellar kinematics, for example, in chapter 5 of the book by Mihalas and Routly (1968).

The peculiar motions, in particular their dispersions, can (together with the dispersion of the radial velocities of the same group of stars) be used for the determination of *statistical parallaxes.*

An important feature of the definition of proper motion given at the beginning of this section is that it is *essentially* (but not exactly) the position change due to the stars' motion with respect to the Sun. The reason for the use of the qualifying "essentially" is, as already pointed out in Section 1.6.2, that stars change their positions not only because of their own motions, but also because of the motion of the coordinate system.

As discussed previously in Section 1.6.3, the changes of the orientation of the equator system with respect to an inertial system are caused by *precession* and *nutation*. The formulas for the calculation of their effects, that is, the matrices **S** and **N** discussed in Section 1.6.5, were established from mechanical theory in which some parameters, especially the constant of nutation and the constant of precession, must be found from appropriate observations (see Sections 2.4.3 and 2.4.4 below).

The analysis of the observations will therefore produce only estimates and not the true values of these parameters, so that these estimators obtained through the adjustment will not completely describe the rotation of the equator system with respect to an inertial one.

Nutation refers to a *periodic* change of the coordinate system's orientation only, and the effect of an inaccurate value of the nutation constant on the proper motions is thus not cumulative, as is that of an inaccurate constant of precession. Proper motion is calculated from the rate at which the mean position of a star changes with respect to an (ideally) nonmoving reference system. Before observations can be used for the calculation of proper motions, they must be reduced to the same coordinate system.

Since, however, the computation of the mean position for equator and equinox at epoch T_0 from the position observed at epoch t involves the precession matrix $\mathbf{S}(t, T_0)$, it is clear that different values of the constant of precession will also lead to different values of the proper motion from the same original observations. Strictly speaking, the precise definition of a proper motion requires that a value of the constant of precession be given, by means of which the relevant observations are reduced from the coordinate system at the epoch of observation to a common coordinate system.

Since in first approximation the influence of precession on equatorial coordinates is given by

$$\begin{aligned} \Delta\alpha &= \Delta t(m + n \sin\alpha \tan\delta) \\ \Delta\delta &= \Delta t\, n \cos\alpha, \end{aligned} \tag{2.21}$$

where m and n are general precession in right ascension and declination, respectively, the systematic differences between the proper motions μ_A and μ_B, referred to precession constants A and B, respectively, are given by

$$\mu_{\alpha A} - \mu_{\alpha B} = (m_A - m_B) + (n_A - n_B) \sin\alpha \tan\delta$$
$$\mu_{\delta A} - \mu_{\delta B} = (n_A - n_B) \cos\alpha, \qquad (2.22)$$

where m_A, n_A and m_B, n_B are the precession parameters derived from the precession constants A and B, respectively.

The second reason why the definition of the proper motions given at the beginning of this section is not exact, is as follows:

The observations from which the proper motions are derived are unavoidably affected by *systematic errors*, so that the proper motions themselves may also contain systematic errors. As we shall see below, a fundamental catalogue defines a system of not only positions but also proper motions. Any such system can, at best, be only an approximation to the true system. In any case, systematic errors in the positions will likely cause systematic errors in the proper motions (unless the positions of all catalogues are affected by the same errors); thus, even when referred to the same constant of precession, they will be uniquely defined only if the system is known in which they have been determined. The proper motions are reduced from one system to the other merely by applying the systematic differences.

The reduction of proper motions μ from the epoch t to the proper motions μ_1 at the epoch t_1 may either be done approximately by means of the formulas (1.11c), since $\partial^2\alpha/\partial t^2 = \partial\mu_\alpha/\partial t$, and $d\mu = (\partial\mu/\partial t)\, dt$. The rigorous formulas were given in a paper by Eichhorn and Rust (1970) and are

$$\mu_{\alpha 1} = \frac{\mu_\alpha}{A^2 + \mu_\alpha{}^2\tau^2}$$

$$\mu_{\delta 1} = \frac{\mu_\delta\left(1 + \frac{\dot{r}}{r}\tau\right) - \left(n^2 - \frac{\dot{r}^2}{r^2}\right)\tau\tan\delta}{\left(1 + 2\frac{\dot{r}}{r}\tau + n^2\tau^2\right)(A + \mu_\alpha{}^2\tau^2)^{1/2}}, \qquad (2.23)$$

where $\tau = t_1 - t$, $A = 1 + (\dot{r}/r - \mu_\delta \tan\delta)\tau$, $\dot{r}$ is the radial velocity and r the distance at t; and $n^2 = \dot{r}^2/r^2 + \mu_\alpha{}^2 \cos^2\delta + \mu_\delta{}^2$. The value for $\dot{r}/r$ at t_1 is given by

$$\frac{\dot{r}_1}{r_1} = \frac{\frac{\dot{r}}{r} + n^2\tau}{1 + 2\frac{\dot{r}}{r}\tau + n^2\tau^2}. \qquad (2.24)$$

In these formulas, all proper motions are understood to be in radians, $\dot{r}$ and r must refer to the same unit of length (for instance, parsecs/yr and parsecs, or kilometers per second and kilometers) and the time unit to which the radial velocity is referred must be the same as that in which τ is counted.

2.4.2 THE DETERMINATION OF PROPER MOTIONS. When achievement of the highest accuracy is not necessary, it is often sufficient to obtain the proper motion μ_α from two positions α_1 and α_2 (the declinations are treated completely analogously) at two epochs t_1 and t_2, respectively, by

$$\mu_\alpha = \frac{\alpha_2 - \alpha_1}{t_2 - t_1}. \tag{2.25}$$

The rms. error ϵ_μ of μ_α, in terms of the errors ϵ_1 and ϵ_2 of α_1 and α_2, respectively, is given by $\epsilon_\mu = (\epsilon_1^2 + \epsilon_2^2)^{1/2}/| t_2 - t_1 |$. Obviously, α_1 and α_2 must be on the same system.

We have seen above that proper motions refer strictly to only one epoch, although their change with epoch is small. If we want to take this into account, we would, instead of using equation (2.25), have to solve equations (1.9a), after $(\dot{x}, \dot{y}, \dot{z})^T$ has been replaced by the expression in (1.10), for μ_α (that is $\dot{\alpha}$), μ_δ and $\dot{r}$, from the known α_0, δ_0, r_0 and α, δ, r. Normally, r_0 and r will not be available. In this case there is no harm in arbitrarily putting them equal to unity; as a matter of fact, it is the only thing one can then do. Equations (1.9a) are linear in μ_α, μ_δ and $\dot{r}$, which greatly facilitates their solution. One could also solve the first two of the equations (1.12) simultaneously for μ_α and μ_δ, where we assume α, α_0, δ, δ_0, r_0 and $\dot{r}_0$ to be known. This cannot be done in closed form. However, a very simple iteration scheme is to proceed as follows: Suppose we aim at finding $\mu_\alpha(t_1)$. Calculate a first approximation $\mu_\alpha^{(1)}$ of μ_α from equation (2.25). From here then, calculate, from equations (1.11d) or (1.12) what α_2 would be if $\mu^{(1)}$ were the proper motion at the epoch t_1. Call this quantity $\alpha_2^{(1)}$. Notice that its computation requires also that we know μ_δ, and if we use (1.12), $\dot{r}/r$. Now we form $\mathrm{d}^{(1)}\alpha_2 = \alpha_2^{(1)} - \alpha_2$, and compute a second approximation $\mu_\alpha^{(2)} = (\alpha_2 - \mathrm{d}^{(1)}\alpha_2 - \alpha_1)/(t_2 - t_1)$. Next we calculate $\alpha_2^{(2)}$ from (1.11d) or (1.12) with $\mu_\alpha^{(2)}$, and so forth. This should now be very close to α_2. If it is not, we continue the iteration until further iterations no longer change the values thus obtained.

If more than two positions are involved in the determination of a proper motion, one has an adjustment problem. Every position $\alpha(t_i)$ generates an equation of condition of the form, previously used as equation (1.11),

$$\alpha(t_i) = \alpha(t_0) + \mu_\alpha(t_i - t_0), \tag{2.26}$$

where $\alpha(t_0)$ and μ_α are the unknowns. In this case, just as in that involving two positions only, μ is a function of the epoch. This can be considered in the same way as above by using an appropriate value of μ for the computation of the just described corrections $\mathrm{d}\alpha$, so that the equations of condition become

$$\alpha(t_i) - \mathrm{d}\alpha = \alpha(t_0) + \mu_\alpha(t_0)(t_i - t_0). \tag{2.26a}$$

Iterations, if they prove necessary, can easily be performed in the same manner as above. Iterative methods are very well suited for electronic computers. It hardly needs to be mentioned that the observed $\alpha(t_i)$ must have been reduced to the same system before they are used in the equations of condition.

Proper motions obtained in this way are on the same system (except for the often-mentioned unavoidable deterioration of systematic accuracy in an adjustment process) as the $\alpha(t_i)$.

A much simpler way of obtaining proper motions, especially of stars fainter than 9^m and within relatively small areas smaller than $5° \times 5°$ is the *photographic method*. This works as follows:

The region in which proper motions are to be determined is photographed twice, with as large as possible an epoch difference, and under circumstances that are as similar as possible, that is, preferably with the same telescope, the same tangential point, the same brand of plates. In such case, one can assume that the image positions of the same star on the two plates differ because of proper motion, and because the two plates were differently oriented and have different scales. In the simplest case, every star will then generate a set of condition equations following the model

$$\epsilon_x + \mu_x + \Delta x = Ax + By + C$$
$$\epsilon_y + \mu_y + \Delta y = Ay - Bx + D, \qquad (2.27)$$

where Δx and Δy are $x(t_2) - x(t_1)$ and $y(t_2) - y(t_1)$, the differences between the measured image coordinates on the plates exposed at epochs t_2 and t_1, respectively. The ϵ_x and ϵ_y are the errors in the Δx and Δy, respectively. These may be regarded as accidental.

The normal equations which follow from (2.27) can be set up in several ways. One may (as is most commonly done in this procedure) regard the proper motion components μ_x and μ_y themselves as random, and use the equations originating from all stars as condition equations. This, of course, deliberately disregards the fact that proper motions have systematic components which originate from the motion of the Sun and from differential galactic rotation. Besides, the dispersion of the proper motions depends on the stars' magnitude and spectral type as well as on their galactic latitude, and on the position angle of the proper motion itself. All this might be taken into account, but seldom is, if ever. This is so because proper motions obtained by regarding them as random are only relative but, nevertheless, serve as well as absolute ones for certain purposes, such as the establishment of a criterion for cluster membership.

Absolute proper motions, that is proper motions in a well-defined system, may be obtained by taking the same model (2.27) for condition equations but setting them up only for "reference stars" whose proper motions in a certain system are known from previous investigations. This procedure was followed by Osvalds (1953) in his determination of proper motions in the region of the Hyades.

A method for the estimation of the systematic components of proper motions was also given by Clube (1966).

It has become quite apparent that the model (2.27) is generally inadequate, even if the plates at the two epochs are obtained under practically the same conditions. Telescopes do not retain their imaging properties over decades, so that one will have to replace the right-hand sides of equations (2.27) by polynomials in x and y, and sometimes even the magnitude, m, and the color equivalent, c.

The popularity of the *differential photographic method* as described above, with investigators, derives probably from the fact that the measurements and calculations can be carried out easily. The plates at the second epoch were, for example, sometimes exposed through the glass, so that the plate pairs could be clamped together film to film with the images only slightly displaced, and the image coordinate differences were then measured directly instead of the image coordinates.

There is no doubt that the differential method has certain advantages, especially when only few stars per plate are observed. The success of the many series of photographic parallax determinations on long focus telescopes testifies to this. However, in this author's opinion, the advantages of the differential method over the direct one become insignificant when many stars on a plate pair are involved, as is usually the case when photographic methods are used for the determination of proper motions for stars in a catalogue. In such a case, it is just as well to use standard methods to derive positions at the various epochs from the different plates, and then to employ these for the determination of proper motions as described above in this section. It is, incidentally, the only possibility if one has to use plates that were taken on different telescopes. An example for carefully carrying out such an investigation by means of the plate overlap method was given by Sofia, Eichhorn, and Gatewood (1969) in their determination of the proper motion of the X-ray source Scorpio X-1.

2.4.3 THE DETERMINATION OF THE CONSTANT OF PRECESSION.

2.4.3.1 THE ORIENTATION OF ECLIPTIC AND EQUATOR IN AN INERTIAL SYSTEM. In Sections 1.6.3 and 1.6.5 above, we have considered the change of star positions due to the motion of the frame of reference (coordinate system) with respect to an inertial system. In this section we shall show how the precession matrix $\mathbf{S}(T, T_0)$ in equation (1.13a) may be established from a combination of theoretical investigations and certain measurements. In connection with *precession*, the term *lunisolar* refers to phenomena produced by the motion of the equator only. The term *planetary* is used for phenomena produced by the (very slow) motion of the ecliptic only. The term *general* is used for phenomena produced by the simultaneous motions of ecliptic and equator.

Consider the system E_0 of the ecliptic (see Section 1.3.3) as oriented at a certain epoch T_0 (say, 1850 or 1900). Ideally, this *system of the fixed ecliptic* would be inertial. As mentioned in Section 1.3.3, the ecliptic constantly changes its orientation with respect to an inertial system as a consequence of the (secular) perturbations which the planets exert on the motion of the Earth.

Let the orientation of the ecliptic at T (which is also called the *ecliptic of date* or the *movable ecliptic*) be expressed in terms of the longitude Π_1 of its ascending node on the *fixed ecliptic* (that is, that at T_0) and the inclination π_1 of the movable with respect to the fixed ecliptic. Note that Π_1 and π_1 give the same information as the angles ϵ and Ω of Section 1.3.3. Since generally the normal unit vector of a plane with the Eulerian angles ϵ and Ω is given by $\hat{\mathfrak{x}}(270° + \Omega, \epsilon)$, where $\hat{\mathfrak{x}}(l, c)$ is defined as in equation (1.1), the normal unit vector $\boldsymbol{n}_e$ of the movable ecliptic in E_0 is given by

$$\boldsymbol{n}_e = \hat{\mathfrak{x}}(270° + \Pi_1, \pi_1) = \begin{pmatrix} p_1 t + p_2 t^2 + \cdots \\ -\, q_1 t - q_2 t^2 + \cdots \\ 1 \end{pmatrix}, \tag{2.28}$$

where $t = T - T_0$ and the numerical values of $p_1, p_2, \ldots, q_1, q_2, \ldots$ may be computed from planetary theory. The coefficient with the largest absolute value, q_1, is about $-47''$ per Julian century. The exact values of the p's and q's depend therefore on the various parameters of the planetary system, that is the elements of the planets' orbits and, in particular, the masses of the planets.

The longitude in E_0 of the ascending node of the moving mean equator (that is, the *mean equator of date*, or the mean equator at T), and its mean inclination to the fixed ecliptic are given by $180° - \Psi_1$ and Θ_1, respectively. Ψ_1 is called the *lunisolar precession in longitude*, Θ_1 has no commonly used name. The normal unit vector $\boldsymbol{n}_q$ of the moving mean equator in E_0 is given, according to what was just said, by

$$\boldsymbol{n}_q = \hat{\mathfrak{x}}(90° - \Psi_1, \Theta_1). \tag{2.29}$$

The dynamical theory of the rotation of the Earth develops Ψ_1 and Θ_1 in the form

$$\Psi_1 = f_1 t + f_2 t^2 + \cdots$$

$$\Theta_1 = \epsilon^0 + \theta_1 t + \theta_2 t^2 + \cdots. \tag{2.30}$$

The numerical values of $f_2, f_3, \ldots$ and $\theta_1, \theta_2, \ldots$ are calculated from this theory which involves some of the parameters of the planetary system, primarily distances and masses. These quantities are also, at most, a few seconds of arc per century, and it so happens that $\theta_1 = 0$, that is, that there is no first order rotation of the movable mean equator on the x-axis of E_0.

The quantities f_1 and ϵ^0 in the expressions (2.30) for Ψ_1 and Θ_1 are called the *constant of lunisolar precession* (in longitude) and the *mean obliquity of the ecliptic at epoch* (that is, at T_0), respectively. The value of ϵ^0 can only be determined from observations, as discussed in Section 2.2.6.1. f_1 can in principle be computed from the parameters of the Earth-Moon system and the density distribution within the Earth, and other parameters describing the Earth. Unfortunately, the density distribution is, at this time, not known with sufficient accuracy, so that f_1 must be determined from observations. f_1 is over 5000 seconds of arc per Julian century.

2.4.3.2 General precession. Let the descending node of the movable equator on the movable ecliptic, which is called the *mean equinox of date*, be denoted by Υ. The direction unit vector $\boldsymbol{n}_\Upsilon$ in E_0 toward Υ is clearly given by

$$\boldsymbol{n}_\Upsilon = \frac{\boldsymbol{n}_q \times \boldsymbol{n}_e}{|\boldsymbol{n}_q \times \boldsymbol{n}_e|} = \hat{\boldsymbol{x}}(-\Psi, 90° - B). \tag{2.31}$$

Equation (2.31) serves also to define B and Ψ, the latitude and the negative longitude, respectively, of Υ. The *mean obliquity of date*, ϵ, is given by

$$\sin \epsilon = |\boldsymbol{n}_q \times \boldsymbol{n}_e|. \tag{2.32}$$

Ψ is called the general precession in longitude and may be expressed as a truncated power series (that is, polynomial) in time by evaluating it from equations (2.31) in which the time polynomials given by equations (2.28) and (2.30) are used to represent the Eulerian orientation angles of the moving mean equator and the moving ecliptic. We thus have

$$\Psi = h_1 t + h_2 t^2 + \cdots, \tag{2.33}$$

where h_1 is called the *constant of precession*.

It can be shown (see, for instance, Woolard and Clemence [1966], p. 242) that

$$h_1 = f_1 - p_1 \cot \epsilon^0. \tag{2.34}$$

We now define the *planetary precession* as the longitude of Υ in a right-handed system A whose x-axis is in the direction of the descending node of the moving mean equator on the fixed ecliptic, and whose x-y plane coincides with that of the moving mean equator. The transformation between the coordinates of a vector $\boldsymbol{x}_A$ in A and $\boldsymbol{x}_E$ in E_0 is given by

$$\boldsymbol{x}_A = \mathbf{R}_1(-\Theta_1)\mathbf{R}_3(-\Psi_1)\boldsymbol{x}_E, \tag{2.35}$$

so that the *planetary precession* a is defined by

$$\hat{\boldsymbol{x}}(a, 90°) = \mathbf{R}_1(-\Theta_1)\mathbf{R}_3(-\Psi_1)\mathbf{R}_1(\epsilon^0)\hat{\boldsymbol{x}}(-\Psi, 90° - B). \tag{2.36}$$

In terms of a, we have approximately (see Woolard and Clemence [1966], p. 243)

$$\Psi = \Psi_1 - a \cos \epsilon^0. \tag{2.37}$$

This equation will be needed later for the discussion of the determination of the constant of precession from the analysis of proper motions. By evaluating equation (2.35), the *planetary precession* a may be expressed as a truncated series (that is, polynomial) in t, so that

$$a = g_1 t + g_2 t^2 + \cdots. \tag{2.38}$$

We now see that the matrix $\mathbf{S}(T_1 T_0)$ which was used in equation (1.13a) may be written

$$\mathbf{S}(T, T_0) = \mathbf{R}_3(a)\mathbf{R}_1(-\Theta_1)\mathbf{R}_3(-\Psi_1)\mathbf{R}_1(\epsilon^0). \tag{2.39}$$

Thus we have, at least in principle, solved the problem of finding an object's mean coordinates at T from its mean coordinates at T_0.

Traditionally, **S** is expressed in terms of usually three rotations on other angles. These developments are as follows.

The unit vector $\boldsymbol{n}_Q$ in the direction of the ascending node Q_0 of the movable mean equator on the fixed mean equator in the system A_0 of the fixed mean equator is given by

$$\boldsymbol{n}_Q = \mathbf{R}_1(-\epsilon^0)\,\frac{\hat{\boldsymbol{x}}(90°,\epsilon^0) \times \boldsymbol{n}_q}{|\,\hat{\boldsymbol{x}}(90°,\epsilon^0) \times \boldsymbol{n}_q\,|}, \tag{2.40}$$

where $\hat{\boldsymbol{x}}(\epsilon^0, 90°)$ is the normal unit vector of the fixed mean equator in E_0. The longitude $90° - \zeta_0$ of Q_0 in the system A_0 is thus defined by

$$\boldsymbol{n}_Q{}^T = (\sin\zeta_0, \cos\zeta_0, 0). \tag{2.41}$$

ζ_0 can be found by equating the right-hand sides of equations (2.40) and (2.41). Let $90° + z$ be the angle (on the movable mean equator) between Q_0 and r. Since the directions $\boldsymbol{n}_Q$ and $\mathbf{R}_1(-\epsilon^0)\boldsymbol{n}_{\Upsilon}$ to these points in the system A_0 are known from equations (2.40) and (2.31), we may calculate $\sin z$ and $\cos z$ in the usual manner by the dot and cross products, respectively, of these two vectors. Again, z may (after some heavy algebra, of course) be represented as a polynomial in time. The angle J between movable and fixed equator, also called the *precession in declination*, is given by

$$\sin J = |\,\hat{\boldsymbol{x}}(90°,\epsilon^0) \times \hat{\boldsymbol{x}}(90° - \Psi_1, \Theta_1)\,|, \tag{2.42}$$

and we see that the description of $\mathbf{S}(T, T_0)$ given above can now be written

$$\mathbf{S}(T, T_0) = \mathbf{R}_3(-90° - z)\mathbf{R}_1(J)\mathbf{R}_3(90° - \zeta_0). \tag{2.43}$$

If all developments have been properly carried out, the equations (2.39) and (2.43) must yield identical matrices. The arguments z, J and ζ_0 of the rotation matrices in equation (2.43) will be given below in Section 3.2.1, equations (3.4).

The *general precession in right ascension*, μ, is now defined by

$$\mu = z + \zeta_0. \tag{2.44}$$

Note that this is, in essence, the increase of the right ascension of a point on the equator due to precession, and thus very nearly the right ascension of the movable mean equinox in A_0.

Remembering that the planetary precession a is an arc on the movable mean equator, we may define ζ by

$$\zeta = z + a, \tag{2.45}$$

so that

$$\mu = (\zeta + \zeta_0) - a. \tag{2.46}$$

$\zeta + \zeta_0$ is the *lunisolar precession in right ascension*.

2.4.3.3 THE INFLUENCE OF PRECESSION ON THE EQUATORIAL COORDINATES OF AN

OBJECT. In the texts on spherical astronomy it is shown that

$$\frac{\partial \alpha}{\partial T} = m + n \sin \alpha \tan \delta$$

and

$$\frac{\partial \delta}{\partial T} = n \cos \alpha, \tag{2.47}$$

where

$$m = \frac{d\mu}{dt} = \frac{d\Psi_1}{dt} \cos \epsilon - \frac{da}{dt}$$

and

$$n = \frac{dJ}{dt} \sec z = \frac{d\Psi_1}{dt} \sin \epsilon. \tag{2.48}$$

m and n are, of course, variable with time, and the derivatives must therefore be evaluated at the epoch T.

From equation (2.37), we have

$$d\Psi_1 = d\Psi + da \cos \epsilon^0, \tag{2.49}$$

and therefore, to the first order,

$$df_1 = dh_1 + dg_1 \cos \epsilon^0. \tag{2.50}$$

From equations (2.48), we get

$$\begin{aligned} dm &= (dh_1 + dg_1 \cos \epsilon^0) \cos \epsilon^0 - dg_1 \\ dn &= (dh_1 + dg_1 \cos \epsilon^0) \sin \epsilon^0. \end{aligned} \tag{2.51}$$

Errors dh_1 and dg_1 in the rates h_1 and g_1 of general and lunisolar precession, respectively, will thus produce the following components in the rates of changes of the coordinates α, δ of an object:

$$\begin{aligned} d\frac{\partial \alpha}{\partial T} &= dm + dn \sin \alpha \tan \delta \\ d\frac{\partial \delta}{\partial T} &= dn \cos \alpha. \end{aligned} \tag{2.52}$$

These will appear added to the proper motion components. In Section 2.4.1 we defined the proper motion components as the rate of change of the coordinates with respect to a certain system. The values of the precession parameters h_1 and g_1 are essential for the definition of the system, and equations (2.52), which are very similar to equations (2.21) which do not provide for a change in g_1, make it clear why the rigorous definition of proper motions requires the knowledge of the precession parameters to which they are referred.

2.4.3.4 THE ESTABLISHMENT OF AN INERTIAL SYSTEM FROM THE ANALYSIS OF PROPER MOTIONS. If a *model for the various systematic components of proper motions* is set up, and if one assumes a *distribution function for the peculiar motions*, one may analyze proper motions to find (1) the parameters in the model for the systematic motions, and (2) the rates at which the coordinate system must be rotated on three noncoplanar axes so that no systematic (uniform) rotation is left in the proper motions. It is very probable that the system obtained in this way is a good approximation to an inertial system. Wayman (1966) has given a very lucid discussion of this subject.

The distribution function of the peculiar motions is usually represented as a random distribution function.[11] The systematic components of the proper motions are (1) the *parallactic motions* (caused by the uniform translation of the sun with respect to the local standard of rest), (2) the later discovered effects of *galactic shearing* or *differential galactic rotation* (described by Oort's constant A), and (3) of *galactic rotation*—a uniform rotation on the axis of the galaxy at a rate given by Oort's constant B.

The parallactic motion of a star will of course depend on its distance r from the Sun, while the effects of galactic shearing and galactic rotation are (to the first order) distance independent. Due to the orientation of these effects along the galactic equator, the models for the systematic parts of the proper motions, apart from those that will reduce the system to an inertial system, are most simply written in terms of proper motions in galactic longitude (μ_l) and galactic latitude (μ_b) as follows:

$$\mu_l = A \cos 2l + B + \frac{1}{kr \cos b} (s_1 \sin l - s_2 \cos l)$$

$$\mu_b = -\tfrac{1}{2}A \sin 2b \sin 2l + \frac{1}{kr} (s_1 \sin b \cos l + s_2 \sin b \sin l - s_3 \cos b), \qquad (2.53)$$

where A and B are *Oort's constants*, l is the galactic longitude referred to the galactic center, b the galactic latitude, k is an inessential conversion factor, r the star's distance from the Sun, and s_1, s_2, and s_3 are the components of the Sun's velocity

11. This is not in accordance with what we know about the distribution function of the residual velocities of the stars in the solar neighborhood. The author believes that significantly improved values for the precession parameters could be obtained from the presently available material, if the known peculiarities of the velocity distribution function were properly accounted for and utilized in the analysis. Also, since here the peculiar proper motions play the role of accidental errors, more accurate values of the parameters would be obtained by weighting the proper motions according to their expected absolute values which could be calculated as functions of the stars' magnitudes, spectra, and so forth, and their positions in the sky.

with respect to the local standard of rest and referred to the system of galactic coordinates. The unknown parameters in the adjustment are A, B, s_1, s_2 and s_3. Note the effect which B has. It represents, as mentioned above, a uniform rotation of the stellar system on the axis of the galaxy which will occur according to the laws of galactic dynamics.

This means, however, that not all of the uniform rotation of the stars contained in the proper motions with respect to the empirical system—nearly, but not perfectly, inertial—will be due to the fact that this system is not inertial. Since we also assume that B must be determined from the analysis of the proper motions, we cannot allow the inertial motion of the system (to which the proper motions are referred) to be modeled in full generality, for instance, by solving for rotation components along three perpendicular axes. The effect of B would then disappear, since its components along the three axes would simply be added to the components of the inertial motion of the system along these axes. In fact, the system of equations for the determination of a general rotation in addition to B would be singular. This was pointed out very clearly in Fricke's (1967) paper.

The only way out of this dilemma is to model the reduction of the system to an inertial system not in full generality, but by allowing only for rotations along two axes. It is imperative for a meaningful analysis that these two axes are correctly chosen.

A selection from the axes of the equator system suggests itself. A rotation on the y-axis (toward the point $\alpha = 6^h$, $\delta = 0°$) corresponds to a change of the precession parameter n (see equation (2.52)). A rotation on the z-axis (toward the north celestial pole) could be interpreted as a change in the rate g_1 of the planetary precession a, or a so-called *motion of the equinox*, denoted by de, which might have its origin in the fact that the right ascensions from which the proper motions were derived are affected by a systematic error in the absolute catalogues. This error is a linear function of time (which seems perhaps a rather artificial assumption) or simply reflects some effect for which a physical model cannot be given at this time, but which must be interpreted in the sense that the system, in order to be rendered inertial, must be rotated on the z-axis.

We have stated above that the only parameters in the precession theory which must be determined from observation are f_1 and ϵ^0. This not only means that we really should not even consider a correction to θ_1, but also that it is unnecessary, and even detrimental, to consider a correction to the precessional rate of change of the obliquity of the ecliptic, that is, a rotation of the system on the x-axis. It is thus that in the analyses of proper motions for finding corrections to the constant of precession, a correction of the rotation rate on the x-axis is traditionally not allowed for. This means, of course, that the uniform rotation in the proper motions will be broken down into components along the y-axis (dn), the z-axis (de) and the axis of the galaxy (B). Only the first two of the components are, however, interpreted as corrections which make the system inertial. Needless to say, the success of this model depends critically on the theory of planetary motion and precession to provide a correct value of the rate of change of the obliquity of the ecliptic, ϵ. If this is not available (or wrong), erroneous values of dn, dm and B will result from the analysis. It was suggested by Aoki (1967) that the available analyses of the observation material show that the presently accepted value of the change of ϵ, as calculated from theory, is indeed not the actual one, and he derives the corresponding value of B, which follows from the dϵ/dt, which he regards as the correct one. How drastically this changes our picture of the galaxy is shown by the fact that Aoki's value of B would result in a mass of the galaxy of 6.5×10^{11} solar masses instead of the 1.8×10^{11} solar masses that result from the value of B associated with the conventional value of the rate of change of the obliquity of the ecliptic. On the other hand, Fricke (1970) explains the discrepancy between the observed and the predicted value of dϵ/dt in part by systematic errors in the early epoch observations of absolute solar declinations which were introduced by inadequate allowance for instrumental flexure.

Fortunately, an error in the rate g_1 of the planetary precession does not have such far-reaching consequences. If we assume the equations (2.51) as the basis of the analysis of the proper motions according to the model given by equations (2.52), dh_1 and dg_1 could be determined; if we, however, introduce de, the above discussed motion of the equinox, the first of equations (2.51) would have to be written

$$\mathrm{d}m = (\mathrm{d}h_1 + \mathrm{d}g_1 \cos \epsilon^0) \cos \epsilon^0 - (\mathrm{d}g_1 + \mathrm{d}e), \tag{2.51a}$$

so that dg_1 and de can no longer be separated. It is then customary to rewrite equations (2.52) to read

$$\mathrm{d}\frac{\partial \alpha}{\partial T} = \mathrm{d}n(\cot \epsilon + \sin \alpha \tan \delta) - (\mathrm{d}a + \mathrm{d}e)$$

$$\mathrm{d}\frac{\partial \delta}{\partial T} = \mathrm{d}n \cos \alpha, \tag{2.52a}$$

where now dn and da + de are the adjustment parameters. Since, according to equations (2.50) and (2.51)

$$\mathrm{d}n = \mathrm{d}f_1 \sin \epsilon^0, \tag{2.51b}$$

it is obvious that an analysis of this type can yield a correction to h_1, the constant of general precession, only if da is either assumed equal to zero, or known from an independent source. Otherwise, only the correction to the constant of lunisolar precession, f_1, can be found.

Böhme and Fricke (1965) give a list of values of df_1 and da + de (in their notation Δp_1 and $\Delta\lambda + \Delta e$, respectively) which were obtained from various investigations; the values obtained for both df_1 and da + de are around 1″/century with uncertainties of about 0.″20/century. Similar values were found by Fricke (1966, 1967, 1967a, 1971) in his investigations which are, so far, the most recent attempt at delineating an inertial system through the analysis of proper motions.

2.4.3.5 THE DETERMINATION OF AN INERTIAL SYSTEM FROM THE ANALYSIS OF PLANETARY MOTIONS. Corrections to the assumed value of the constant of precession can be obtained independently from investigations of stellar proper motions by investigating the motions of the planets; see, for instance, the article by Clemence (1966). This is plausible when one considers the definition of an inertial system as one in which Newton's laws of motion are valid.

In order to actually build an inertial system in this way, one must have a *system*, realized by observations, to which the positions (or the locations) of the planets are referred, and a *theory of the planets*, which consists of precepts for computing their coordinates as functions of time, certain parameters, such as their masses, and the constants of integration, which in this case are the elements of their orbits.

In analyzing the observations, the empirical system will then be subjected to a rotation to render it inertial. The other unknown parameters in the adjustment are the orbital elements of certain planets and the masses of certain others. Only a rotation on the y-axis of the equator system is allowed for; that means it is assumed that the empirical system differs from an inertial system only by a rotation on the y-axis.

According to Clemence (1966), to whom we refer the reader for further details, only the observations of Mercury, Venus and the Sun can contribute significantly to the solution of the problem, and only the determination of a correction to the precession, and the masses of Mercury and of Mars is feasible without overloading the system of normal equations. The value of dh_1 results to $+1.''86 \pm 1.''59$ (rms.) which is a much larger error than what is obtained from the analysis of proper motions as just described in Section 2.4.2.4. Within their errors, the two results do, however, agree.

2.4.3.6 DETERMINATION OF THE CONSTANT OF PRECESSION FROM POLAR TELESCOPES. Suppose a camera is directed toward a Celestial Pole, whose image will then be close to the tangential points of the plates. Also assume that at least five (and usually more) exposures are put on each plate. After every exposure, the plate is reset so that it occupies the same position with respect to the horizon system at the beginning of each exposure. In this way, a situation is simulated in which the exposures will produce images on the plate in the same way as if the plate had not been moved with respect to the Earth at all, and as if the images had been generated by exposures made at very regular intervals of such short duration that they show no elongation due to their diurnal motion, and appear as circular dots for all purposes.

The images produced by the same star will lie on almost circular ellipses, from the parameters of which the location of the celestial pole with respect to the stars can be computed and is thus monitored. If such a series of observations is continued over a sufficiently long period, the constants of precession and nutation can be derived from the data gathered.

Efforts toward this end are under way at the Yale Observatory (see, for instance, Williams [1967]), at the Pulkova Observatory and in Mizusawa, Japan. No definitive results are, however, yet available.

2.4.3.7 DETERMINATION OF THE CONSTANT OF PRECESSION FROM PROPER MOTIONS OBTAINED BY REFERENCE TO DISTANT GALAXIES. The method of using distant galaxies to define an inertial frame of reference was discussed above in Section 2.4.1; the proper motions which will be obtained in this way will (except for unavoidable errors in the reduction constants) be directly referred to an inertial frame, provided that the system of galaxies does not rotate. One will thus be able to analyze them using only the model described by equations (2.53) without the added terms $d(\partial\alpha/\partial T)$ and $d(\partial\delta/\partial T)$ given by equation (2.52a). A, B, s_1, s_2 and s_3 will therefore become available from these proper motions completely independently of df_1 and $da + de$. These can be found by analyzing the systematic differences between the proper motions μ_g obtained by reference to galaxies, and the proper motions μ_c obtained from meridian (or conventional) astrometry according to the model described by equations (2.52a). This is so since

$$\mu_{\alpha c} - \mu_{\alpha g} = d\frac{\partial\alpha}{\partial T}$$

and

$$\mu_{\delta c} - \mu_{\delta g} = d\frac{\partial\delta}{\partial T} \tag{2.52b}$$

represent those components that appear in the conventional proper motions because the system to which they are referred is not inertial.

Once the μ_g is available, the value of the constant of precession can become determined with an accuracy much higher than that attainable with present-day means. The first results of these efforts have already been published (Fatchikhin 1969, Vasilevskis 1970) and more definitive results should become available in the 1970s.

2.4.4 THE CONSTANT OF NUTATION AND ITS DETERMINATION. The orientation of the mean equator of date is given by the angles Ψ_1 and Θ_1, introduced by equation (2.29) and discussion in Section 2.4.2.1.

The plane defined in E_0 by the orientation angles $\psi = \Psi_1 + \Delta\Psi_1$ and $\epsilon_1 = \Theta_1 + \Delta\Theta_1$ is that of the *true equator of date*. The theory of the rotation of the Earth on its axis shows that

$$\Delta\Psi_1 = -17.''233 \sin\Omega + 0.''209 \sin 2\Omega$$
$$-0.''204 \sin 2L_{☾} - 1.''273 \sin 2L_{\odot} + \cdots,$$

and

$$\Delta\Theta_1 = 9.''210 \cos\Omega - 0.''090 \cos 2\Omega$$
$$+0.''088 \cos 2L_{☾} + 0.''552 \cos 2L_{\odot} + \cdots, \tag{2.54}$$

where Ω is the mean longitude of the ascending node of the lunar orbit on the ecliptic, and $L_{\odot}$ and $L_{☾}$ are the mean longitudes of the Sun and Moon, respectively.

The most accurate expressions so far for $\Delta\Psi_1$ and $\Delta\Theta_1$ were derived by Woolard (1953). At this time, Melchior (1970) is introducing further sophistications into the theory.

For the reduction of coordinates from A to the system of the true equator and equinox, it is more advantageous to know the orientation angles of the true equator not with respect to E_0 but with respect to E_1, the system of the ecliptic of date. These are defined by $\psi = \Psi + \Delta\Psi$ and $\epsilon = \Theta + \Delta\epsilon$. Formulas for $\Delta\Psi$ and $\Delta\epsilon$ are calculated by reducing the formulas (2.54) from the ecliptic of epoch to the ecliptic of date, and are found, for instance, on pp. 44 and 45 of the Explanatory Supplement (Nautical Almanac Offices, 1961).

In terms of $\Delta\Psi$ and $\Delta\epsilon$, the nutation matrix **N** in formulas (1.13a) and (1.13b) is given by

$$\mathbf{N} = \mathbf{R}_1(-\Theta - \Delta\epsilon)\mathbf{R}_3(-\Delta\Psi)\mathbf{R}_1(\Theta), \tag{2.55}$$

or in terms of $\Delta\Psi_1$ and $\Delta\Theta_1$ by

$$\mathbf{N} = \mathbf{R}_3(a_1)\mathbf{R}_1(-\Theta_1 - \Delta\Theta_1)\mathbf{R}_3(\Delta\Psi_1)\mathbf{R}_1(\Theta_1)\mathbf{R}_3(-a), \tag{2.55a}$$

where a_1 can be calculated from a. It is given by

$$a_1 = a + \Delta a$$

with $\Delta a = (0.''00983 \sin \Omega - 0.''00128 \cos \Omega + 0.''00073 \sin 2L_\odot + \cdots)T$, T being measured in Julian centuries since 1850.

The coefficient of $\cos \Omega$ in $\Delta\Theta_1$ (equations 2.54) is called the *constant of nutation*. It is here given as 9.″210, which is the value presently accepted by international agreement.

The constant of nutation, N, may be determined in many ways: by observing meridian zenith distances, prime vertical transits and by means of zenith telescopes as they are used by the International Latitude Service. Fortunately, the angle Ω has a period of 18.6 years, so that undetected annual and diurnal systematic errors will have little influence on the value for this constant as obtained from the observations. It is worth noting that there is a *rigorous relationship between the constants of lunisolar precession* f_1 (for 1900) *and nutation*, N, and the Moon's mass μ (in terms of the Earth's mass). See, for instance, Kulikov (1964), p. 135, namely,

$$N = \frac{47.6237 f_1}{\dfrac{1}{\mu} + 178.822},$$

where the value $(C - A)/C = 0.00327893$ was used for the dynamical flattening of the Earth. (C and A are the Earth's principal moments of inertia.)

Kulikov's book, the original Russian version of which was published in 1956, also gives a fairly detailed discussion of the methods used for the determination of the constant of nutation, and a list of the individual determinations up to then.

Böhme and Fricke (1965) list the determinations of the constant of nutation between 1920 and 1965.

Fortunately, the effects of nutation on the frame of reference are, by definition, periodic, so that they concern catalogue astronomy only insofar as the mean catalogued position must be computed from the observed true ones. This is also true—mutatis mutandis—for the effects of aberration and the determination of their constant to be discussed next.

2.4.5 THE CONSTANT OF ABERRATION AND ITS DETERMINATION. In the analysis in Section 1.6.4 of the velocity components of the Earth which cause aberration effects, we have seen in equations (1.15) that there is, in essence, a variation of the Earth's velocity vector with a period of one year. There we called it v_1''. Because of this, the apparent positions of the stars describe, to a high degree of approximation, ellipses around their mean positions whose semimajor axes all have the value k (measured on a great circle) and are parallel to the ecliptic, and whose semiminor axes have the value $k \sin \beta$ (with β the star's ecliptic latitude). In first approximation (if we assume the Earth's motion to be circular), the equation of this ellipse in parametric representation is

$$\lambda' - \lambda = -k \cos(\odot - \lambda) \sec \beta$$

$$\beta' - \beta = -k \sin(\odot - \lambda) \sin \beta,$$

where (λ, β) and (λ', β') are the star's true and apparent longitude and latitude respectively and $\odot$ is the ecliptic longitude of the Sun. Note that the factor $\sec \beta$ appears in the expression for $\lambda' - \lambda$, since this angle is reckoned on a small circle. k, in seconds of arc, is rigorously given by

$$k = \frac{n \rho_0 a}{c \pi_\odot \omega \sqrt{1 - e^2}}, \tag{2.56}$$

which is the mean value of $v_1'' \csc 1''/c$. $\omega = 206264.8$ (the number of seconds per radian), v_1'' is as defined in Section 1.6.4, and c is the speed of light in the same units as v_1''. n is the mean motion of the Earth in seconds of arc, a the semimajor axis of the Earth's orbit (which is very close to the astronomical unit), $\pi_\odot$ the solar parallax, and e the eccentricity of the Earth's orbit.

A discussion of the determinations of the value of the constant of aberration up to 1956 is given by Kulikov (1964) on pp. 74–98; a listing of values determined later is given by Böhme and Fricke (1965). Although it is, in principle, possible to determine the value of k generally from the observation and analysis of apparent positions, other methods have been used in preference to this.

It is also possible to determine the value of k from the analysis of radial velocities observed throughout a year.

The preferred method has, however, been to determine the constant of aberration from series of latitude determinations by the Horrebow-Talcott method from the effect of annual aberration on declination. One of the advantages of this method is that one need not be concerned about diurnal aberration, which has no influence on declinations.

The values obtained at different places and by different methods vary considerably. This was discussed by Kulikov, and may have various sources, since the observing stations' latitudes and the declinations of the stars enter critically in the equations of condition. Any undetected annual or diurnal effects of precession or polar motion will influence the value of k obtained from a particular series and cause the adjustment to show an error of k that is too small. In fact, the various values of k obtained by different authors disagree much worse than they should according to their nominal errors. Perhaps it may sometime be possible to analyze from a critical discussion of these determinations the effects which falsify the individual determinations of k.

The currently adopted value, which is compatible with the other astronomical constants presently adopted, and which satisfies equation (2.56), is $k = 20.''496$. Before 1968, the value $20.''47$ was generally adopted for the computation of apparent places from true (or mean) ones, although the system of astronomical constants in use before that date did not satisfy equation (2.56).

The importance of the consistency of the systems of astronomical constants was discussed by Clemence (1948).

Chapter 3

GENERAL DISCUSSION OF STAR CATALOGUES

3.1 Introduction

3.1.1 HISTORICAL CATALOGUES. The oldest record of star positions is given in Aratus' astronomical poem, *Phaenomena*, in which are quoted those declinations of twenty-five of the brightest stars as determined by Eudoxus, a student of Plato. Other catalogues were observed in ancient Greece by Aristillus and Timocharis, by Menelaus, and by Hipparchus.

The best known of the ancient catalogues is found in Ptolemaios' (Ptolemy's) *Almagest*. It is possibly just Hipparchus' catalogue precessed from 128 B.C. to 138 A.D., with an inaccurate constant of precession at that, but it is still the principal source for ancient stellar positions. The rms. errors of ecliptic latitudes and longitudes in it are 0.°58 and 0.°37, respectively. New editions of it were issued as late as 1843 (by F. Baily).

It was apparently not until about 1534 that positions of 1018 stars were newly observed and compiled into a catalogue by Ulugh-Beg in Samarkand. Around 1594, the positions of 1004 stars were observed and published by Christof Rothmann and Wilhelm IV, Landgrave of Hessia. The accuracy of positions in this catalogue corresponds to a rms. error of about 6′. Rothmann and Wilhelm IV established the connection between the Sun and the stars by taking Venus, and not the Moon, as an intermediary point. The advantage of this procedure is quite obvious, considering that settings on Venus can be made much more accurately than on the Moon.

Essentially the same technique was applied by Tyge (or, in its usually hellenized form, Tycho) Brahe, whose catalogue contains 1005 stars for the epoch 1601. He measured the declinations with a mural quadrant, and the distances between the stars with a sextant, and used α Arietis as reference point. His star positions are less accurate than his measurements of planetary positions.

The last catalogue which was observed (at Danzig) without a magnifying device is that of Hevelius, containing 1563 stars for the epochs 1661 and 1701. The positions of his 335 southern stars were, however, observed by means of a telescope. Hevelius' catalogue was, by the way, the first which gave right ascensions and declinations of stars as well as their ecliptic longitudes and latitudes. Included in Hevelius' catalogue are the above-mentioned positions of 335 stars (*Catalogus Stellarum Australium*) which had been observed 1676–1678 by Edmond Halley (by means of a telescope) at St. Helena; Hevelius' naked-eye positions are, however, just as accurate as Halley's telescopic ones. In 1725, the *Historia Coelestis Britannica* was published, containing the positions of 3310 stars to 7^m which had been observed by Flamsteed. The rms. error of these positions is about $2''$. In 1750–62, James Bradley observed the positions of 3268 stars on an iron quadrant made by Graham.

The positions of 9766 southern stars were observed from 1751 to 1752 by Lacaille (*Coelum Australe Stelliferum*); those of nearly 50,000 northern ones were published by Lalande under the title *Histoire Celeste* and are based on observations around 1760–1800. Piazzi observed the positions of 7646 stars at Palermo from 1792 to 1813.

The *British Association Catalogue* of the positions of 47,390 stars, compiled by Fr. Baily from observations by Lalande and others, appeared in 1847.

3.1.2 THE CLASSIFICATION OF STAR CATALOGUES. According to the definition in Section 1.1.1, a star catalogue is essentially a list of accurate star positions.

An *independent catalogue* (in the terminology of Newcomb [1906], p. 380) or *observation catalogue* is a list of positions that were observed for the construction of the catalogue at hand, and which are not based on previously published and fully-reduced positions. In contrast to this, a *compilation catalogue* lists positions that were computed by combining positions from more than one independent catalogue. Since these independent catalogues generally list positions for different epochs, compilation catalogues frequently contain as primary information not only the positions but also the proper motion components calculated from the incorporated independent catalogue positions; that is, they are complete catalogues (see below). Not all catalogues that give proper motions are compilation catalogues, however.

The positions given in an independent catalogue may have been computed from relative observations, that is, observations which were carried out with respect to a coordinate system that was realized by referring to reference stars whose positions were assumed known. (See Sections 2.2.7 and 2.3.) In this case, the catalogue will be a *relative catalogue*. If, on the other hand, the positions are absolute, that is, if the positions were determined by methods which in no way depend on previously known star positions, the catalogue will be an *absolute catalogue*. (See Section 2.2.5 and 2.2.6, especially for the different types of absolute observations and qualifications.) Since absolute positions are often called fundamental positions, the term fundamental catalogue is sometimes used for a catalogue of absolute positions. This should be avoided since a fundamental catalogue is something else, see below.

The positions in an independent relative catalogue may have been determined by direct observations on a transit circle (or other instrument), or by measuring the relative positions of star images on photographic plates. Accordingly, we speak of *visual* and of *photographic catalogues.*

When a (usually independent relative) catalogue lists positions around the sky within a narrow zone of declination, say 5° wide, it is called a *zone catalogue.* Zone catalogues will be discussed below in Chapter V. All photographic catalogues are (of necessity) zone catalogues. Also, the faintest stars whose positions are catalogued usually appear only in zone catalogues.

A clear-cut division of compilation catalogues into absolute and relative ones is not possible and, besides, would be of little value. In fact, most important compilation catalogues are the results of combining absolute and relative catalogues.

A compilation catalogue is sometimes constructed with the purpose of defining that frame of reference (that is, coordinate system) which is almost the best approximation to an inertial system that can be achieved at the time. This is done by incorporating primarily a selection from the available absolute catalogues (see Sections 4.1 and 4.2). Such a catalogue is then called a *fundamental catalogue.* High-accuracy relative positions are, as a rule, also used for deriving the positions and proper motions listed in a fundamental catalogue, but only after they have been carefully reduced to the system which is defined only by the absolute catalogues. A fundamental catalogue defines a *fundamental system,* and consists essentially of two sets of positions for all stars in the catalogue at two different epochs, but may also be given in the form of a list of positions and proper motion components. In this case, however, the constant of precession utilized for the computation of the proper motion components must be known. This does not mean that a fundamental system of positions will depend on the choice for the constant of precession, although the associated system of proper motions will, as was pointed out in Section 2.4.1. If a star catalogue is to provide the means for the prediction of the mean positions of all stars listed in it for all epochs, it must contain positions as well as proper motions for each star, and the value assumed for the constant of precession must also be fixed. In all rigor, the ratios of radial velocity to distance for each star would have to be available as well, but these are almost always assumed to be equal to zero. A catalogue giving proper motions in additions to the positions may be termed a *complete catalogue.*

Some lists of observed star positions, although termed "catalogues" are not really catalogues in the sense discussed here since they were not fully reduced. They should more appropriately be called "observations," and usually are. In this category are the *annual catalogues* of several observatories, from which an "N Year Catalogue" is later compiled and published as a true independent catalogue.

3.1.3 THE CHARACTERISTIC FEATURES OF STAR CATALOGUES.

The positions published in the various catalogues were obtained from the raw observed position material by means of a more or less tedious and sophisticated reduction process that involves a number of constants.

As described above in Chapter II, the readings obtained on the instruments must first be converted to refracted apparent positions for the epoch of the observations by applying the instrumental corrections such as level error, azimuth, collimation, circle division errors, and so forth. The mean positions with respect to equator and equinox of the catalogue are obtained by removing the effects of refraction, daily and annual aberration, parallax (in the very rare cases when indicated), nutation, and precession between the equator and equinox to which the positions in the catalogue are to be referred, and the epoch of observation. Although this is not good practice, available proper motions were in a few cases used to reduce the positions to the same epoch as that of the equator and equinox. This step should always be restricted to compilation catalogues.

The values used for the reduction constants, notably the values for the constant of precession, have undergone appreciable changes over the period that precision catalogues have been observed. It is easy to see that a wrong value for the constant of aberration will produce a small annual periodic error in the right ascensions if most observations are made, say, before midnight, as, for obvious reasons, they actually were.

One of the most difficult problems is the accurate computation of the refraction (see Section 2.2.5.2), which is indispensable for the absolute determination of declinations from meridian observations. In order to evaluate properly the information given in a star catalogue, it is desirable to know exactly how the refraction was computed.

Most important, the constant of precession is intimately connected with the proper motion system of the catalogue. The total change of the mean position of a star is a combination of the effects of proper motion and of precession. The effect of an error in the assumed value for the constant of precession will be compensated by a systematic effect in the proper motions, see equations (2.22). The assumption of a distribution function of the velocities of the stars in the galactic system is one way to arrive at a determination of the constant of precession. With regard to the discussion of proper motion given above in Sections 2.4.1 and 2.4.3, it should be pointed out that the changes in the various assumed values of the constant of precession, used during the past two centuries for the reduction of position observations, resulted partly from both an increase of the number of observed star positions and an increase in the average time interval available for determining proper motion. Thus, more and more accurate proper motions became available for the determination of the constant of precession. This resulted, partly, from the use of improved models for the distribution function of the proper motions, namely, the consideration of the systematic effect which the motion of the Sun among the other stars has upon the proper motions of the stars. While most early determinations of the constant of precession were made under the assumption of complete randomness of the proper motions (even those by Struve and Peters whose results were used until the twentieth century), Newcomb determined the constant of precession allowing for the parallactic motions, that is, the systematic part of the proper motions produced by the motion of the Sun. Later, still better determinations of the constant of precession were made allowing for the effects produced by differential galactic rotation, see, for instance Oort (1943b), Williams and Vyssotsky (1948), Fricke (1967, 1967a), Fatchikhin (1968), but Newcomb's value for the constant of precession has been used since the beginning of the twentieth century by international agreement. Likewise, the value for the constant of aberration used by international agreement was only recently changed to the most accurate one available. However, this change to the better value will hardly have a significant effect on the mean star positions calculated from the apparent ones.

Because of all this, it is mandatory to carefully read the introduction before one uses positions from any star catalogue. In particular, one may want to ascertain the following:

1) Are the positions given for the epoch of observation, or were they reduced to the epoch of orientation of the coordinate system by the application of proper motions? If so (and this is only rarely the case), which proper motions were used?

2) In the case of absolute observations, how was the refraction computed? Which systematic instrumental corrections were applied?

3) Which constants were used for the reduction from apparent place referred to the coordinate system at the epoch of observation to the mean place for the coordinate system of the catalogue?

4) To which system (fundamental or other) are the catalogue positions referred?

Only if these questions are answered, can the originally observed values which may be needed for a more careful reduction be reconstructed.

The precise meaning of the rest of the data given in a star catalogue must also be completely understood before the data can be used in critical investigations. Questions that will come up concerning the peripheral data are:

How were the magnitudes obtained? On which system are they? Were they newly observed for this catalogue or taken from some previously existing source? How accurate are they?

How were the spectra (if given) obtained? On which system, and how accurate, are they?

What is the exact meaning of the precession terms (annual and secular variation and third term when given)? Do these terms incorporate only the effects of precession on the mean positions, or also those of proper motion? Which value of the constant of precession was used to compute them?

How were the proper motions (when given) obtained? What systematic corrections were applied to the catalogue(s) that supplied the early epoch(s) material? Were any systematic corrections applied to the proper motions themselves?

In that case where the proper motions are so large that the prediction of positions at epochs sufficiently far from the average epoch of the catalogue requires the ratio of radial velocity to distance to be known, is this ratio available, and if so, from what sources?

Finally, the sky coverage provided by the catalogue is important. Virtually all catalogues provide positions around the sky (that is, from 0^h through 24^h) in their declination range. The magnitude range of the catalogued stars is also of interest. Some catalogues provide complete magnitude coverage up to a certain magnitude, and list stars fainter than the limit of completeness only sporadically, and, of course, stars fainter than a certain magnitude not at all. Other catalogues leave out the brightest stars and cover a certain magnitude interval either selectively or completely. Usually, the zone catalogues aim at completeness up to a certain magnitude, and absolute catalogues, usually, at a selection of bright stars (say, less than 6^m).

3.1.4 THE ORGANIZATION OF A CATALOGUE. Most catalogues follow a general pattern for the publication of the data recorded in them.

Serial numbers and *magnitudes* are almost always given. The magnitudes are sometimes independently estimated, frequently at the same time that the stars' positions are observed on the instrument; sometimes taken from other, previously existing sources. With few exceptions, the magnitudes are rather inaccurate and show considerable systematic differences from one catalogue to the next. They are usually not considered suitable for photometric investigations requiring high precision.

The *most important* (and quite often the only original) *data* in a catalogue are the *mean positions* referred to a coordinate system as oriented at the beginning of a Besselian year. For older catalogues, the epoch of orientation is usually chosen close to the average of the median observing epochs, but in catalogues observed in the twentieth century it is almost always either 1925.0, 1950.0, 1975.0 or 2000.0. The *right ascensions* are usually given in terms of hours, and minutes and seconds of time to $0.^s001$ or $0.^s01$, and the *declinations* in terms of degrees, and minutes and seconds of arcs to $0.''1$ or $0.''01$. In rare cases, north polar distances (NPD) are given instead of declinations (NPD $= 90° - \delta$) and expressed in the same units. The stars are usually listed in the order of increasing right ascension. In zone catalogues, the positions are often given in the order of increasing right ascension within the several zones, and the serial numbers will start new in each zone. For a number of very accurate precession operations, especially in high declinations, the fact that the mean positions were not completely freed from all aberration effects (namely, the perigee terms, see Section 1.6.4) must be properly considered.

Frequently, the catalogues, especially the more recent ones, indicate the *fundamental* (*or other*) *system to which the positions are referred.* Experience, however, has shown that one must not always take the observer's or compiler's word regarding the relationship between the system actually defined by a certain catalogue and the system the author claims it to be on. As a rule, it is very difficult to completely avoid systematic differences between the positions of the catalogue stars and those of the primary stars that define the system.

Until recently, most catalogues gave values for the computation of the effect of precession of the positions according to the formula

$$\alpha(T) = \alpha_0 + p_\alpha T + \frac{v_s}{200} T^2 + \frac{v'}{6}\left(\frac{T}{100}\right)^3 \tag{3.1}$$

(and analogously for the declination) where α_0 is the published mean right ascension, $\alpha(T)$ the mean right ascension at the time T usually reckoned in tropical years from the epoch of orientation. p_α and v_s are *annual* and *secular variations*, respectively. v' is the so-called third term of precession which can usually, depending on the accuracy requirements, be neglected except in high declinations and for time intervals of, say, over 50 years. In order to be certain as to how the precessional data given in a catalog should be used, one should consult the introduction to the catalogue.

The precession coefficients p_α, v_s, and v' were usually computed individually for every star in the catalogue, assuming a certain value for the constant of precession. Before these precession coefficients are used, one must ascertain from the introduction or preface to the catalogue which value of the constant of precession is the basis of their computation and whether or not the secular variations were corrected for proper motion.

In those catalogues in which the star positions were not only precessed to refer to the coordinate system, but also, by application of the proper motion effects, reduced to the epoch of the coordinate system, the proper motions were usually added to the annual precession; their sum is sometimes called annual variation. In this case, equation (3.1) not only precesses, but also reduces the positions to the epoch of the new coordinate system. Ignorance concerning the accurate meaning of the precession coefficients which are communicated with the positions may lead to small but appreciable errors, which are very often extremely hard to trace.

Modern catalogues frequently do not give annual and secular variations but terms usually called I, II, III; that is, first, second and third term of precession, respectively. The formulas to be used with these are typically

$$\alpha(T) = \alpha_0 + \mathrm{I} \cdot T + \mathrm{II}\left(\frac{T}{100}\right)^2 + \mathrm{III}\left(\frac{T}{100}\right)^3. \tag{3.1a}$$

The computation of precession, by using one of these truncated series, always carries the implicit danger that the series may have been broken off too early or that it may not converge. In high declinations and for precessing positions over long time intervals, formulas of type (3.1) or (3.1a) are inaccurate.

The only reason why they were used to begin with is that their application involves less computing than that of the rigorous formulas (3.5) below. This is, or was, an important consideration in the age of logarithms or desk calculators, but is of no consequence when electronic computers are available. There is actually no good reason to continue the publication of precession terms in catalogues, since the precession effects can be computed without danger of truncation errors by the formulas (3.5).

Remember, that the use of formulas established for logarithmic or desk calculator computations on an electronic computer is comparable to using a newly-acquired desk calculator only for adding logarithms. The calculation of precession by truncated series is a perfect example of a procedure that was very useful in the logarithm and the desk calculator eras, but which has become superfluous and even detrimental in the computer age.

At least one new catalogue (the zone $-80°$ to $-90°$ of the Cape Photographic Catalogue for 1950.0), see Section 5.3.4, lists, besides mean right ascensions and declinations, also the components of the *position unit vector* $\hat{\mathfrak{x}}(\alpha, 90° - \delta) = (\cos\delta \cos\alpha, \cos\delta \sin\alpha, \sin\delta)^T$, which are the rectangular coordinates of the star on the unit sphere, in the system of mean coordinates. This avoids the singularity at the poles (where the right ascension is not defined) and one should seriously consider to adopt this practice for all catalogues. The components of $\hat{\mathfrak{x}}(\alpha, 90° - \delta)$, when given to eight figures, define a position to better than $0.''01$ and avoid our clumsy system of measuring angles in terms of hours, minutes, and seconds of time in one coordinate, and in degrees, and minutes and seconds of arc in the other. Star positions might as well be given in terms of the components of $\hat{\mathfrak{x}}$ (which are their direction cosines in the system to which they are referred) as in terms of α and δ.

One should also realize that α and δ are in no way "natural" for the definition of star positions. There is no good reason why *star positions* should not be given in terms of direction cosines *with respect to the galactic system*, which is much more "natural" in connection with the stars than with the equatorial system which is essentially determined by the kinematics of the Earth alone. It might, however, be argued that right ascensions and declinations are the direct products of observations and that the transformations of the positions to any other system should be performed by the user, especially those transformations to the galactic system whose definition has not remained uniform over the last few decades; besides, the precise orientation of the galactic equator and the direction to the galactic center have not yet been rigorously defined. To these objections it could be replied that the published mean positions of the stars are themselves already the results of rather involved reductions of directly observed quantities, so that another transformation, namely that from the equator system to the galactic system, would not complicate matters very much more, especially since there are no conceptual difficulties associated with it and it could be easily and routinely performed by a computer. As to the objection concerning the "accurate" values of the transformation parameters, one could follow the present practice of arbitrarily defining the orientation of the galactic equator and the direction to the galactic center, and leaving it at that. Whenever sufficient new material accumulates for a meaningful revision of the assumed values of the transformation parameters to the galactic system, these revisions can, from time to time, be made. It is in this way revisions are made to the adopted value of the constant of precession, which is, after all, also one of the parameters used to compute the catalogues' positions from the directly observed quantities.

The suggestion of changing the customary form of publishing star positions in terms of α and δ to publishing them in terms of the components of their direction unit vector with respect to the galactic system is also supported by the fact that most astronomers who are interested in proper motions are interested in their galactic components. The responsibility of transforming them (back) to equatorial components may very well be left to the minority of specialists whose object of interest is the investigation of the Earth's inertial motion rather than that of the stars in the galactic system.

The remaining columns of a catalogue usually give the following information:

The *number of single observations* incorporated in the published positions. This information is important in determining the rms. error of a given position and may differ for right ascension and declination of the same star.

The *average of the epochs of observation.* They may be different in right ascension and declination for some stars. The serial *number of the "zones"* in which the star was observed. From these, the original individual observations can sometimes be reconstructed or simply looked up.

The serial *numbers of the star in other star lists* (BD CPD CoD) of other catalogues. *Proper motions* (where available or desired), *spectral types, color indices,* and *remarks.*

The proper motions given in all of the old and most of the new nonfundamental catalogues are neither sufficiently accurate nor sufficiently free from systematic errors for sophisticated investigations. They should, as a rule, be used with extreme caution. Also, always note to which constant of precession the proper motions refer.

3.2 The Handling of Star Catalogues

3.2.1 THE TRANSFORMATION OF POSITIONS AND PROPER MOTIONS (POSITION UPDATING). A star position component, say, α, refers to an epoch of place t and a coordinate system (equator and equinox) as oriented at a different epoch of orientation, T. This was pointed out before in Section 1.6.3. Thus we write symbolically

$$\alpha = \alpha(t, T); \qquad \delta = \delta(t, T). \tag{3.2}$$

Frequently, it is necessary to change both t and T, that is, compute the position for a different epoch (of place) and also change the orientation of the coordinate system. The most general transformation will calculate α and δ for t and T from α and δ for t_0 to T_0. This transformation is expressed most succinctly in terms of direction unit vectors associated with a position. We introduce $\hat{\mathbf{y}}(t, T) = \hat{\mathbf{x}}[\alpha(t, T);$

$90° - \delta(t, T)]$ and have, according to equations (1.9) and (1.13)

$$\begin{aligned} \hat{\mathbf{y}}(t, T) &= \mathbf{S}(T, T_0)\left[\hat{\mathbf{y}}(t_0, T_0) + (t - t_0)\frac{\dot{\mathbf{x}}}{r}(t_0, T_0)\right] \\ &= \mathbf{S}(T, T_0)\hat{\mathbf{y}}(t, T_0) \\ &= \hat{\mathbf{y}}(t_0, T) + (t - t_0)\frac{\dot{\mathbf{x}}}{r}(t_0, T), \end{aligned} \tag{3.3}$$

where r is the distance to the star and $[\dot{\mathbf{x}}/r]$ (t, T) is the velocity vector divided by the distance at the epoch t referred to the coordinate system as oriented at T. For $t - t_0 = 0$, the equations (3.3) perform the precession of the coordinates α and δ from the orientation epoch T_0 to the orientation epoch T.

The matrix $\mathbf{S}$ is calculated from equation (2.43), where the arguments of the rotation matrices are given as follows:

$$\begin{aligned} \xi_0 &= (23042.''53 + 139.''75\tau + 0.''06\tau^2)\mathrm{T} + (30.''23 - 0.''27\tau)\mathrm{T}^2 + 18.''00\mathrm{T}^3 \\ z &= \xi_0 + (79.''27 + 0.''66\tau)\mathrm{T}^2 + 0.''32\mathrm{T}^3 \\ J &= (20046.''85 - 85.''33\tau - 0.''37\tau^2)\mathrm{T} + (-42.''67 - 0.''37\tau)\mathrm{T}^2 - 41.''80\mathrm{T}^3, \end{aligned} \tag{3.4}$$

where τ is $(T_0 - 1.900)$ and $\mathrm{T} = T - \tau$, both reckoned in tropical millenia.

A further simplification results from the consideration that $\mathbf{R}_3(\phi)\hat{\mathbf{x}}(\alpha, \delta) = \hat{\mathbf{x}}(\alpha - \phi, \delta)$. For $t = t_0$, equation (3.3) thus becomes, considering equation (2.43)

$$\hat{\mathbf{x}}[\alpha(T) - z - 90°;\ \ 90° - \delta(T)] = \mathbf{R}_1(J)\hat{\mathbf{x}}[\alpha(T_0) + \zeta_0 - 90°;\ \ 90° - \delta(T_0)]. \tag{3.5}$$

Equations (3.5) are rigorous and simple, and should always be used (at least when a computer is available) to calculate the effects of precession.

According to the second equation (3.3) the transformation would be performed by referring, in the coordinate system as oriented at T_0, the position to the epoch t by applying the proper motions as referred to T_0, and then transforming (by the rotation matrix $\mathbf{S}$) this position to the coordinate system as oriented at T.

According to the third equation (3.3), the position for epoch t_0 would first be transformed to the new epoch of orientation, T, and then referred to the epoch of place, t, by adding the influence of the proper motion. Note that in this case, however, the change of the proper motion components through precession must be considered before the proper motions are applied. This apparently involves more work than the application of the second equation (3.3), which will thus be the preferred one.

Since proper motion components are also part of the information given in star catalogues, especially fundamental ones, we will also have to give the solution to the following problem: Given are μ_α, μ_δ and $\dot{r}/r$ for the epoch of place t_0 and the epoch of orientation T_0. Find from this, and whatever other information is necessary, the value of these quantities for the epoch of place t and the epoch of orientation, T. If the epoch of orientation is to remain unchanged (that is $T = T_0$), this problem has already been solved by the equations (2.23) and (2.24). We are thus faced with the special problem of calculating the influence of precession on the proper motion components only, since the radial velocity is invariant with respect to rotations of the coordinate system. This problem can be solved rigorously as follows:

We obtain from considering equations (3.3), since $r(t_0, T) = r(t_0, T_0)$,

$$\dot{\mathbf{x}}(t_0, T) = \mathbf{S}(T, T_0)\dot{\mathbf{x}}(t_0, T_0). \tag{3.6}$$

In this equation, imagine the velocity vectors $\dot{\mathbf{x}}$ being expressed by α, δ, r and their time derivatives as in equation (1.10). Equation (3.6) then constitutes a linear system of equations in the unknowns $\dot{r}/r$, μ_α and μ_δ at T. Their coefficients are functions of α and δ at T. The right-hand side involves only $\mathbf{S}$ and position and proper motion components at T_0, all of which are regarded as knowns. Before the system can be solved for the unknowns, α and δ must be available for (that is, precessed to) the new epoch of orientation, T. This is accomplished by equations (3.5).

The procedure just outlined, although rigorous, is quite cumbersome and will have to be applied only if $T - T_0$ is large. Besides, whenever $\dot{r}$ and/or r are unknown, one will have to assume $\dot{r}/r = 0$. In practice, the approximate formula

$$\mu_\alpha(T) = \mu_\alpha(T_0) + (T - T_0)\frac{\partial\mu_\alpha}{\partial T} \tag{3.7}$$

and an analogous one for $\mu_\delta(T)$ will give results that are, to the required accuracy, identical with those obtained by the rigorous method. The expressions for $\partial\mu_\alpha/\partial T$ and $\partial\mu_\delta/\partial T$ were given above in equations (1.14).

The transformation of proper motions from the epochs (t_0, T_0) to the epochs (t, T) is accomplished by first computing the motions for (t, T_0) through application of the formulas (2.23) and (2.24), and then referring these to (t, T) by the process just outlined.

3.2.2 THE IMPROVEMENT OF COMPLETE CATALOGUES BY THE INCORPORATION OF NEW POSITIONS. Consider the situation that a catalogue contains positions and proper motions for different epochs of place, but one common epoch of orientation. The typical user of a star catalogue is, however, not interested in that position of a star which is recorded in a particular catalogue, but rather interested in exerting some reasonable effort to obtain a more accurate position and proper motion for a particular star than he will find in any catalogue. A typical nonspecialist user will not aim at the best position and proper motion of a star which can be derived from the entirety of the available material; this would rather be the aim of the compiler of a fundamental catalogue. He will, however, be interested in improving a position (and proper motion) published in an available compilation catalogue, such as the GC (see Section 4.2.4) or the Catalog of the

Smithsonian Astrophysical Observatory (see Section 4.3.1), by combining the information in the catalogue with a selection from the material which was not used in the compilation of the catalogue and may have only recently become available. Procedures toward this purpose were described in the past by various authors, for example by Newcomb (1906, pp. 369–376), Boss (1937, pp. 49–51), and most recently by Eichhorn and Googe (1969). The developments in this section closely follow the procedures given in their 1969 paper.

The simplest case, namely the calculation of proper motions μ_α from right ascensions $\alpha(t_1)$ and $\alpha(t_2)$ at epochs (of place) t_1 and t_2, respectively, was described above in Section 2.4.2 by equation (2.25). In the same section, an iterative procedure was suggested for taking account of the fact that right ascensions and declinations are not strictly linear functions of the epoch of place. We shall, however, assume in the present section, that $\alpha(t) = \alpha(t_0) + (t - t_0)\mu_\alpha$ with the understanding that the true, nonlinear relationship between α, t and μ may be accounted for by the same process that was suggested in Section 2.4.2.

If $\epsilon_1 = \epsilon_\alpha(t_1)$ and $\epsilon_2 = \epsilon_\alpha(t_2)$ are the rms. errors of $\alpha(t_1)$ and $\alpha(t_2)$, respectively, the error of μ_α computed from them is given by

$$\epsilon_\mu = \frac{\sqrt{\epsilon_1^2 + \epsilon_2^2}}{|t_1 - t_2|} \tag{3.8}$$

and the error of an inter- or extrapolated α at the epoch t is

$$\epsilon_\alpha(t) = \frac{\sqrt{\epsilon_1^2(t_2 - t)^2 + \epsilon_2^2(t_1 - t)^2}}{|t_1 - t_2|}. \tag{3.9}$$

By solving the equation $d\epsilon_\alpha(t)/dt = 0$ for t, one obtains

$$t_c = \frac{\epsilon_1^2 t_2 + \epsilon_2^2 t_1}{\epsilon_1^2 + \epsilon_2^2} \tag{3.10}$$

as the epoch for which the position interpolated between $\alpha(t_1)$ and $\alpha(t_2)$ is most accurate.

In the terminology of Newcomb, t_c is called the *central date.*

Clearly, the central date cannot be calculated unless the rms. errors of the positions incorporated are known. By inserting the expression for t_c given in equations (3.10) for t and in equations (3.9), we obtain for the rms. error of the position at the central date

$$\epsilon_\alpha(t_c) = \frac{\epsilon_1 \epsilon_2}{\sqrt{\epsilon_1^2 + \epsilon_2^2}}. \tag{3.11}$$

If $\epsilon_\alpha(t_c)$ is used in equation (3.9), we obtain the simple expression

$$\epsilon_\alpha(t) = [\epsilon_\alpha^2(t_c) + (t - t_c)^2 \epsilon_\mu^2]^{1/2}. \tag{3.12}$$

We now assume that a catalogue is available which contains estimates for the position $\alpha(t_1)$ for the epoch t_1, and proper motion μ, of a star, together with $\epsilon(t_c)$, the rms. error of the position at the central date t_c, ϵ_μ, the rms. error of μ, and t_c itself.

Also assume that another independent catalogue is available which contains an estimate of $\alpha(t_2)$ of the star's position at the epoch t_2, whose rms. error is ϵ_2.

Our aim is to derive from all these data improved estimates $\alpha_i(t_1)$ and μ_i of the star's position at t_1, its proper motion, the new central date t_{ci} referring to α_i and μ_i, and the rms. errors $\epsilon(t_{ci})$ and $\epsilon_{\mu i}$, respectively.

According to the developments in the above quoted paper by Eichhorn and Googe (1969), we have

$$\alpha_i(t_1) = \frac{1}{e}\{[\epsilon_2^2 + (t_2 - t_1)(t_2 - t_c)\epsilon_\mu^2]\alpha(t_1) + [\epsilon^2(t_c) + (t_1 - t_c)(t_2 - t_c)\epsilon_\mu^2][\alpha(t_2) - (t_2 - t_1)\mu]\}, \quad (3.13)$$

and

$$\mu_i = \frac{1}{e}\{[\epsilon_2^2 + \epsilon^2(t_c) + (t_1 - t_c)(t_2 - t_c)\epsilon_\mu^2]\mu + (t_2 - t_c)\epsilon_\mu^2[\alpha(t_2) - \alpha(t_1)]\}, \quad (3.14)$$

where

$$e = \epsilon^2(t_c) + \epsilon_2^2 + \epsilon_\mu^2(t_2 - t_c)^2. \quad (3.15)$$

The new central date t_{ci} is given by

$$t_{ci} = \frac{t_2\epsilon^2(t_c) + t_c\epsilon_2^2}{\epsilon^2(t_c) + \epsilon_2^2} \quad (3.16)$$

the rms. error $\epsilon(t_{ci})$ of $\alpha(t_{ci})$ is

$$\epsilon(t_{ci}) = \frac{\epsilon(t_c)\epsilon_2}{\sqrt{\epsilon^2(t_c) + \epsilon_2^2}}, \quad (3.17)$$

and the rms. error $\epsilon_{\mu i}$ of the improved estimate μ_i of the proper motion by

$$\epsilon_{\mu i} = \left(\frac{1}{\epsilon_\mu^2} + \frac{(t_2 - t_c)^2}{\epsilon^2(t_c) + \epsilon_2^2}\right)^{-1/2}. \quad (3.18)$$

The advantage of these formulas is that after the incorporation of each additional position estimate, the results will appear in the same form as before, so that the formulas (3.13) through (3.18) form an iteration scheme, the results of which are identical with what would be obtained if all the available data (that is, positions in dependence on their epochs) were subjected to a least squares solution for $\alpha_i(t_1)$ and μ_i on the basis of condition equations of the form (2.26) as described above in Section 2.4.2. Needless to say, the individual positions $\alpha(t_i)$ would have to be assigned weights proportional to $1/\epsilon_i^2$.

Another advantage of the iteration scheme proposed in this section is that not only the $\alpha_i(t_c)$ and μ_i, and perhaps also their rms. errors will be calculated and recorded, but also t_{ci}. Most users of the least squares adjustment technique, unfortunately, habitually neglect to communicate in addition to the unknowns and their mean errors also the *covariance matrix* of the unknowns, which they must have calculated anyway in the course of their computations if they quote the rms. errors of the unknowns.

Only if the covariance matrix of the "unknowns" (or, more correctly, the adjustment parameters) is known, will it be possible to combine newly available observation material with the results based on all previously incorporated observations. In the case at hand, the covariance matrix depends only on $\epsilon(t_c)$, ϵ_μ and t_c—see Eichhorn and Googe (1969), their formulas (6) or (16)—and all three of these quantities are needed in the formulas (3.13) through (3.18). They ought to be listed routinely in all modern compilation catalogues which give positions and proper motions, so that new observations of a star position may be used to improve the data given in the compilation catalogue without it being necessary to refer to the observations that were originally used for the derivation of the data in the compilation catalogue.

3.2.3 SYSTEMATIC CORRECTIONS TO CATALOGUE POSITIONS.

3.2.3.1 DEFINITIONS; LIMITATIONS TO THEIR DETERMINACY. In Sections 1.6.6 and 1.6.7 above it was pointed out that the set of star positions listed in a catalogue defines a coordinate system, the system of this catalogue. A subset of positions also defines a system. Of particular interest are those subsets which contain all catalogued positions in a relatively small coherent area of the sky which is a part (subset) of the area covered by the whole catalogue. This may be the entire sky, but most catalogues contain positions only within a zone. In the above-mentioned sections it was pointed out that the catalogued positions define a coordinate system in each such region. If this system stays the same as the region changes, the catalogue was termed *self-consistent.* Most catalogues are not self-consistent and the positions listed in them, therefore, require *systematic corrections* to achieve self-consistency. We must also remember that the true system, that is the inertial system which the catalogue is intended to represent, is rigorously not available but may only be achieved by closer and closer approximation. Systematic corrections will thus not reduce a catalogue to the true system but to some, and not always the best, approximation. Another reason why systematic corrections will not transform the catalogue system (as it varies from region to region) to an ideal, self-consistent system is as follows. The corrections are essentially obtained by investigating the deviations of the difference between certain standard positions and the catalogue positions in a region. The regional averages of these deviations are then either immediately regarded as the systematic corrections, or first subjected to some smoothing process in the transition from region to region and then regarded as corrections. This is an adjustment process and necessarily involves only a finite number of stars. The systematic corrections will thus essentially be the results of an adjustment process, regardless whether they were regarded as completely independent from region to region, or whether they were represented by some mathematical model. Now, it is common knowledge that no adjustment process will yield the true values of the parameters. *All results of any adjustment process are statistical variates with an associated covariance matrix and thus uncertain.* The uncertainty depends, among other things, on the *particular form of the mathematical model* chosen to describe the variations of the corrections from region to region. More complicated models will usually result in smaller residuals but in larger uncertainties of the resulting parameters. A sufficiently complicated reduction model will produce adjustment residuals that are very small, but the effects and even the presence of some parameters will have no physical justification, and thus may produce rather than remove systematic errors. All this illustrates a fact which sophisticated investigators have known for a long time, namely, that the accuracy with which the systematic errors of a series of observation can be determined is limited by the accuracy of these observations themselves.

Another condition necessary for a successful determination of corrections is that the catalogue in question contain a sufficiently large number of stars whose positions are also listed in other catalogues whose systematic errors are known. Otherwise it would be impossible to separate the accidental from the systematic errors.

A large number of star lists contain only relatively small numbers of occasionally observed stars. These lists are thus of rather limited value for more critical work, because their systematic errors cannot be determined with sufficient accuracy.

Very important for the determination of the systematic corrections to the positions in a catalogue is that these corrections be essentially the same for stars of equal magnitude in (about) the same region of the sky, and that the variation of the corrections with magnitude and possibly color is fairly well-behaved. To satisfy this condition, in which case the system is called *homogeneous*, the observations must be made with great care. (See also Section 3.2.5.) It is obviously impossible to observe all the stars which are to be incorporated into a catalogue in one and the same observing session. The systematic errors, by which the stars observed on any particular evening will be affected, depend to a greater or lesser degree on the meteorological and other peculiarities of the occasion. Therefore, if star positions observed on different nights, but with the same instruments, are later to be combined into a homogeneous catalogue, the influence of the peculiarities of the occasion must be eliminated as far as feasible. To achieve this is the principal aim of the observation of zone catalogues described above in Section 2.2.7.

One will, therefore, have little confidence in catalogues which are the result of a compilation of unsystematic and random observations.

The most important and most valuable of the relative catalogues are therefore those which are the result of homogeneous and systematic observations in zones. Even there, the user often need not accept the results of the reductions by the observer (or anyone else) as they were published in the catalogue. Frequently, especially up to the first quarter of this century, the observations themselves were published zone by zone and can be reduced again if the user believes that he has reasons to mistrust the original and published reductions. New reductions of older catalogues were frequently published. Some important catalogues, especially early epoch absolute ones, were even re-reduced more than once.

3.2.3.2 The arguments in the catalogue corrections. We shall now discuss various models that have been used for the representation of the systematic corrections which reduce the system of one catalogue to that of another. Only very few star positions have been observed completely absolute, that is with reference to a coordinate system whose realization was an integral part of the observations themselves. Most star positions have been obtained with respect to a set of reference stars whose absolute positions were assumed known for the reason that relative observations are much less laborious than absolute observations. Only thus could the meridian positions of a relatively large number of stars have been established.

It is obvious that this mode of observing will cause (see Section 1.6.1) the positions to be affected by essentially two kinds of systematic errors: (1) The systematic errors of the reference stars, and (2) systematic deviations from the system of the reference stars.

It is immediately evident that a large part of the systematic errors in the reference star positions will be transferred to the positions observed relative to them. As explained in Section 1.6.6, a fundamental system can be established only as a more or less accurate approximation to the "true physical system." In evaluating the systematic errors of a given catalogue, one will therefore not aim at establishing the systematic corrections to the physical system, but at learning the systematic corrections to one of the fundamental systems currently in use (see Sections 4.1 and 4.2).

The systematic corrections which have to be added to positions published in a certain catalogue to reduce them to a given fundamental system can depend on all observational integral characteristics of a star, namely, position, (apparent) magnitude and color. If we denote by α_f the fundamental and by α_c the catalogue's right ascension of a star, we get $\alpha_f - \alpha_c = \Delta\alpha(\alpha, \delta, m, c) + \epsilon_\alpha$ where $\Delta\alpha$ is the systematic correction for reduction to the fundamental system, and ϵ_α the accidental error. The arguments in the function $\Delta\alpha$ are right ascension, declination, magnitude m, and color equivalent c. The latter, in the absence of color excess may be regarded as a numerical representation of the spectrum of the star. (The same principles apply, of course, to declination also.) $\Delta\alpha$ can be determined (either analytically or in tabular form) only by comparing a sufficiently large number of α_c with the corresponding α_f and by making a suitable assumption about the distribution function of the ϵ_α. Usually, they are assumed to be random.

From comparisons with fundamental positions, $\Delta\alpha$ *can be determined properly as a function of its arguments only if their range in the fundamental positions covers that in the catalogue positions.* This condition is satisfied as far as the positions are concerned. They are, after all, practically identical in the fundamental and the relative catalogues. The faintest stars whose positions have been determined fundamentally (that is, absolutely) are, however, about three magnitudes brighter than the faintest zone stars. The range of colors is practically the same for reference (fundamental), and zone stars.

This situation makes it theoretically impossible to use comparison of positions for achieving a reliable reduction of the system of any catalogue containing faint stars to a fundamental system, without making certain restricting assumptions regarding the magnitude dependence of the systematic errors, or without independently determining the way in which the corrections depend on magnitude.

In order to appreciate this situation, it is necessary to consider its historical development. At this time, investigations on the *color dependence of visually observed star positions* are very sparse. Considering δ only, there is a possibility of a color dependence error due to the general lack of adequate color corrections for use in the computation of refraction. Tables for color correction for use in the computation of refraction exist, but are little used. The appropriate color correction would need to depend not only on the spectral type of the star, but also on the characteristics of the objective and "possibly" the observer. A color dependent error in computing the refraction would, of course, introduce a "color-zenith distance" dependent error in the resulting declinations. Van Herk (1952) found a slight error correlated with spectral type by comparing various Northern Observatory Catalogues with Southern Observatory Catalogues. This process works because a large zenith distance star for a northern observatory would be a small zenith distance star for a southern observatory and vice versa. Scott (1970) computed, but did not publish, the results of such a comparison of a northern catalogue (Pub. U.S.N.O. Vol. 15) with a Cape catalogue observed at about the same epoch. His results were similar to those obtained by van Herk. They both found a few hundredths of a second of arc difference between early and late-type stars. More investigations of this type would probably show that the color equations in the declinations would very likely vary from one observer to the next, and might even vary for one observer in the course of time. We may probably assume that the dependence of the corrections on color may safely be neglected when we deal with visual star positions.

Also, the realization came relatively late that the time of transit recorded by an observer might be early or late depending on the magnitude of the star. This is due to psychologically conditioned differences in the observer's perception. Traditionally, the *magnitude equation* in meridian astronomy is understood to be a correction to the right ascensions which usually varies linearly with the magnitudes. For catalogues that were observed by the eye-ear method without screens, a mean value for the correction of $-0.^{s}0077$ per magnitude is apparent, indicating that the transit of a faint star was usually perceived too late. As described in Section 2.1.6, efforts to avoid the introduction of a magnitude equation started soon after its discovery, but even in observations made now, there is still the chance of a magnitude equation.

In some catalogues, magnitude-dependent errors in the declinations have been noticed. However, corrections compensating for them are seldom available. This is probably so because errors of this nature were unsuspected until recently since no one could think of a good reason why they should occur at all.

3.2.3.3 Corrections involving complete catalogues. According to the definition above in Section 3.1.2, a complete catalogue is one that contains proper motions as well as positions. In most independent catalogues, the epoch of observation differs from star to star; the positions in complete catalogues can all be reduced to the same epoch of observation, say, the average of the epochs of the positions of all stars, by application of the proper motions. In all catalogues that are not complete catalogues, it is, strictly speaking, meaningless to refer to the "epoch of observation of the catalogue," although one may refer to the average epoch (of observation) of the catalogue. In any case, the individual epochs of the stars in an independent catalogue will usually not differ from the average by more than a few (say, five) years. Suppose now we want to establish the "system" of an independent catalogue, C, by comparing it with a complete catalogue whose system is well established, say, the fundamental catalogue F. For this purpose, the positions in both catalogues must, of course, refer to the same epoch of orientation. In addition, each position from F must be reduced to the epoch of observation of the corresponding star in C. If there were enough stars common to F and C, correction functions $\Delta\alpha(\alpha, \delta, t)$ and $\Delta\delta(\alpha, \delta, t)$ could be established which, when added to the right ascensions and declinations in C, reduce them to the system of F. Since not all epochs of the stars in C are the same, it must be expected that, generally, the functions $\Delta\alpha$ and $\Delta\delta$ will depend on the epoch t itself, as was indicated above. When the interval of observation epochs in C is not too large, $\Delta\alpha$ and $\Delta\delta$ will not be appreciably different at the beginning and at the end of this interval; besides, in most cases the catalogues simply don't contain a sufficiently large number of stars to allow one to establish the form of the time dependence of $\Delta\alpha$ and $\Delta\delta$ reliably. The correction tables found in the literature, which give the reductions of individual independent catalogues to fundamental systems, are assumed to be independent of the range in epochs in the independent catalogue. The tables represent the difference between the independent catalogue and the fundamental system at the mean epoch of the independent catalogue (see Section 3.2.3.5).

On the other hand, this assumption cannot be made about the systematic differences between the positions in two complete catalogues, K_1 and K_2. Assume $\Delta\alpha(t_0)$ to be the systematic difference (in the sense $K_2 - K_1$) between the right ascensions in K_2 and K_1, respectively, when all positions refer to the epoch t_0. These will, of course, depend on α, δ and m as discussed above; and quite analogous considerations are valid for the declinations. The systematic differences $\Delta\mu$ between the proper motions (in right ascension) in K_2 and K_1 will depend on the same arguments as the $\Delta\alpha$, except that they will not be time dependent. In order to obtain $\Delta\alpha(t)$, the systematic difference between the positions in K_1 and K_2 when the positions are all reduced to the epoch t, we have

$$\Delta\alpha(t) = \Delta\alpha(t_0) + (t - t_0)\,\Delta\mu, \tag{3.19}$$

and an analogous formula in declination.

Suppose now we have the $\Delta\alpha$'s (in the form of tables, or formulas) to reduce the positions in an independent catalogue C to the system of a fundamental catalogue F_1, but we need to reduce them to the system of another fundamental catalogue F_2. And, in addition, we have (in the sense $F_2 - F_1$) the systematic difference $\Delta_F\alpha(t_0)$ and $\Delta_F\mu$ between the positions in F_2 and F_1 at the epoch t_0, and the proper motion in F_2 and F_1, respectively. If t is the epoch to which that star refers which is to be reduced to the system of F_2, we will have to add to its right ascension the quantity $\Delta\alpha + \Delta_F\alpha(t_0) + (t - t_0)\Delta_F\mu$. Since most stars were observed at different epochs t, this quantity will differ from star to star. Therefore, the reduction of the positions in C from the system of C to that of F_2 cannot be accomplished by a simple table of $\Delta\alpha$'s. If, however, the $\Delta_F\mu$ are reasonably small and the $t - t_0$ are within a certain range, then one may adopt $\Delta\alpha + \Delta_F\alpha(t_0) + (\langle t\rangle - t_0)\Delta_F\mu$ as correction, which now no longer depends on the epoch of observation of the individual star in C. $\langle t\rangle$ is, of course, the average of the epochs of the stars' positions in C. The maximum error committed by using the time independent instead of the time dependent correction is, of course, given by $|(t_{\max} - \langle t\rangle)\Delta_{F\max}\mu|$, where $t_{\max}$ is that individual epoch which deviates most from $\langle t\rangle$, and $\Delta_{F\max}\mu$ the largest value of $\Delta\mu$ for any realistic combination of its arguments. If this error is intolerably high, the rigorous, time dependent form of the corrections must be used.

3.2.3.4 ANALYTICAL MODELS FOR CATALOGUE CORRECTIONS. Models for the functions $\Delta\alpha(\alpha, \delta, m, c)$ and $\Delta\delta(\alpha, \delta, m, c)$ which were mentioned in the previous section must be adopted before the corrections can be applied. By analyzing the process of meridian observations, one may argue that the forms

$$\Delta\alpha = \Delta_1\alpha(\alpha) + \Delta_2\alpha(\delta) + \Delta_3\alpha(m)$$

$$\Delta\delta = \Delta_1\delta(\alpha) + \Delta_2\delta(\delta) + \Delta_3\delta(m)$$

will properly represent these functions. The familiar notation for these equations is

$$\Delta\alpha = \Delta\alpha_\alpha + \Delta\alpha_\delta + \Delta\alpha_m$$

$$\Delta\delta = \Delta\delta_\alpha + \Delta\delta_\delta + \Delta\delta_m \tag{3.20}$$

where $\Delta\alpha_\delta$, for instance, denotes a correction to the right ascension which is a function of the declination only. $\Delta\delta_m$ is usually considered insignificant and thus left out, so that the standard form for the $\Delta\alpha$ and $\Delta\delta$ is usually written as

$$\Delta\alpha = \Delta\alpha_\alpha + \Delta\alpha_\delta + \Delta\alpha_m$$

$$\Delta\delta = \Delta\delta_\alpha + \Delta\delta_\delta. \tag{3.20a}$$

Most series of meridian observations are made with the instrument sometimes pointing to the north, sometimes to the south of the zenith. Experience has shown that the *system defined by an instrument often undergoes a discontinuity when the instrument passes through the zenith.* In part, this is certainly so because the observer's feet always point north or south as the telescope points north or south. Unless the observations are entirely objective (and only the most modern ones are) the observer, through his judgment, is an integral part of the apparatus through which the positions are obtained. The discontinuity in the observer's position, which accompanies the instrument's passing the zenith, will at least in part be responsible for the resulting *zenith discontinuity* in the observations. Thus, one must always watch for a zenith discontinuity when establishing the $\Delta\alpha_\delta$ and $\Delta\delta_\delta$ functions for any catalogues.

The zenith discontinuity manifests itself as follows: while $\Delta\alpha_\delta$ and $\Delta\delta_\delta$, characteristic for a certain catalogue, are generally continuous functions of declination, a discontinuity will occur at that declination which corresponds to the zenith of the observer's position.

It is obvious that this classical model (equations (3.20a)) is a considerably specialized form and must be regarded as an approximation only. In many cases it will not even be suitable to remove in their entirety the systematic errors depending on position only. This particular model for the corrections was originally chosen by catalogue astronomers because it was perhaps believed that the arrangement of the observations was such that no additional parameters could be obtained. Furthermore, the observing instruments (transit circles) were assumed to behave in accordance with ideal models and, therefore, no one recognized a good reason why the various effects that produce declination dependent systematic errors (errors in refraction, circle division, and so forth) should also depend on right ascension. It was realized only recently that, in practice, they do.

It is, on the other hand, also obvious that the form given in equations (3.20) suffices to represent errors at least for zone catalogues that give positions of stars in fairly narrow strips of declination. For catalogues that cover a more extended region of the sky, a more adequate representation of the corrections can be given by providing several sets of, say, $\Delta\alpha_\alpha$, for various more or less narrow declination zones.

In evaluating the traditional, time-tested procedures of positional astronomy, one must consider two all-too-easily forgotten things:

First, every correction table or formula (giving the Δ with the various arguments) is the result of many hundreds of individual comparisons, in which the plots of the residuals vs. right ascension or declination show a considerable scattering of the individual points; that is, the accidental errors are, more often than not, considerably larger on the average, than the systematic errors. If a formula of the form (3.20a) is found to remove all systematic errors that can be "seen," one will, as a rule, be satisfied, for the reason given below.

Second, the processing of catalogue comparisons without electronic computing aids is extremely tedious. Until the efforts at the Heidelberg Astronomisches Rechen-Institut (Brosche 1966 and 1970, see next Section), correction tables were derived graphically. Of course, the material would have provided an opportunity par excellence for the application of a least squares adjustment, but the formation of the many hundreds of normal equations and their solution on a desk calculator—let alone logarithms—would have been enough to scare even the most enthusiastic data reduction expert.

One might also raise doubts as to whether more rigorous determinations (in the sense of mathematical statistics) would have given "better" results. The numerical values of the corrections depend to some extent on the relative weights assigned the individual position comparisons, and weighting beyond a certain point is still arbitrary.

3.2.3.5 BROSCHE'S MODEL. The traditions of the logarithmic and desk calculator periods and the limitations in computing and analyzing power need not, of course, be carried into the electronic computing age. Therefore, Brosche (1966) devised a method to carry out catalogue comparisons (in effect, the derivation of corrections for the reduction of the positions and/or proper motions in one catalogue to the system of another one) which abandons the specialized form as given in equations (3.20). Instead, the corrections are postulated as depending on position only and written in the form

$$\Delta(\alpha, \delta) = \sum_{n,m} P_{nm}(\delta)\,(S_{nm} \sin m\alpha + C_{nm} \cos m\alpha) \tag{3.21}$$

where $P_{nm}(\delta)$ are the associated Legendre polynomials, and S_{nm} and C_{nm} numerical coefficients. The representation given by formula (3.21) is equivalent to a sum of spherical harmonics.

The $P_{nm} \sin m\alpha$ and $P_{nm} \cos m\alpha$ are orthogonal functions. If the stars were distributed homogeneously over the sphere, the coefficients S_{nm} and C_{nm} could be obtained by the well-known integrals, or to a high degree of approximation, their corresponding Riemann sums, to which every star contributes one term. This means that the S_{nm} and C_{nm} could be calculated without the use of least squares. This orthogonality is lost since the stars in any catalogue are not distributed homogeneously over the sphere. Brosche gets around this difficulty by making use of the fact that a system of linear combinations of the "weighted" spherical harmonics that is, the vectors Φ_{nm} whose components are $S_{nm}P_{nm}$ $(\cos\delta_i)$ $\sin m\alpha_i$—and an analogous one with C_{nm}—with every star responsible for one of the i will form an orthogonal system, so that the then uncorrelated coefficients of these new linear combinations can be determined independently of each other by the evaluation of certain (Riemann) sums without a least squares adjustment. This is particularly valuable since the addition or removal of one of the orthogonal functions leaves the values of the coefficients of the others unchanged. The coefficients S and C of the original spherical harmonics can then be obtained relatively easily from the coefficients of the linear combinations.

The number of terms in the series given by equation (3.21) that are needed to adequately represent a particular systematic error is determined by an objective statistical criterion. This makes the results reproducible and independent of the arbitrary judgment of the investigators, and thus constitutes an important step toward the unique definition of the concept of catalogue systems.

Brosche's investigation combines the best points of the German and the American schools for the construction of a fundamental system, which advocate finding the Δ by graphs and by simple formulas, respectively. Brosche (1970) has written a computer program for the application of his method, and it is to be hoped that this or equivalent procedures will always be used in the future for catalogue comparisons.

Even the magnitude equation can be explored in this way. By assuming that

$$\Delta(\alpha, \delta, m) = \Delta_1(\alpha, \delta) + m\Delta_2(\alpha, \delta), \tag{3.22}$$

$\Delta(\alpha, \delta, m)$ can be established if the material is divided into two magnitude groups, with average magnitudes m_1 and m_2 respectively, since from the equations

$$\Delta(\alpha, \delta, m_1) = \Delta_1(\alpha, \delta) + m_1\Delta_2(\alpha, \delta)$$

$$\Delta(\alpha, \delta, m_2) = \Delta_1(\alpha, \delta) + m_2\Delta_2(\alpha, \delta)$$

the functions Δ_1 and Δ_2 can generally be determined. Equation (3.22) is, of course, equivalent to postulating a linear magnitude equation whose coefficient is, however, a function of the position.

3.2.3.6 SOURCES OF SYSTEMATIC CORRECTIONS TO THE POSITIONS IN VARIOUS CATALOGUES. The usefulness of a catalogue of star positions is severely restricted if the systematic errors affecting the positions contained in it are not known, or not sufficiently well known, and also if the standard deviation of the positions is not known. So far, only two catalogue comparisons have been published by Brosche (1966, 1970) that make use of his method. Most of the tables of catalogue corrections follow the form in equations (3.20a).

The establishment of correction tables, which give the $\Delta\alpha_\alpha$ and so forth, necessary to reduce the positions to some standard system, and the determination of the standard deviation of the positions, require fairly extensive comparisons of the positions from the catalogue under investigation with the positions in the standard system because apart from a zero point error, the accidental errors are usually considerably larger than the systematic error. The classical way to establish these corrections is as follows: the residuals "standard minus catalogue" are freed from their all-over average and sorted with respect to right ascension. Their systematic run with right ascension is then removed (that is, the $\Delta\alpha_\alpha$ and/or $\Delta\delta_\alpha$). If the catalogue covers a wide range in declination, the $\Delta\alpha_\alpha$ and $\Delta\delta_\alpha$ are sometimes determined for narrower declination ranges when the results indicate that this would bring about a better representation of the residuals. The residuals minus the $\Delta\alpha_\alpha$ or $\Delta\delta_\alpha$, as the case may be, are then sorted according to declination, and the $\Delta\alpha_\delta$ and $\Delta\delta_\delta$ are determined. Many investigators, however, for instance those who compiled the FK3, prefer to determine $\Delta\delta_\delta$ and $\Delta\alpha_\delta$ first. A discussion of the causes for various terms (that is, $\Delta\alpha_\alpha$, $\Delta\alpha_\delta$, $\Delta\alpha_m$, $\Delta\delta_\alpha$, $\Delta\delta_\delta$) is found on pp. 156–166 of Vol. I of the GC (Boss 1937) where reference is also made to more complicated models for the systematic catalogue errors. For instance, it is explained there that $\Delta\alpha_\alpha$ is usually periodic in α, but that it may also vary with declination. In these cases, the correction will be given as the product of a periodic function in α times a function of δ which is specified in every single case where its application is warranted. An annual periodic error in determining the polar point would, for instance, introduce an error of the form $\epsilon(\alpha)\tan\delta$.

There are several sources for the systematic corrections of the several catalogues to a fundamental system. The most important are:

(1) Appendix III of GC. I, that is the first volume of the GC (Boss 1937).
(2) The tables by Gyllenberg (1948).
(3) The paper by Nowacki and Strobel (1968).
(4) The tables in the publication by Kopff, Nowacki and Strobel (1964).
(5) The tables in Wahl's (1937) dissertation.

There exist also several other collections of tables, see for example, Auwers (1903), but we shall not list these here since they all provide the corrections to catalogue systems that are of only historical interest. All corrections are referred to the average of the epochs of the positions in the catalogue in question.

The items (1) and (2) above give the catalogue corrections to the system of the GC. Boss' and Gyllenberg's tables are complementary. Boss gives corrections only to those catalogues which appear before the Compilation of the GC; Gyllenberg's tables cover the post-GC catalogues and some of those pre-GC catalogues which Boss had left out. For the convenience of arrangement, the systematic corrections for zone catalogues are given in (1) after the corrections to the other catalogues. Also note that the $\Delta\alpha_m$ in (1) have m on the Harvard system of magnitude as argument. See Section 4.2.4 concerning the consequences of this and the difficulties thus caused.

The tables in the items (3), (4), and (5) give corrections to the system of the FK3. The tables (3) encompass those catalogues which were used for the derivation of individual corrections to the stars in the FK4 but had not already been used in the work toward the FK3. The epochs of the most recent catalogues considered there are 1956. Comprehensive collections of correction tables for catalogues with epochs after 1956 have at this time (1970) not appeared anywhere. The tables in (3) differ but little from those given in (4); as a matter of fact, they are based on them, and are, in some instances, identical to them. There are, however, in (4) some correction tables for catalogues that had already been incorporated in the FK3. Wahl's dissertation contains the systematic corrections of nine catalogues to the FK3 system.

Except for (1) and (2), there is considerable overlap between these tables. At this time, investigations have not been published which would decide which of the sources are most reliable, and the user will probably do best to take averages. However, since Gyllenberg derived all corrections listed in (2) graphically, the other sources will probably give the better results.

In (3), (4), and (5), only the fundamental stars, that is, the stars in the FK3 and FK4, were utilized for the derivation of the corrections. In the presence of significant magnitude effects, these corrections will therefore not be valid in the faint stars. In (1) and (2), the corrections are based on comparisons with the stars in the GC which goes to considerably fainter magnitudes than does the FK4. These are, therefore, perhaps preferable for fainter magnitudes. A disadvantage of the α corrections in (1) and (2) is, that the $\Delta\alpha_\alpha$ were forced to be of the form $\Delta\alpha_\alpha = a + b \sin(\alpha - \alpha_1) + c \sin 2(\alpha - \alpha_1)$. The universal validity of this restriction is certainly not assured.

At this time, reliable systematic catalogue corrections are simply not available for the entire magnitude range of all catalogues. The reason for this is, that computers were only becoming easily available after the termination of the work toward the establishment of the FK4. The catalogue comparisons necessary for the computation of the correction tables are very tedious, even now that computers are generally available, because the data in the catalogues must still be transferred to either punched cards or magnetic tape.

Also, the derivation of reliable corrections would presume the existence of a standard of comparison, that is, a true General Catalogue. Such a catalogue does not exist at this time, as will be pointed out below (Section 3.2.5.). It is clear that there is much left for industrious investigators to do.

3.2.4 ACCURACY (WEIGHT) OF CATALOGUED STAR POSITIONS.

3.2.4.1 THE ACCIDENTAL RMS. ERROR OF STAR POSITIONS. All star positions that are listed in any catalogue are in error. There are two causes contributing to this, namely, the (systematic) errors of the catalogue system, that is, the deviations of the catalogue system from a fundamental, or an inertial, system, and the accidental errors of the positions within the system of the catalogue. In this section we shall deal only with the latter.

Obviously, not all star positions contained in the same catalogue will have the same accidental standard deviation. For one, most catalogues carry a column showing that different numbers of individual observations of the catalogued stars were combined into the positions finally printed. Obviously, the accuracy will increase with the number of observations incorporated into position. Unless some device was employed to compensate for the stars' differences in magnitude (as for instance a filter), it is plausible that the observations of very bright stars will not have been as accurate as those of stars of average magnitude, while very faint stars, too, will have proved difficult to observe accurately. The accuracy of the observations will depend on the instrument used, the care with which the observations were made and reduced, and the skill of the observer. It may also depend on the season, the time of the night, the position of the observer's head (zenith distance) and even the star's color. The establishment of a formula giving the standard deviation of the positions as a function of all these quantities would require not only a tremendous amount of numerical work, but also considerable effort in determining an entirely adequate analytical form for representing the standard deviations. Performing all this work without an electronic computer is out of the question.

Weights rather than standard deviations for catalogued positions have been established and published in the past. The compilations of weights for catalogues which have been published so far (Auwers 1900, 1903; L. Boss 1903e; B. Boss 1937; Wahl 1937; Kopff 1938; Gyllenberg 1948; Nowacki and Strobel 1968) usually give the weights of the $\alpha\cos\delta$ and δ with only the number of observations as argument. Only in a few cases are different weights given for separate magnitude intervals and, in rare cases, as functions of the stars' coordinates. Auwers (1903) warns that the weights refer to the bright stars in the catalogues referred to in his particular paper, and that the weights of the positions of the "telescopic" stars (that is those with a magnitude fainter than $6.^m0$) might differ considerably from these.

No investigations have so far been published that will, for a large number of catalogues, give a precept on how the rms. error of a position published in them can be found as a function of all arguments (as enumerated above) which this error may depend on. As a matter of fact, the tables giving systematic corrections and weights quoted above give the weights usually as a function of the number n of observations incorporated into the position. B. Boss and Gyllenberg give, for every catalogue which they include, the weights p (referred to a unit weight which corresponds to a *probable* error of $0.''30$) separately for $\alpha\cos\delta$ and δ. They postulate the probable error to depend only on n and be of the form $\sqrt{\epsilon_0 + \epsilon_1/n}$, that is, essentially equation (3.24) below; ϵ_0 and ϵ_1 are formed from a type of trial and error solution, see L. Boss (1903e).[1] The standard deviations (rms. errors) $\epsilon_\alpha \cos\delta$ and ϵ_δ of a position are thus computed from the weights given by Boss and by Gyllenberg by the formulas

$$\epsilon_\alpha = 0.''455/\sqrt{p}\cos\delta, \qquad \epsilon_\delta = 0.''455/\sqrt{p}. \tag{3.23}$$

These weights were found empirically from the residuals obtained by comparing the catalogued positions with those in the GC, after the systematic corrections to the GC system had been applied. Boss gives details in GC.I, pp. 166 & 168, while Gyllenberg does not give any further details. Their weights would correspond to *external* weights p_e in the terminology of the investigators of the German School.

The investigators of the German School follow the procedures originally established by Auwers. A distinction is made between external weights, p_e and internal weights p_i. The latter are those calculated from the rms. errors quoted in the introductions to the individual catalogues by the various observers; the p_e are obtained from the residuals which result from comparing the catalogue in question to a very accurate fundamental catalogue (which is considered as practically error-free) after it has been corrected to the system of the fundamental catalogue. The actual procedure used for finding the weights is described in great detail by Kopff, Nowacki and Strobel (1964, pp. 2–5). They found, by and large, that the ratios p_e/p_i for the various catalogues were somewhat less than, but nearly equal, to 1. The smallest value found was $\frac{1}{2}$. p_e was preferred over p_i in all applications; for example, the

1. It thus is inexact to speak of "the weight of a catalogue," although $p(\bar{n})$ could be so termed where $\bar{n}$ is the average number of observations incorporated into a position listed in that particular catalogue.

calculation of the individual corrections to the FK3 stars for the formation of the FK3R (see Section 4.1.5.2). The weights given by the German School investigators refer to rms. errors (mean errors) of unit weight of $0.^s0381$ in $\alpha\cos\delta$ and $0.''396$ in δ respectively. The formula

$$\epsilon^2(n) = a^2 + \frac{b^2}{n} \tag{3.24}$$

is postulated for the dependence of the square of the rms. error (the variance) of a position on the number n of observations incorporated. Kopff, Nowacki and Strobel are of the opinion that the values for a^2 are nearly identical for all modern catalogues and quote $a_\alpha = \sqrt{0.0000314}$ and $a_\delta = \sqrt{0.00785}$ in seconds of time and of arc, respectively. b^2, according to these authors, varies from catalogue to catalogue and must therefore be determined individually. They recommend that this be done by using the average error, d, which is defined as the arithmetic mean of the absolute values of the residuals and very easy to obtain. For Gaussian error distributions (and it is assumed that the accidental rms. errors of the catalogued positions follow this distribution), the relationship $\epsilon = \sqrt{\pi/2}d$ holds between rms. errors ϵ and average errors d. The relationship (3.24) is thus equivalent to

$$d^2(n) = \bar{a}^2 + \bar{b}^2/n, \tag{3.24a}$$

with $\bar{a} = \sqrt{0.000020}$ seconds of time for $\alpha\cos\delta$, and $\bar{a} = \sqrt{0.0050}$ seconds of arc for δ. With these values for $\bar{a}$ and after finding, from the comparison residuals the different values of d for various n, the value of $\bar{b}$ which characterizes the catalogue in question can be calculated for every n. If the formula (3.24a) and the assumption of a constant $\bar{a}$ for all (modern) catalogues are realistic, the values for $\bar{b}$ found from investigating residuals generated by positions with the same number n, in any one catalogue, should essentially be all the same. However, the example given by Kopff, Nowacki and Strobel on p. 5 of their above quoted paper shows that this is not always so. There also appears no good reason why $\bar{a}$ should not vary, even for all modern catalogues.

After $\bar{b}$ has become known—and frequently an average of the values found from the various n-groups is adopted—the average error $d(1)$ as a function of n can be computed from equation (3.24a). The rms. errors of unit weight quoted above correspond to average errors of unit weight of $d_0 = \sqrt{0.000924^s}$ in $\alpha\cos\delta$, and $d_0 = \sqrt{0.0998''}$ in δ. The weight p_n of a position which is the mean of n observations is given by $p_n = d_0^2/d_n^2$. Since, with $\bar{a}$ and $\bar{b}$ known, d_n^2 can be calculated, a table of p_n vs. n can now be established. Alternatively, the d_n^2 can be established directly from groups of residuals with the same n and used for calculating p_n without going the way of finding $\bar{b}$.

3.2.4.2 Critique and suggestions. Anyone following the classical precepts for finding $p(n)$, as outlined in the previous section, must be aware of their imperfections. For one, there is no real justification for the continued use of weights, which offer by themselves no information about the accuracy of the material they are intended to describe, unless the rms. (or any other well-defined) error of unit weight is also known. Rms. errors, however, would give this information immediately. In order to appreciate the practices of the past it is well to remember that the investigators had a huge amount of material in their hands, and could hope to manage this only if they used the simplest methods and procedures that gave them perhaps not the best but still a usable result. Thus is explained the use of the easily obtained average errors in preference to the rms. errors whose computation requires much more labor. Investigators who had to work with logarithms and desk calculators preferred weights which were, for simplicity's sake, almost always rounded to the nearest integer numbers to rms. errors, and tables to formulas. There is, however, no reason why these primitive, and, by present standards, crude procedures should not be changed, considering that the availability of electronic computers has made the amount of computing labor a secondary consideration.

At the present time, one ought to proceed like this: Assume that, for any one catalogue, the comparison residuals v can be represented by the following model, obtained after removal of the systematic corrections,

$$v^2 = a + b/n + cm + dm^2 + ef(z), \tag{3.25}$$

and make a standard least squares adjustment, regarding the v^2 as the observations, and a, b, c, d and e as the unknown parameters. n is, as before, the number of individual observations incorporated into the position, m the magnitude, and $f(z)$ some function, not necessarily monotonically increasing, of the zenith distance.

Formula (3.25) considers the fact that variances will be the sum of the component variances. After $a, \ldots, e$ have been found, equation (3.25) can be used to compute $\epsilon = \sqrt{v^2}$, the expected rms. error, from v^2.

This method will need a considerable amount of investigation before it can be used. It is, for instance, particularly embarrassing when certain, actually-occurring combinations of n, m and z yield negative computed values for v^2. Should this occur, the model must be carefully examined and, when necessary, replaced by a more realistic one. A method very analogous to this one was used by Eichhorn, Googe, Lukac and Murphy (1969) for estimating the rms. errors of input material (plate measurements) in a photographic astrometric problem. Their experience was that the magnitude dependence of the variance had to be represented in the form $a(m - m_0)^4$ with m_0 being a zero magnitude which was also an unknown adjustment parameter. If this was not done, the "bottom" of the $v^2(m)$ curve was not flat enough, and negative computed values for v^2 resulted. Actually, the best form of equation (3.25) will have to be found by trial and error, probably with the help of a lot of plotting and graphing.

3.2.4.3 THE TOTAL ERROR EXPECTANCY OF A STAR POSITION. As pointed out before (see Section 3.2.4.1), the total error of a star position, with respect to any system, is composed of the deviation of the catalogue system from the target system and the error of the position within the catalogue system.

The most important case is the error ϵ_i of a star position with respect to the actual physical inertial system which the catalogue is intended to represent; for instance, the error of a star in a fundamental catalogue with respect to the intended inertial system. In order to compute this error, the error ϵ_s of the system and the error of the star positions ϵ_p with respect to the catalogue system must be calculated separately. Then, we have

$$\epsilon_i^2 = \epsilon_p^2 + \epsilon_s^2. \tag{3.26}$$

ϵ_p (in α and in δ) at the epoch t is calculated with the use of equation (3.12); the data necessary for this are taken from the various catalogues: they are the central epoch t_c, and the rms. errors $\epsilon(t_c)$ and ϵ_μ of position and proper motion (within the catalogue system), respectively.

ϵ_s is calculated from the same formula, but with different data. For the calculation of ϵ_s, t_c is the mean epoch of the catalogues that contributed to the establishlishment of the *system* of the catalogue in question; for example, about 1913 for the FK3 in both coordinates, and 1935 for the FK4 declinations. $\epsilon(t_c)$ and ϵ_μ are the accidental error of the system of positions at its mean epoch, and the accidental error of the system of proper motions, respectively. These data are given for the FK3 in Table IV-2, Section 4.1.4, for the FK4 in Table IV-3 (right ascension system Section 4.1.5.3), and IV-4 and IV-5 (declination system Section 4.1.5.4), respectively.

Finally, it should be noted that the ϵ_s calculated with the use of the aforementioned data measures the errors within the ideal FK4 (or FK3) system, which is known not to be an inertial system because it is based on a constant of precession which is not the most accurate one that is currently available.

For very critical investigations, the contribution of the uncertainty of the reduction from the catalogue system to the fundamental system in question will also have to be considered.

3.2.5 THE CONSTRUCTION OF COMPILATION CATALOGUES. The individual compilation catalogues and the details of the procedures that were actually employed in their construction will be discussed below in Chapter IV. In this section we shall suggest that most general principles that underly the compiling of a star catalogue.

In planning a compilation catalogue, serious consideration must be given to (1) the definition of the system on which the catalogue is to be based, (2) the selection of the catalogues to be incorporated, and (3) the choice of a model for the representation of the systematic corrections of each of the individual catalogues to the chosen fundamental system. This means that a pair of model functions $\Delta_k\alpha(\alpha, \delta, m; a_{k1}, \ldots, a_{km})$ and $\Delta_k\delta(\alpha, \delta, m; b_{k1}, \ldots, b_{kl})$ when added to the positions in a certain catalogue, will reduce them to the system adopted in (1).

Certain conditions must be satisfied before the construction of a compilation catalogue can actually be undertaken. The catalogue positions must all refer to the same epoch of orientation, and the systems represented by the individual catalogues should be homogeneous (see Section 3.2.3.1). A detailed explanation of the last requirement is as follows: assume that the function $\Delta\alpha(\alpha, \delta, m; a_{k1}, \ldots, a_{kl})$ is determined for the k-th catalogue at the point (α_0, δ_0) from the stars in the catalogue in the neighborhood of this point. If the function $\Delta\alpha$ (and also, or course, $\Delta\delta$) shows no significant change no matter which subset of the stars around (α_0, δ_0) in the catalogue are employed in its determination, the system of the catalogue is termed *homogeneous* at (α_0, δ_0). This definition could be made more rigorous by the specification of significance criteria, a required confidence level, closer delineation of the size of the neighborhood of the point, and so forth. It says, in essence, that the system of a catalogue is homogeneous at a point, if the corrections required to put the positions at this point into some other well-defined system, are of essentially the same functional form, no matter from which group of stars they were determined. Since magnitude m is a parameter in $\Delta\alpha$ and $\Delta\delta$, these functions will generally be different for different magnitudes, that is $\Delta\alpha(m_1) \neq \Delta\alpha(m_2)$ when $m_1 \neq m_2$. This does not mean that the system of the catalogue under consideration is not homogeneous. If, however, two sets S_1 and S_2 of catalogue stars within a narrow magnitude interval around m_0 could be chosen such that the corresponding correction functions Δ_{S1} and Δ_{S2} would be significantly different, the system of the catalogue would be *nonhomogeneous*. The most important source of nonhomogeneity in older catalogues is the variation of latitude, which was recognized only relatively late, since about the turn of the century. The same considerations can also be applied with respect to any other of the arguments in the correction functions $\Delta\alpha$ and $\Delta\delta$. It is also conceivable that the correction functions have the star color equivalents as arguments. This may sometimes have to be taken into consideration, but does not change the principles outlined in this section.

The material at hand for the construction of a compilation catalogue are the right ascensions α_{ik} (declinations are treated completely analogously) of star number i in the catalogue number k, the magnitudes m_i (and colors c_i) of the i-th star (which obviously, except in the rare cases of variables, do not depend on the catalogue) and the observation epochs t_{ik}. Also available must be a sufficient number of α_{i0} and μ_{i0}, that is positions and proper motions of stars on the fundamental system which is intended to be the basis of the whole compilation catalogue. The α_{i0} and μ_{i0} are, of course, available only for a selection of the stars and will usually come from an already existing fundamental catalogue. All positions α_i will refer to the same epoch of place, t_0. If, however, the compilation catalogue to be established is the fundamental catalogue itself, the situation is somewhat different and will be discussed further in this section.

The adjustment parameters or star parameters, that is the unknowns in the problem, are obviously the positions α_i that is, $\alpha_i(t_0)$, and proper motions μ_i of all stars involved, and also, although not of primary interest, the catalogue parameters a_{ik} in the correction functions. The equations of condition are (with the appropriately weighted observations on the right-hand side):

For fundamental stars

$$\alpha_i = \alpha_{i0}$$

$$\mu_i = \mu_{i0} \tag{3.27}$$

and for all stars

$$\alpha_i + \mu_i(t_{ik} - t_0) - \Delta_k(a_{k1}, \ldots, a_{kn}) = \alpha_{ik}. \tag{3.28}$$

The equations (3.27) are generally the best way to set up a determinate system of normal equations whose solution are the star and catalogue parameters. If the problem is, however, the establishment of a fundamental catalogue itself, there are usually no observations α_{i0} that come from catalogues for which no corrections Δ_k are required. In this case, equations of the form (3.27) will be absent. In order to avoid getting a singular system of normal equations, the Δ_k from information available from absolutely observed catalogues, must be carefully chosen and some of the a_{kj} subjected to equations of conditions. If, for example, all Δ_k are of the form $\Delta_k = \Delta_k' + a_{kn}$, that is if a bias term a_{kn} is allowed for in all catalogue corrections, the normal equations will be singular. This could be prevented, for instance, by enforcing the condition $\sum_j a_{k_j n} = 0$ where the summation goes over all j that denote a number k_j of an absolutely observed catalogue.

The simultaneous determination of the star and catalogue parameters as proposed here is a problem that is completely analogous to the determination of photographic star positions by the plate overlap method which was discussed above in Section 2.3.8.2. In order to carry out a rigorous adjustment, one could follow the algorithms suggested by Googe, Eichhorn and Lukac (1970). The formulas given in their paper could be used almost as they are by treating the catalogue parameters in the same way the plate constants are treated there. The knowledgable investigator will have no difficulty in setting up the system of normal equations from which all star and catalogue parameters are simultaneously determined as unknowns. The matrix is a banded-bordered matrix, typical for overlap problems. A suggestion for the inversion of such matrices was given by Brown (1968).

The matrix of the complete system of normal equations will be of the order $2s + c$ in each coordinate, where s is the number of stars involved, and c the total number of catalogue parameters. For an undertaking of the size of the General Catalogue, this matrix will be in the order of $10^5 \times 10^5$, which is huge by any standards. Even if one follows the procedure which is standard in plate overlapping (or, in photogrammetric terminology, block adjusting) and eliminates the star parameters, the matrix of the remaining system still is of the order of some $10^3 \times 10^3$, and any manipulation of matrices of this size is hopeless without computers.

If one evaluates the processes used in the past for the construction of compilation catalogues, it is apparent that the procedures suggested above in this section have in principle been followed universally, although the details always deviate rather drastically from those outlined above. Most important of all, computers appeared on the scene rather late and the determination of the unknowns was accomplished by some process of successive iteration that could be accomplished on desk calculators. The astronomers in charge of compiling large catalogues could not be expected to suddenly adjust their thinking to the availability of computers, and therefore, computers were initially used only to perform faster the very same calculations that had previously been carried out with the aid of logarithms and desk calculators. Brosche's (1966) investigations are the first genuine progress and break with a long established procedure, and the principles for the construction of compilation catalogues outlined in this section still have to be applied in full generality.

The most recent detailed suggestions (Brouwer 1960) for the construction of compilation catalogues still reflect the steps generally followed in the past. These were essentially (1) the establishment of a fundamental system, (2) the determination of very simple correction functions of the various catalogues, and (3) the computation of the star constants. The system was essentially determined by averaging the available absolute positions, and the correction functions were essentially determined by comparison of the catalogue positions with the previously determined positions in the fundamental system only. In this way, a very powerful condition which helps the determination of the Δ functions is not utilized, namely, that position and proper motion of a star are unique, and that there must not be any significant systematic differences between the sets of positions that are extracted from different catalogues after application of the appropriate corrections to the same system, even for the positions of nonfundamental stars, and from differentially observed catalogues.

The suggestions in this section for the construction of compilation catalogues are analogous to the plate overlap method in photographic astrometry, while the traditional methods which were actually applied, correspond to the determination of photographic star positions using the classical method.

3.3 Lists of Star Catalogues

3.3.1 RISTENPART'S LISTS. Ristenpart (1901) wrote an article "Sterncataloge und Karten" for the "Handwörterbuch der Astronomie," a five-volume encyclopedic work edited by W. Valentiner. In the appendix to it he gives a list of all star catalogues that appeared up to 1900 which are of any value, except merely historical, for the recording of accurate star positions. This list was reissued as a booklet (Ristenpart 1901a) in a somewhat expanded form.

Ristenpart's list gives the following information: Column I: Author; Column II: Abbreviation (code) for the catalogue, usually consisting of the first letters of the name of either the author or the place of observation and a subscript when necessary to avoid ambiguities; Column III: The number of star positions given in the catalogue; Column IV: Sundry information (which stars observed, limits of coverage in region and magnitude, and so forth); Column V: Epoch of orientation; Column VI: Title of the catalogue.

This list is based in part on the previous similar and in-parts-more-extensive list by Knobel (1877) which contains some useful information concerning catalogues that are now only of historical interest.

3.3.2 GESCHICHTE DES FIXSTERNHIMMELS (GFH). Even toward the end of the last century, the number of catalogues had grown tremendously. The point was reached where even experienced scientists had to go through a time-consuming search of catalogues and references in order to find all independently observed positions of a certain star, if they needed this information for their own astrometric work and for the determination of a star's proper motion.

In 1897 F. Ristenpart, then Assistant Astronomer at the Heidelberg Observatory, wrote to A. Auwers, the leading authority on star catalogues in Germany at that time, and suggested the establishment of a card catalogue of star positions. There was to be a card for every star whose position was observed at least once. On this card, all published positions of the star and other pertinent information were to be entered. The card catalogue would have been kept up-to-date by entering new positions as they appeared in newly issued catalogues. Ristenpart's original idea would, of course, have been only of limited value. First, it might have been difficult to find an institution which would have taken on the arduous task of keeping the card catalogue up-to-date, and secondly, it would have been difficult for any astronomer not connected with this institution to obtain the information contained in this card catalogue. Auwers, who basically liked the idea, regarded it therefore as necessary to limit the card catalogue to position catalogues that had appeared before a certain date, and to eventually issue it in book form. Thus originated the "Geschichte des Fixsternhimmels," a forty-eight volume work, which was only recently completed (see Table III-1). The original proposal was submitted in 1898 to the Royal Prussian Academy of Sciences by Auwers and Ristenpart. The funds were appropriated in the same year and the work started immediately. The original plan called for the following:

(1) Collection of all available (through publication or otherwise) star positions which had been determined between 1750 and 1900 by transit observations or equivalent methods, and which had been fully reduced.

(2) The reduction of these positions to the coordinate system for 1875, (using Struve's constant of precession) and finally the reduction of these positions to a uniform system.

(3) The construction of a General Catalogue of star positions for 1875 on the basis of this collection of positions observed between 1750 and 1900, with precession terms and with proper motion components where these can be determined.

(4) The publication of the collected positions for every single star, and of the general catalogue.

The supervision of the work was initially in the hands of Ristenpart who mainly employed people who were not astronomically trained. In 1908 he went to Santiago de Chile to become the Director of the Observatory. Upon Auwers' death in 1915, the gigantic undertaking had lost both of the men who had started it and taken an active and personal interest in it.

One of the most important by-products of the initial phases of the work on the GFH is Ristenpart's (1909) comprehensive list of errata in the star catalogues of the 18th and 19th centuries which, in itself, is a list of star catalogues with short descriptions.

The original plan was not carried out in its entirety. Only point (1), the collection of the positions, was completed. Of point (2), only the first part (that is reduction to orientation 1875) was achieved and even this did not encompass all of the material. The work on points (3) and (4) was never even started.

Yet, much can be said in favor of Ristenpart's and Auwers' original plan. The need for accurate positions and proper motions of stars, in a uniform system, is now greater than ever and is expected to grow. There is no telling how much of the very-much-needed accuracy in star positions is not realized because every user is,as an individual or as a team, limited in what he can accomplish. The availability of electronic data processing equipment, together with advances in the field of theoretical numerical analysis, now makes it possible to consider again the construction of a true General Catalogue, similar to that envisioned by Ristenpart and Auwers. It would still be a huge enterprise, but there are astronomical institutions who have the personnel and the equipment to handle the task.

The period between the commencement of the work and publication of the first volume in 1922 saw the first world war. The difficulties of the war and postwar periods were mainly responsible for the curtailment of the original plans.

The collection of the observations reduced to 1875.0 was published separately for the northern and the southern sky. (Abteilung I and II, respectively.) One volume is assigned to every hour of right ascension, Volume 1 to 0^h R.A., Volume 2 to 1^h R.A., and so forth.

From Table III-1 below one can see that at the end of World War II, the Abteilung I had appeared completely, and so had Volumes 1 through 7 of the Abteilung II.

After Ristenpart's departure, H. Paetsch became astronomer-in-charge. His signature appears under the introduction of all volumes of Abteilung I, although Johannes Haas, who signs as being responsible for the first seventeen volumes that cover the southern sky (that is, Abteilung II), informs us that he and Paetsch shared in the work almost to equal parts (Haas 1954).

World War II retarded the completion of the GFH very considerably. Due to frequent bombings, the work had come to a virtual standstill in 1943; the material, much of it on cards, had been packed in boxes and stored in a bombproof room.

Due, obviously, to the difficult situation in postwar Berlin, the work was not immediately resumed. Late in 1947, Haas moved to Bonn and with him went the material for Volumes 8 through 17 of the Abteilung II. The Volumes 11 through 17, however, contain the positions as given in the original catalogues; freed from errors where such have been detected, but not reduced to the coordinate system of 1875. This change in plan was mainly due to a lack of funds, and perhaps, also, because other influential astronomers in West Germany had expressed doubts as to whether the material from old star catalogues of comparatively low accuracy was really worth the effort.

Table III-1

The Volumes of the Geschichte des Fixsternhimmels

ABETELUNG I

Vol.	*Year*	*Corrections in*	*Remarks*
1	1922	I: 2, 4, 9, 10, 13, 18, 20, 21 II: 4	V
2	1923	I: 3, 4, 9, 10, 12, 16, 18, 21, 24 II: 4, 5, 7	
3	1924	I: 4, 6, 8, 9, 12, 13, 16, 21, 23 II: 4, 9	N
4	1925	I: 6, 8, 9, 10, 14, 16, 21, 22, 23 II: 4	
5	1926	I: 5, 6, 8, 9, 10, 15, 16, 21, 23 II: 2, 5	
6	1927	I: 6, 7, 8, 9, 11, 14, 15, 16, 21, 23	
7	1927	I: 8, 9, 13, 15, 16, 21, 23, 24 II: 4	
8	1928	I: 8, 9, 16, 23	
9	1928	I: 9, 14, 23	
10	1929	I: 10, 15 II: 8	
11	1929	I: 11, 14, 21 II: 3	

ABTEILUNG II

Vol.	*Year*	*Corrections in*	*Remarks*	
1	1937	II: 1, 2, 3, 5, 8		
2	1938	II: 3, 5, 8, 10		
3	1938	II: 5, 8		
4	1939	II: 5, 8, 9		
5	1940	II: 5, 8, 9	N	
6	1941	II: 8, 10		
7	1943	II: 7, 8, 9		
8	1948	II: 8, 9, 10		
9	1951	II: 9, 10		
10	1954	II: 10, 12	a: $9^h - 10^h4^m$.	V
11	1957	II: 11	S: $10^h5^m - 10^h59^m$, n.s.	V

12	1930	I: 13, 20		12	1956		n.s.	
		II: 1, 5						
13	1930	I: 21, 23		13	1954	II: 12	n.s.	
		II: 3						
14	1930	I: 15, 21		14	1954		V, n.s.	
		II: 9, 10						
15	1931	I: 21		15	1953	II: 14, 15	n.s.	
16	1931	I: 21		16	1953	II: 14, 15, 16	V, n.s.	
17	1932	I: 17, 21		17	1958		V, N, n.s.	
18	1932	I: 21		18	1964			
		II: 10						
19	1933	I: 19		19	1963			
20	1934	I: 20		20	1960			
		II: 5						
21	1934	I: 22		21	1959			
		II: 5, 10						
22	1935	I: 22		22	1957			
		II: 1, 5, 9, 12						
23	1936	I: 23		23	1956			
		II: 5, 10						
24	1936			24	1952			V
		II: 3, 5	N					

The failure to precess the positions to 1875 in the Volumes 11 through 17 of Abteilung II means that certain errors in the catalogued positions will remain undetected since the unprecessed positions are very dissimilar. It has, on the other hand, the advantage that the precession can be computed according to the user's wishes. After all, the Struve-Peters precession constant used for the GFH is not the one internationally in use these days. Furthermore, since the advent of electronic computers, the computation of precession has become a very trivial matter and is best performed by a rigorous formula, thus removing any doubt as to how many terms in the series development should be carried.

Volumes 18 through 24 of the Abteilung II were issued by a team at the "Deutsche Akademie der Wissenschaften" at (East) Berlin, namely, Julius Dick, Martin Lange and Gerhard Felsmann. The Berlin group adhered to the original plan, that is, precessing the stars to 1875 with the Struve-Peters constant.

The latest catalogues included in the GFH are those which contain positions for a coordinate system of not later than 1900.0, if the observations of these positions started not later than during the year 1900. One exception is the AGK1 Berlin C, whose coordinate system is oriented to 1905 and for which all observations were made after 1900.

When the first volume appeared, positions from 492 catalogues had been included in the GFH. The volumes of the GFH contain, for every star, a small table at the head of which are listed its number in the BD or the CoD (Bonner Durchmusterung or Cordoba Durchmusterung), or the number of this star in some other catalogue (usually AGK1) if neither BD nor CoD contain it, its approximate position for 1875 (R. A. to 1^m, declination to $1'$,) and its BD or CoD magnitude. Occasionally, magnitudes are from different sources and are shown in parentheses.

The columns of the tables contain the symbol for the catalogue and the number of the star in this catalogue (Autorität), the epoch of orientation of the original catalogue (Äq). Columns α and δ contain the seconds of time and arc, respectively, of the originally-given position, precessed to 1875. Column B gives the numbers of observations, column Ep the epochs of observation. In columns Äq and Ep, italics and bold type refer to the eighteenth and the twentieth centuries, respectively.

The figures under α and δ must be added to the approximate positions given at the head of every table for a star to get the 1875 position computed from the original catalogue position.

Known blunders, in particular those in Ristenpart's already-mentioned list of errors, were corrected in the GFH listings. Whenever necessary, that is whenever proper motions had been applied to the position published originally, the position given in the GFH was reduced to the epoch of the observation by removing the previously applied proper motion. *No systematic corrections of any kind were applied* except that, to some catalogues which are mentioned in the introductions (all observed at Cordoba: Gou, GouCl, Co81, Co82), declination dependent corrections $\Delta\alpha_\delta$, $\Delta\delta_\delta$ to the system of the FC (see Section 3.5.3) had been applied in both coordinates. The positions were precessed (except as pointed out before, in Volumes 11 through 17 of Abteilung II), using the annual and secular variations and the third order term of the well-known series, for stars with declinations between −80° and +80°; for those within ten degrees of a pole, the precession was computed by means of the rigorous trigonometric formulas. The influence of large proper motions on the precession was considered by modification of the secular variation whenever necessary. The values for the annual and secular variation for 1875.0 (computed with the Struve-Peters constant without the proper motion correction to the secular variation) are given at the bottom of every table listing the positions of a star. The actually-used values are found, in brackets, below the values for the secular variations in the volumes concerning the southern celestial hemisphere (that is Abteilung II), whenever these differ from the "regular" ones because of a modification due to proper motion.

During the decades that the GFH was worked on, a few more catalogues turned up which should have been included at the beginning. The positions in these are listed in the various "Nachtrag" of volumes which have appeared later. Table III-1 indicates the volumes which contain a "Nachtrag" of positions. In this table, column 1 gives the number of the volume, column 2 the year of its appearance, column 3 the volume numbers in which corrections to this volume are given, and column 4 contains notes. N means that the volume contains a "Nachtrag," V means that the preface (the "Vorwort") contains information about the setup of the GFH or information concerning the history and organization, n.s. means that the positions refer to the equinox of their original catalogue. Some supplementary information to Abteilung I is also given by Felsmann (1961).

Stars whose positions were included in the FC, NFK or the FK3 were either omitted or are just referred to by number because their positions had already undergone a thorough investigation in the above-mentioned fundamental catalogues. This feature is regarded by this author as a defect.

3.3.3 THE "INDEX DER STERNÖRTER." The example of the GFH, which took more than half a century to finish, has shown the difficulties which stand in the way of such an enterprise. On the other hand, it is necessary to keep track of the many published star positions if they are not to be lost to science.

In 1924, R. Schorr, then Director of the Hamburg-Bergedorf Observatory, organized the establishment of an index to star positions which was to indicate in which catalogues positions of stars observed between about 1900 and 1925 could be found. The aim of this enterprise was more modest than that of the GFH, since the "Index" was to give only the serial numbers assigned to the individual stars in the various catalogues.

Even a limited list like this would have required the effort of several people for many years. In 1926, however, there was an abundance of intelligent and highly-trained clerical personnel available due to the high rate of unemployment in Germany at that time. Some of these were engaged for work on the Index within the framework of the State Unemployment Relief. Due to this circumstance, the manuscript was finished in the first half of 1927.

The Index (Schorr 1925) refers to star positions contained in the following catalogues:

(1) All lists and catalogues of stars not considered in the GFH which appeared in print before completion of the manuscript for the Index.

(2) Lists of star positions which were being worked on at various observatories at the time of the publication of the Index, which had been scheduled for publication within the next few years. These lists were put at the disposal of the Hamburg Observatory before they were actually published. (As it happens, some of these have not yet been published even as this is written.)

Only meridian positions or those of equivalent or higher accuracy were considered; in particular no relative observations on refractors were included. Only those photographic positions were indexed which are based on comparison star positions of the same epoch, as for instance, the stars which had been photographically observed to serve as reference stars of the asteroid Eros, and those of the Yale catalogues which had then appeared. The Astrographic Catalogue, however, (see Section 5.5) was not included.

The first volume of the Index refers to the northern, the second to the southern hemisphere (equator 1855).

The references are given in zones of 1° width according to the BD from +90° to −23°, the CoD from −23° to −62° and the CPD from −62° to −90°. Within every zone, the Index lists (against the BD, CoD or CPD number) the number of the star list or catalogue in which the star occurs, and the number of the star in this catalogue (if one was assigned). The 281 lists and catalogues which were searched are listed alphabetically according to the place of the observatory, at the end of the volumes. There are gaps in the numbers, assigned to the catalogues, so that they go up to 401.

Observed stars which have no BD (CoD, CPD) number (Anonymae) are listed with their approximate 1855.0 positions in the region covered by CoD and CPD, respectively, at the end of each zone.

About 365,000 stars are referenced in the Index, 185,000 on the northern and 180,000 on the southern hemisphere.

A continuation of the Hamburg-Bergedorf Index is the "Index der Sternörter 1925–1960 (Index II)" by A. Kahrstedt (1961–1966).

As indicated by the title, the Index II is a reference catalogue for positions of fixed stars that were determined by transit observations and published between 1925 and 1960. According to the definition by its author, this includes also those photographic positions which were tied to a system of reference stars that had been observed on transit circles. Essentially, the format of the original Index (which Kahrstedt calls Index I) was retained. This means that positions are referenced by the assigned catalogue numbers and the serial number of the star in that catalogue.

Table III-2 gives a survey of the nine volumes of Index II. The increase from the two volumes of Index I to nine volumes of Index II is an indication of the increased activity in the field, even though Index II covers 35 years as compared to the 25 years of activity covered by Index I.

Table III-2 **Index II**

Vol.	*Area Covered*		*Appeared*
IX	BD+51° to BD+89°		1966
VIII	+31	+50	1965
VII	+21	+30	1964
VI	+11	+20	1963
V	+0	+10	1963
II	−22	−0	1962
III	CoD−40 to CoD−22		1964
IV	{CoD−61	CoD−41	1965
	CPD−89	CPD−62	1965
I	total sky (anonymae)		1961

Pages III through XXVII of all nine volumes are identical and give the preface (III–IV), instructions for the use of the Index II (V–VI). The table of 589 catalogues and lists which are newly referenced in Index II is roughly arranged alphabetically according to the name of the place of observation (p. VI–p. XVII).It gives the number of the catalogue in Index II, the name of place of observation, the epoch of orientation and the title of the catalogue. Pages XVIII–XX give a list of 71 catalogues referenced in Index II which were already referenced in Index I. Their three-digit Index II numbers start with 6, and their numbers in Index I are also given. Next follows a list of 172 lists and catalogues (pp. XX–XXIV) that are referenced in Index II and were also referenced in the GFH; their GFH numbers are given, and their three-digit Index II numbers start with 7 or 8. Finally, a list of corrections to Index I occupies pages XXIV to XXVII.

The anonymae, whose number had substantially increased (to about 57000) were treated in the first volume. There they are listed according to increasing right ascension in 1° wide zones. Their approximate (1950) positions are given to 1^s and $1'$ in right ascension and declination, respectively. Volume I of Index II contains also the anonymae of Index I, and Kahrstedt suggests that the numbers now assigned in Index II, Volume I should from now on be used for these stars in the same way as BD, CoD or CPD numbers. The positions of the anonymae from Index I were precessed to 1950. They are easily recognized by the fact that the numbers assigned to the catalogues from which they were taken (found under the column "Ort") all begin with "6."

The anonymae of the GFH were listed in an appendix to Index II, Volume I, Appendix A, in one degree wide zones with serial numbers, lists the GFH anonymae missing in Index II, and appendix B is a list of those anonymae which are referenced in Index II. Appendices C and D refer to stars which were not originally in the BD, and double star components, respectively.

The material for Index II had been processed on punched cards, which were, incidentally, also used for the preparation of the main part of the manuscript.

3.3.4 OTHER LISTS. The sources treated so far list all star catalogues published up to 1960, and should first be referred to when looking for published catalogues or even small lists that give the positions of only a few stars, as long as they are either directly or indirectly based on transit circle observations. Beside these, there are many star positions which were observed relative to transit circle stars at visual refractors (indeed, some of the stars in the southern extension of the AGK1 at La Plata were observed that way), quite in the same way that comets or asteroids are observed at visual refractors, and there are the catalogues published after 1960. These have been listed in the "Astronomischer Jahresbericht," issued from 1899 to 1968 annually by the "Astronomisches Rechen-Institut" formerly at Berlin, but at Heidelberg since after World War II.

The Astronomischer Jahresbericht lists all astronomical publications of a certain year and gives short abstracts of the articles. It has been, for the entire twentieth century, an indispensable help for all those engaged in astronomical research. Included are, of course, references to newly published lists of star positions under the heading "Sternkataloge, Sternkarten."

From 1969 on, the Heidelberg Rechen-Institut has been preparing twice a year, the "Astronomy and Astrophysics Abstracts" (published by the Springer-Verlag), which are essentially a continuation of the Jahresbericht.

The catalogues with coordinate systems 1900–1932, which were partly included in Index I, are given in an important list by Kopff (1933) together with titles more complete and accurate than listed in the Index. The special importance of this list (which is not complete) lies in the subsequently and widely used chiffres, or codes, suggested by Kopff for these catalogues. He lists the catalogues in two groups: independent catalogues and compilation catalogues, since the latter are neither in the GFH nor in the Indices.

A continuation of Kopff's list was given by Heinemann (1964). The first part (pp. 7–18) of his paper is a reprint of Kopff's (1933) list, the second (pp. 19–22) an appendix to the lists of Auwers (1907) and Kopff. Listed in it are those catalogues which were not included in the lists of Auwers and Kopff, but were used for the construction of the FK3.

The third part (pp. 23–39) is the continuation proper of Kopff's list. The catalogues listed in it appeared usually between 1932 and 1964, and are in two groups: 514 independent catalogues, and 91 compilation (and other nonindependent) catalogues. The numbers of the independent catalogues are marked by an asterisk. The chiffres, which were one of the main purposes of the publication of Auwers' and Kopff's lists, are not continued in Heinemann's, since he feels that they are no longer necessary in the computer age. The catalogues are listed in alphabetical order of the places (or names) of the observatory or origin. Further listed are the authors (or principal calculators), the epoch of orientation, the number in Index II (as compared to Index I from Kopff's list) and the title of the catalogue. Information concerning the average epoch, the number of stars listed in each catalogue, the mean error of a typical listed position, and so forth, the average number of observations that were combined into one published position, the coverage of sky area, the star density within the area covered, and so forth, are available only if the title gives a clue to it.

Heinemann concludes with a key to the chiffres of those catalogues that were incorporated into the FK4.

Finally, it is worth noting that the FK4 (see Section 4.1.5) carries a list of catalogues that were used in its construction.

3.4 A Table of the More Important Independent Catalogues

Table III-3 is a list of independent star catalogues which was compiled from the other, more complete lists discussed in Section 3.3. It contains catalogues of relative as well as of absolute positions. Since some of the older catalogues contain not only absolute, that is fundamentally determined positions of the bright stars, but also relative positions of some faint stars, a strict and always applicable division of the catalogues into absolute and relative catalogues is not always possible. Also, as pointed out before, some star catalogues selected their entries rather randomly, while others, the so-called zone catalogues, aimed at complete coverage within a more or less small declination zone down to a certain magnitude.

A complete list of all independent catalogues published to date, similar to that which will be given below (Table III-3) for the meridian catalogues of the Astronomische Gesellschaft, would not only exceed the purpose of this monograph, but would also require a tremendous amount of labor for its compilation without really yielding relevant and indispensable information. One would have to go to most of the original catalogues for the required data, and many of these catalogues are not easily accessible. A complete list of published catalogues could be established from the sources discussed in detail in Section 3.3.

In order to point out the more important catalogues, Table III-3 lists, with a few exceptions and without any claim to completeness, those meridian catalogues which contain the precise relative positions of at least two thousand stars. AG type catalogues are also excluded because they will be dealt with in Chapter V.

The table illustrates very well the tremendous increase of the weight of relative meridian positions between about 1800 and the present, reflected in a weight ratio of approximately one to forty between bad 1800 and good modern catalogues. One may say that around 1900, after the introduction of the Repsold-type impersonal micrometers into meridian work, a mean error of about 0.″30 in either coordinate was the standard for a competently determined relative meridian position.

As discussed in Chapter II, essential progress in observing technique has been achieved between 1900 and now. Due to a considerable degree of sophistication in measuring technique, reduction technique and the technological aspects of the observing instruments, the accidental mean errors of relative positions obtained from a single observation on a first class transit circle nowadays (1970) are about 0.″15 in right ascension and 0.″20 in declination.

These improvements were brought about mainly by automatic (objective) reading of the declination circle and the developments of techniques for observing transit times that are (almost) independent of the judgment of the observer. For a description of one of the most modern operating transit circles see, for instance, Laustsen (1967).

At the present time, there is no more of the "wild" observing which was fairly common in the nineteenth century and the first decade of the twentieth. Most modern catalogues were and are being observed with a very definite purpose: As, for example, the repetition of an AGK1 zone to obtain proper motions (for instance, by Gyllenberg [1926] at Lund), but this procedure has for some time been completely replaced by photography. The chief purposes now in observing star positions visually on a transit instrument are the establishment of reference star catalogues for the various photographic zone catalogues, and the observation of absolute star positions in order to supply material for the continuous improvement of the fundamental system.

The information for each catalogue in Table III-3, which is given in detail below, is presented in a block in the following format:

<table>
<tr><td rowspan="4">1</td><td rowspan="3">2</td><td>4</td><td>7</td><td>11</td></tr>
<tr><td>5</td><td>8</td><td>12</td></tr>
<tr><td>6</td><td>9</td><td>13</td></tr>
<tr><td colspan="2">3</td><td>10</td><td></td></tr>
</table>

The information given in the various fields of this scheme is described below. Data that could have been obtained only with considerable effort have frequently been left out.

1. The name of the author(s) and the (abbreviated) title of the catalogue. Sometimes no author is given when several astronomers have shared in making and reducing the pertinent observations, and frequently no author's name appears on the title page of the book. Alternately, it may be indicated that the work on the catalogue was done "under the direction of N.N., Director" or a similar phrase.

   In doubtful cases, no effort was made to establish any rigorous criteria for who should be listed as author. In general, it will be the person listed as such on the title page.

   The title has sometimes been abbreviated so that it is still recognizable. An effort has been made to retain (if quoted) the place of observation and the period of observation.
2. Medium and place of publication.
3. Year of publication.
4. Epoch of orientation of the coordinate system.
5. Year the observations began.
6. Year the observations ended.
7. Number of stars for which the catalogue contains positions.

8. and 9. The accidental rms. error in right ascension and in declination, respectively, in units of seconds of arc, computed from the weight tables in the first column of the GC (GC.I) or Gyllenberg (1948). It is assumed that two observations were combined to give the published position. As explained above, the rms. errors of the positions may vary within any particular catalogue as a function of position and magnitude; in particular, the right ascensions frequently become more accurate for higher declinations. In Table III-3 average values are recorded. In order to determine the mean errors of the positions more accurately, in particular as functions of the number of incorporated observations, the original tables (GC.I, Gyllenberg) must be consulted.

10. The average number of observations combined into a star position system, either GC.I or the Gyllenberg paper.
11. The code used for the catalogue in GC.I or when preceded by an asterisk, in Gyllenberg's paper.
12. The chiffre assigned in German sources (Ristenpart, GFH, Kopff, Heinemann).
13. Number in Heinemann's list.

Table III-3 **A Selected List of Meridian Catalogues and Data Referring to Them**

I. Fedorenko:	St. Petersburg	1790	4673	*Fed
Pos. moy. des étoiles			2.6	Fdr
circompolaires . . .	1854		2.6	
B. A. Gould:	Mem. Nat. Acad.	1800	6497	*d'Ag
Red. of the obs. of fixed	of A. and Sci. Vol.	1783	3.6	D'Ag
stars made by Joseph	I Washington	1785	2.2	
Lepaute D'Agelet at Paris				
in 1783–1785.	1866			
A cat. of those stars in the		1800	47390	*Lal
Histoire Celeste . . . for			3.1	Lal
which Tabs. or Red. have			2.2	
been published by Prof.				
Schumacher, red. by *Bailey*	1847			
J. Bossert:		1800	3950	*LBo
Supplement à l'Hist.			3.1	LBo
Cel . . .			2.2	
Piazzi:	Panormi	1800	7646	Pi 1800
Praec. stell. inerr. pos.		1792	3.5	Pi
med. . . . ab anno 1792 ad a.		1813	2.0	
1813	1914			
F. W. Dyson and	Edinburgh,	1810	4243	Groo 10
W. G. Thackeray:	Neill & Co.		1.4	Grb
New red. of Groombridge's	Ltd.		1.0	
circumpolar cat. for the ep.				
1810 under the dir. of Sir				
W. H. M. Christie	1905			
J. E. de Vos van Steenwijk:	Astr. Abh.	1825	2840	Bris 25
Cat. of the dec's. of 2840 so.	4 No. 7	1823		Ptta
st. mostly btw. $-30°$ and		1828	1.5	170
the so. pole. . .				
Paramatta NSW	1923			

Table III-3 (*cont'd*)

M. Weisse: Pos. med. st. fix. in zonis Regiom. a Besselio inter $-15°$ et $+15°$ declin. . . .	St. Petersburg 1846	1825	31085 2.6 1.8	*W W_1
M. Weisse: Pos. med. st. fix. inter $+15°$ et $+45°$	St. Petersburg 1863	1825	31445 2.6 2.0	$*W_2$ W_2
F. G. W. Struve: Stell. fix. impr. dupl. et. mult. pos. med. . . . annis 1822 and 1843 in sp. Dorpatensis	St. Petersburg 1852	1830 1822 1843	2874 1.1 0.5	Dpt 30 StrPM
Th. G. Taylor: A gen. cat. . . . at Madras in the years 1830–1843	Madras 1844 New ed. by Downing Edinburgh 1901	1835 1830 1843	11015 1.4 1.1	Madr 35 TayD
F. R. Robinson: Places of 5345 st. obs. from 1828 to 1854 at Armagh Obs.	Dublin 1859	1840 1828 1854	5345 1.4 1.1	Arm 40 Rob
J. Halm: New Red. of Henderson's Cat. for the ep. 1840	Edinburgh Ann. **2** 1906	1840	3595 1.4 0.8	Edin 40
G. Santini: Descrizione del circ. mer. dell I.R. Oss. di Padova	dal Nuovi Saggi dell Accad. di Padova **5** (1840) and **6** (1847) (see also Mem. R.A.S. **12,** 273) 1840, 1847	1840	4092 2.0 2.2 1.5 2.2	$*San_1$ $*San_2$ San_1 San_2

Table III-3 (*cont'd*)

The Cape Cat. of St. ded. from obs. . . Royal Obs. Cape of Good Hope 1834–1840	Capetown 1878	1840 1834 1840	2892 1.4 0.8	Cape 40 CP(40)
F. W. Argelander: Arg's Zon. Beob. v. 45° −80° nördl. Decl. . . . v.W. Oeltzen. 1841–1844	Ann. Vienna Obs. (3) 1 and 2	1842 1841 1844	26425 1.1 1.1	*AOe AOe
Cat. of 2156 stars. . . from 1836 to 1847 at Royal Obs. Greenwich	Greenwich Observations 1847, app. 1847	1845 1836 1847	2156 1.0 0.7	Gch 45 12yI 12yII
C. Rümker's Hamburger Stern,-verzeichnis 1845.0	Hamburger Sternwarte Bergedorf 1922	1845 1836 1856	17724 1.2 1.2	*RüH 253
The Radcliffe Cat. of 6317 st. chiefly circumpolar . . . made at the Radcliffe Obs. . . . 1840–1853	Oxford 1860	1845 1840 1853	6317 1.0 0.8	Rad 45 RC
Thomas Mclear: Cat. . . Royal Obs. Cape of Good Hope during. . . 1849–1852	Cape	1850 1849 1852	4810 1.15 0.71	Cape 50 Cp(50)
W. S. Eichelberger and F. B. Littell: Cat. of 20521 st. btw. 13° 35′ and 45° 25′ so. decl. f. the equ. 1850	Wash. Publ. (2) 7 1911	1850 1846 1852	23521 1.0 1.2	—

Table III-3 (*cont'd*)

Gilliss:	Washington Obs'ns	1850	16600	Gilliss 50
A cat. . . . southern stars	for 1890, App. 1	1849	1.00	GiZ
deduced from. . . obs. made		1852	1.40	
at Santiago de Chile . . .				
1849–1852	1892			
F. W. Argelander:	Wien	1850	18276	
Cat. d. Arg. Zonen v. 15			1.7	AWe
bis 31 gr. südl. Dekl. hg. v.			1.1	
Dr. E. Weiss	1890			
F. W. Argelander:	Bonner Beob. **6**	1855	33811	*BoVI
Mittl. Oe . . . Mer. Kreis		1845	0.9	BoVI
der Bonner Stw. . . . 1845–		1867	0.9	
1867 angest. Beob.				
Carrington:	London	1855	3716	Carr 55
A cat. of 3735 Circumpolar		1854	0.75	Carr
St. obs'd. at Redhill 1854–56	1857	1856	0.53	
Pos. moyennes de 3542	Pulkova Obs. Pub.	1855	3542	Pulk 55
étoiles. . .cercle me. de	8, sect. 2, p. 227	1840	.81	Pu M
Pulkova dans les années		1869	.70	
1840–1869				
Klinkerfues, W. Schur:	Astr. Mitt.	1860	6900	*Kli
Stern-Cat. . . . i.d. Jahren	G/ttingen, 2. Teil	1858	1.6	Kli
1858–63 angest. Zonenbeob.	1891	1863	1.4	
7-yr. cat. . . . ded'd from	Greenwich Obs'ns.	1860	2022	Gch 60
ob'ns. . . from 1854 to 1860	1862, app. 1	1854	.70	7y
at R. Obs. Greenwich		1860	.53	
Second Radcliffe Cat.	Oxford	1860	2386	Rad 60
. . . from 1854–1861		1854	1.14	RC_2
	1870	1861	1.14	

Table III-3 (*cont'd*)

Moesta:	Observationes astr.	1860	2309	Stgo 60
Ascens. rectas i distancias	de Santiago de	1856	1.14	Moe_2
polares. . . 1856 a 1860 con	Chile **2,** Dresden	1860	1.14	
el circulo meridiano				
Santiago de Chile	1875			
G. Santini:	Memorie dell I.R.	1860	2706	*San_3
Pos. med. . . di st. . . nella	Instituto Veneto	1856	1.8	San_3
zona fra 10° e 12°30′ di	di Sci. Lettere e	1858	1.6	
decl. austr. 1856–58	Arti **7,** pt. II, 311			
	1858			
G. Santini:	Mem. dell Inst.	1860	2246	*San_4
Pos. med. . . di st. nella	Ven. **10,** pt. II, 231	1857	1.4	San_4
zona fra −12°30′ e li −15°		1861	1.4	
di decl. austr. 1857–61	1862			
Trettenero:	Mem. dell Inst.	1860	1425	
Pos. med. dist. 1452. . .fra	Ven. di. Sci. Lett.	1861	1.3	San_5
0° e −3° di decl. austr.	e arti **15,** 329	1863	1.4	
	1870			
M. Yarnall:	Washington	1860	10964	Wash 60
Cat. of st. obs'd. at	Obs'ns.	1845	1.00	Ya
USNO during years 1845–		1877	1.00	
1877				
Kat. v. 26060 St. zw.	Ann. Wien **24**	1860	26060	
14°57′ und 20°44′ n. Dek. . .		1856		Oe_{60}
n. Zonenbeob. v. *W.*		1858		230
Oeltzen	1928			
New 7 year Cat. . . . ded'd.	Greenwich Obs.	1864	2760	Gch 64
from obs'ns. . .1861 to 1867	1868, app. II	1861	0.70	N7y
at Royal Obs. Greenwich		1867	0.53	
H. Struve:	Beob. Erg. K. Stw.	1865	2338	Berl 65
Mittl. Örter v. 2338	Berlin No. 16	1855	1.0	
Vergleichsternen für 1865.0	1914	1868	1.0	243

Table III-3 (*cont'd*)

E. Quetelet: Cat. de 10792 étoiles obs. a l'Obs. Roy. de Bruxelles de 1857–1878	Bruxelles Annals, n.s. **6,** 1887	1865 1857 1878	10791 1.00 0.63	Brus 65 Que
Schjellerup: Stjerne-Fortegnelse ind. 10,000 Pos. of tel. Fixst. im. −15 or +15 Gr. Dekl. Kjobenhavns Meridian-kreis. . .1861–63	Copenhagen 1864	1865 1861 1863	8946 1.4 0.81	*Sj Sj
Robert Grant: Cat. of 6415 stars from obs'ns. made at Glasgow Univ. Obs. during. . .1860 to 1881	Glasgow 1883	1870 1860 1881	6415 1.1 1.0	*Gl Gl
Nine year Cat. of 2263 stars. . .1868–1876. . .at Royal Obs. Greenwich	Greenwich Obs'ns 1876, app. 1	1872 1868 1876	2263 0.81 0.70	Gch 72 9y
B. A. Gould: Catálogo general Argentino 1872–80	Cordoba resultados **14** 1886	1875 1872 1880	32448 0.81 0.70	Cord 75 Gou
F. G. Robinson: 2nd Armagh Cat. of 3300 stars. . .at Armagh Obs. . . 1859–1883	Dublin 1886	1875 1859 1883	3300 1.6 1.1	$*Arm_2$ Arm_2
Copeland and Börgen: Mittl. Oerter der i.d. Zonen −0° u. −1° der B.D. enth. St. bis zu 9^m0. . .	Astron. Mitt. d Göttinger Stw. I. Teil 1869	1875	6595 1.4 0.75	*CB CB
H. Osten: Dritter Radcliffe Cat.	Nova Acta, Halle **92,** No. 1 (1910)	1875 1862 1876	5839 1.4 1.0	Rad 75 265

Table III-3 (*cont'd*)

D. J. Dubyago: Cat. de l'obs. de Kasan. Pos. isolées des ét.	Trudy Kasan **13, 14** 1903 & 1904	1875 1869 1892	4304	
C. M. Smith: Res. of Obs. . . .w.t. Madras Mer. Circle under direction of the late N. R. Pogson	Res. of Obs'ns of the fixed stars made with the Madras Mer. Circle vol. IX, Genl. Cat., 1899	1875	5303 2.0 1.4	Madr 75 MaP
Romberg: Cat. v. 5634 Sternen. . .am Pulkowaer Mer. Kr. . . 1874–1880	Suppl. III aux Obs. de Poulkowa St. Petersburg 1891	1875 1874 1880	5634 1.00 0.53	Pulk 75 Rbg
J. R. Eastman: Second Wash. Cat. of stars. . .USNO w. the 8.5-in. transit circle 1866–91	Washington Obs'ns. for 1892, App. 1	1875 1866 1891	5151 0.81 0.70	Wash 75 Wa_2
J. Palisa: Kat. v. 3458 St. f. . . .auf Grund d.i.d. Bdn. 11–29 d. Wiener Ann. 3. Folge Mer. Kr. Beob. (1843–1879)	Vienna Annals **19** 1908	1875 1843 1879	3458 1.6 1.5	*WPal WPal
Cat. de l'Obs. de Paris Etoiles obs. aux Instr. Meridiens de 1837–1881	Paris tom. **1** 1887; **2**, 1891; **3**, 1896; **4**, 1903. 1887–1903	1845, 1860, 1875 1837 1881	I–III 34733 0.80 1.0 1.0 0.8 0.8 0.8	Paris 45 Paris 60 Paris 75 Par_1 Par_2 Par_3

Table III-3 (*cont'd*)

Ten-year-Cat. of 4059 stars. . .from 1877 to 1886 at Roy. Obs. Greenwich under dir. of *W. H. M. Christie*. . .	Greenwich obs'ns 1887, App. II	1880 1877 1886	4059 0.63 0.63	Gch 80 10y
E. J. Stone: Cat. of 12441 stars. . . at Cape of Good Hope. . .1871–79	London 1881	1880 1871 1879	12441 0.89 0.58	Cape 80 Cp80
H. Seeliger and J. Bauschinger: Erstes Münchener Sternverz. 1840–1872	Neue Ann- der K. Stw. in Bogenhausen b. München **1** 1890	1880 1840 1872	33082 1.7 1.7	*$Mü_1$ $Mü_1$
J. Bauschinger: Zweites Münchener Sternverz, 1884–89	Neue Ann. d.K. Stw. in Bogenhausen bei München **2** 1891	1880 1884 1889	13200 1.4 1.0	*$Mü_2$ $Mü_2$
J. Kowalczyk: Cat. v. 6041 St. zw. 1°50′ u. 7°10′ südl. Dekl. . . Reichenbach and Ertel Mk. d. U. Stw. zu Warschau (1876–1886)	Warschau 1904	1880 1876 1886	6041 1.0 1.4	Wars 80 War
J. Seyboth: Kat. v. 6943 St. f.d. Ep. 1885,0 aus d. Beob. v. H. Romberg	Poulk. Pub. (2) **7** II 1909	1885.0 1881 1894	6943 .63 .53	Pulk R 85 267
F. Bidschof: Kat. v. 2417 Sternen auf Gr. d.i.d. Bdn. III, V, VIII, u. IX der Ann. d. . .zu Wien erh. Merkr. Beob. (1881–1889)	Ann. d.k.k. Univ. Stw., Wien (3) **15** 1902	1885 1881 1889	2417 1.4 1.2	Wien 85 WiB

Table III-3 (*cont'd*)

J. G. Porter:	Pub. Cincinnati	1885	4050	*CiZ
Zone cat. of 4050 stars. . .	Obs. **9**	1885	1.6	CiZ
obs'd with the three-inch		1887	1.8	
transit of Cincinnati Obs.				
1885–87	1887			
N. Herz:	Anh. Abh. Akad.	1890	3310	
St. kat. f.d. Zone v. 6° bis	Berlin	1888	.65	
10° südl. Dek. . . .auf der v.		1891	1.15	
Kuffner'schen Stw. zu Wien				
(Ott.) 1. Abt.: Kat. v. 3310				
i.d. Zonen mehrfach beob.				
St.	1906			
J. G. Porter:	Cincinnati Obs.	1890	2000	Cin 90
A. cat. of 2000 stars	Pub. **13**		0.81	Ci_1
	1893		0.63	
Robert Grant:	Glasgow	1890	2156	*Gl_2
Second Glasgow Cat. of		1886	1.4	Gl_2
2156 stars. . .Glasgow Univ.		1892	1.1	
Obs. . .1886–92.	1892			
W. H. M. Christie:	Greenwich Obs'ns	1890	6892	Gch 90
Second Ten-Year Catalogue	1898, App. II		0.63	II 10y
			0.58	
D. Gill:	London	1890	3007	Cape 90
A Cat. of 3007 stars. . .at		1885	0.81	Cp_{90}
the Roy. Obs. C. of Good		1895	0.63	
Hope. . .1885–1895	1898			
Third Melbourne General	Melbourne	1890	3068	Melb 90
Cat. of 3068 st. f. the equ.	A. J. Mullett	1884	.9	Me_3
1890	1917	1894	.7	151
Edw. J. Stone:	Oxford	1890	6424	Rad 90
Cat. of 6424 stars. . .		1880	0.89	RC90
Radcliffe Obs., Oxford		1893	0.70	
during. . .1880–93	1894			

Table III-3 (*cont'd*)

Cat. de l'Obs. de Bordeaux. Réobs. des ét. comprises dans les zones d'Argelander entre 15° et 20° de dec.	Paris, Gauthier- Villars 1909	1890 1881 1895	6999 .60 .70	Bord 90
W. Valentiner: Kat. der St. zw. d. Aequ. u.d. 8 Grad. südl. Dekl. 1855 b.z.8. Grössenkl. am Mer. kr. d. . .Stw. zu. Karlsruhe i.d. Jahren 1882–94	Veröff. der Grossherz. Stw. zu Heidelberg (Astrometr. Inst.) **2** 1903	1890 1882 1894	6892 0.45 0.58	Karls 90 Val
Battermann: Res. aus Beob. v. 379 Anh. St. 1640 d. Anschl. best. St. . . .1892–97 am grossen Berliner Meridiankreise	Beob. Erg. d. Kg. Stw. zu Berlin, H. No. 8	1895 1892 1897	2019 0.31 0.31	Berl 95 Bm_1
J. G. Porter: A Cat. of 2030 stars. . . w. an app. giving. . .prop. motions for 971 stars	Cincinnati Obs. Pub. No. **14**	1895	2030 0.70 0.70	Cin 95 Ci_2
Obs. d'Abbadia. Ob'ns. faites au cercle mer. . .par M. l'abbé Verschaffel	Impr. de l' obs. d'Abbadia **3** 1905	1900 1902 1903	6284	
Observatoire d'Abbadia Cat. de 13532 ét. comprises entre +5°15′ et −3°15′	Abbadia Obs. Hendaye 1914	1900 1906 1912	13532 .70 .63	Abb 0°00 Abb_2 3
Observatoire d'Abbadia Cat. de 14263 ét. compr. entre +16° et +24°	Abbadia Obs. Hendaye 1915	1900 1899 1906	14263 .63 .58	Abb +20°00 Abb_1 1

Table III-3 (*cont'd*)

Abbé Verschaffel:	Abbadia Obs.	1900	7443	Abb
Cat. de 7443 ét. compr.	Hendaye			−6°00
entre −2°45′ et −9°15′ app.		1912	.63	Abb_3
a la zone phot. de San		1916	.58	4
Fernando.	1917			
Second cata'ogue de l'Obs	Hendaye	1900		Bord 00
de Bordeaux publ. sous la			.70	2Bord
dir. de L. Picart. Obs. des			.53	45
ét. de rep. du cat. phot.				
Zone +10° a +18°	1924			
C. Mönnichmeyer:	Bonn Veröff. **16**	1900	2199	
Kat. v. 2199 St. f. 1900.0		1901		Bo *16* Mö
	1921	1911		40
J. G. Porter:	Cincinnati Obs.	1900	4280	Cin 00
A cat. of 4280 st. f. the	Pub. No. **15**	1897	0.70	Ci_3
epoch 1900 (1897–1904)		1904	0.70	
E. Smith, J. G. Porter:	Cincinnati Pub. **19**	1900	4683	$Cin_{13}00$
A Cat. of 4683 st. for the			.75	Ci_4
ep. 1900	1922		.75	67
C. D. Perinne:	Res. Cordoba **20**	1900	5791	Cord 00
Cat. de 5791 estr. II Cat.			.8	CGA_2
Gen. Argent.	1911		.7	73
L. C. Guerin:	Cordoba, Res. del	1900	6429	
Cat. de 6429 estr. de. rep.	Obs. Nac. Argent.	1917		
del cat. astr. red. a 1900	**34,** entr. 2	1918		329
sin. m.p. entre −23°0′ y				
−32°0′	1935			
D. Gill:	Edinburgh	1900	3365 +	Cape 00
Cat. of stars from Obs'ns. . .			995 +	
during 1900–1904			63 +	
			41	
		1900	.44	Cp00
		1904	.44	
	1906			

Table III-3 (*cont'd*)

Second Nine Year Cat. of St. f. the ep. 1900.0 from obs'ns. . . .at the Roy. Obs. Gch. und. the dir. of Sir W. H. M. Christie	Edinburgh Neill & Co. Ltd. 1909	1900	11668 .73 .58	Gch 00
Cat. of 2713 zod. st. for the equin. 1900.0 from obs'ns made at the Roy. Obs. Edinburgh	Edinburgh Annals **3** 1910	1900 1898 1908	2713 1.5 1.0	Edin 00 249
W. Doberck: Cat. of R.A.'s of 2120 southern stars. . .from obs'ns. made at Hong Kong Obs.	App. to Hong Kong Obs'ns. 1903 1905	1900 1898 1904	 .7	HK00
Fritz Cohn: R. A. Beob. v. 4066 St.	Konigsberg Beob. Abt. **42,** 355 1909	1900 1901 1907	4066 .31 —	Kön 00
R. H. Tucker: Cat. of Southern Stars of Piazzi for the epoch 1900 (1898–1900)	Lick Obs. Pub. **6** 1903	1900 1896 1901	3113 0.44 0.36	Lick 00 TuPi
D. Gill: A cat. of 8560 Astrographic Standard St. betw. Decl. −40° and −52°. . .Cape of Good Hope 1896–99	London 1906	1900 1896 1899	8560 0.44 0.44	Cape 00 CpAs
Cat. of 20554 faint stars in in the Cape Astrographic Zone −40° to −52° for the equ. 1900	London 1939	1900	20554	
W. E. Cooke: Cat, of 1625 st. btw. −39° and −41°	Ann. Perth Merid. Obs. **2**	1900	1625 .7 .6	Perth 00 Perth **2** 179

Table III-3 (*cont'd*)

W. E. Cooke:	Ann. Perth Merid.	1900	1846	Perth 00
Cat. of 1846 st. btw. −33°	Obs. **3**		.7	Perth **3**
and −35°			.6	176
W. E. Cooke:	Ann. Perth Merid.	1900	2047	Perth 00
Cat. of 2047 st. btw. −31°	Obs. **4**		.7	Perth **4**
and −33° for 1900.0			.6	175
	1910			
W. E. Cooke:	Ann. Perth Merid.	1900	2043	Perth 00
Cat. of 2043 stars btw.	Obs. **5**		.7	Perth **5**
−35° and −37°	1912		.6	177
W. E. Cooke:	Ann. Perth Merid.	1900	2025	Perth 00
Cat. of 2025 st. btw. −37°	Obs. **6**		.7	Perth **6**
and −39°	1912		.6	178
Cat. de 10656 étoiles de	Orleans	1900	10656	Paris 00
repère de la cart du ciel			.8	$Par_{00}Rep$
(zone +16° a +24°) obs. a			.6	172
l'obs. de Paris	1929			
V. Seraphimoff:	Pulkovo Publ. (2)	1900	8820	*Pu56
A cat. of 8820 st. btw. 15°	**56**	1896	.7	
s. and 15° n. dec. from obs.		1908	.5	473
by M. Morin and A.				
Kondratieff made with the				
Pulkovo M. C.	1940			
A. de Legge, F. Giaconelli:	Roma	1900	4012	
Cat. di st. compil. sulle oss.		1875		LG_3
fatte al oss. del Campidoglio	1911	1903		206
D. Saint Blancat:	Toulouse Annals **4**	1900	3719	Toul 00
Premier catalogue de			0.54	Tou_1
Toulouse	1901		0.36	

Table III-3 (*cont'd*)

N. E. Gosserat: Renfermant une partie des travaux exécutés de 1898 à 1905 sous la direction de M. B. Baillaud (+3° to +12°)	Toulouse Annals **8,** C 1912	1900 1898 1905	6447 .53 .36	Toul 00
F. Küstner: Katalog von 10663 Sternen zw. 0° u. 51° n. Dek.	Veröff. Bonner Stw. **10** 1908	1900	10663 0.36 0.32	Bonn K 00 Kü
H. R. Morgan: A Cat. of 4526 stars.	Wash. Pub. (2) **9**/1	1900	4526 .31 .31	Wash F 00 Wash **9** Morg 223
R. H. Tucker, A. J. Roy, W. B. Varnum: San Luis Cat. of 15333 st. f. the ep. 1910, prep. at the Dudley Obs, Albany, N.Y.	Carnegie Inst. of Wash. 1928	1910	15333 .36 .32	SL10 San Luis 211
B. Boss, A. J. Roy, W. B. Varnum: Albany Cat. of 20811 st. f. the ep. 1910	Carnegie Inst. Wash. 1932	1910	20811 .32 .32	$Alb_{14}10$ Alb_{10} 12
A. S. Flint, A. J. Roy: Madison catalogue of 2786 for the ep. 1910 from Washburn obs.	Washington, Carnegie Inst. Pub. No. 515 1939	1910	2786 .36 .45	*Mad F1 409
N. Morin: I. Kat. v. 3396 St. zw. 39° u. 46° n. Dek. f.d. Aequin. 1910 aus Beob. am Pulkowaer Mk. von A. Kondratjeff u. M. Morin II. Geleg. beob. St.	Pulkovo Pub. (2), **41** 1933	1910 1909 1925	3369+ .36 .36	*Pu41 470

Table III-3 (*cont'd*)

Greenwich Cat. of St. for	R. Obs. Gch.	1910	12368	Gch Z 10
1910.0 Pt. II: St. in the		1906	.7	$Grw_{10}Zo$
Zone +24° to +32°	1920	1914	.5	77
J. C. Hammond and	Wash. Pub. (2) **11**	1910	2714	$\left\{\begin{matrix}Wash_{10}10\\Wash_{14}10\end{matrix}\right\}$
C. B. Watts:				
Res. of obs'ns w. the 6″ tr.		1909	.31	$Wash_{10}$
cir. 1909–1918	1927	1918	.31	228–9
H. Philippot et E. Delporte:	Ann. Roy. Obs. de		3553	Bruss 00
Cat. de 3553 ét. de repère	de Belgigue N.S.		.6	Uc00 Rep_1
de la zone +21°, +22° de la	**13**/1, 1		.7	50
carte phot. du ciel.	1911			
M. Harwood and	Harvard Ann.		~14000	
E. C. Pickering:	**75**/1			
Bond zones of faint equat.				
st. in the zone 1° n. of the				
equ. (refractor observations)	1913			
S. Chevalier:	Ann. Zô-Sè **15**	1920	14268	Zô-Sè 15
Cat. de al zone −0°50′ a		1917	.36	
+0°50′ d'apres les phot. du		1920	.45	
tour de l'équateur	1928			
				1920 .4
R. Prager:	Veröff. Berlin-	1920	8803	Berl 20
Kat. v. 8803 St. zw. 31° &	Babelsberg **4**	~1919	.53	Bab 4 Prg
40° n. Dek. nach gem. m.			.45	24
K. F. Bottlinger. . .Berl.				
Bab.	1923			
H. R. Morgan:	Wash. Pub. (2)	1920	9989	
Cat. of 9989 st'd. & inter-	**13,** 161	1913		
med. st. +90° to −36°		1926		505
dec. . .red'd. without p.m.	1933			
C. Mönnichmeyer und	Bonn Veröff. **20**	1925		$Bonn_{18}25$
J. Hopmann:			.45	Bo**20**Int
Kat. d. intermed. St. v.			.45	42
+50° Dek. bis z. Nordpol	1927			

Table III-3 (*cont'd*)

First Greenwich Cat. of	London	1925	2643	$Gch_{18}25$
stars for 1925.0		1915	.5	1.Grw_{25}
	1924	1921	.6	79
W. Gyllenberg:	Malmö	1925	11800	Lund 25
Kat. v. 11800 St. d. Zone		1920	.5	LuGyll
+35° bis +40° A. G. Lund	1926	1925	.6	144
F. W. Dyson:	Greenwich	1925	2111	
Second Greenwich Cat. of		1922		
stars for 1925.0. 1st. pt.		1930		339
Fund. st. obs'd. dur'g. the				
yrs. 1922–1930, red. with				
p.m. to the ep. 1925.0	1935			
F. W. Dyson:	Greenwich	1925	10584	*2.Grw_{25}
Second Greenwich Cat. of		1922	.45	
Stars for 1925.0, 2nd pt:		1930	.45	339
cat. of st. in the zone +32°				
to +64°. obs'd d'g. the yrs.				
1922–1930.	1935			
F. Dolberg:	Hamburg. Stw. in	1925	4983	Hamb 25
Erstes Bergedorfer	Bergedorf	1913	.32	Bgd_{25}
Sternverzeichnis 1925.0		1926	.32	33
	1928			
P. Calot:	Hendaye	1925	1272	*4.Abb
Quatrième cat. comprenant		1926	.36	
1272 ét. dont 662 fond. de		1927	.45	275
+45° a −26° obs 1926–27	1934			
E. Jost:	Königsberg Astr.	1925	2043	
Kat. v. 2043 St. zw. 50° u.	Beob. auf. d. Univ.			
55° n. Dek. f.d. Äquin. 1925	Stw. z. Kbg. Abt.			365
n. Beob. am Repsold'schen	**46,** 1938			
Mk.d. Univ. Stw. zu				
Königsberg	1938			

Table III-3 (*cont'd*)

J. Jackson:	London, HMSO	1925		
Second Cape Cat. of stars		1925		
for the equ. 1925.0 cons. of		1933		**316**
pt. I., st. south of dec.				
−30°, pt. II., zodiacal stars.	1949			
J. Jackson:	London, HMSO	1925	6597	
Third Cape Cat. of stars		1932		
for the equ. of 1925.0 Cat of		1936		317
6597 st. btw. the equ. and				
dec. −30° Compiled from				
obs. with the RTC.	1950			
G. Peisino, M. Campa:	Milano, Consiglio	1925	2390	*Tri$_{25}$
Cat. di 2390 stelle oss. al	Naz. delle Ric.,	1931	.5	
cerc. mer. del R. Oss. astr.	Com. per l'astr	1933	.7	**501**
di Trieste negli anni 1931–				
32–33 e ridotte al 1925.0	1942			
P. Herget:	Pub. Cincinnati	1925	2300	
Cat. of 2300 st. for the equ.	Obs. **22**			
1925.0	1946			328
W. S. McClenahan:	Publ. Dom. Obs.	1925	2436	
Cat. of 2436 st. from obs.	Ottawa **15,** No. 1	1911		
w.t. rev. mer. cir. made at		1923		451
the Dom. Obs. Ottawa				
during the years 1911–1923	1949			
C. B. Watts, A. N. Adams:	Pub. USNO (2)	1925	2383	
Res. of obs'ns made w.t. 6″	**16** pt. 1, 205	1930		
T.C. 1925–1941 Cat. of 2383		1935		**511**
st. fr. 1925.0 from obs. i. th.				
yrs. 1930 to 1935. Red.				
without p.m. to 1925.0	1949			
L. Courvoisier:	Veröff Berlin-	1925	2261	Berl VC 25
Obs. Dek. v. 2261 Fund. st.	Babelsberg **7,** 4	1916	—	Bab **7**
f.d. Aquin. 1925 n. Beob.				Cour
am Vert. Kr.	1929	1927	.29	30

Table III-3 (*cont'd*)

H. R. Morgan: R.A. and dec's of the 2438 stand. st. red. with prop. motions to 1925.0 & comp'd. w. other catalogues	Wash. Pub. (2) **13** 1933	1925	2438	— 505
J. C. Hammond: Cat. of 3520 st. bs'd. on obs'ns. with the 6″ T.C. 1928–1930, red. without p.m. to the equ. 1925.0	Wash. Pub. USNO **14**/1	1925 1928 1930	 .32 .36 4.5	*Wash 14 (1) **506**
Cat. de 2251 faibles reduites a 1930.0	Ann. Obs. Strasbourg **4,** fsc. 1 1950	1930	2251	 482
H. A. Martinez: 2123 estrellas del catalogo Boss Comprendidas entre −15° y −80°	La Plata Obs. Astron. de la Univ. Nacional Pub. **12** 1936	1932	2123 .45 .41	*LaPl 12 **374**
H. R. Morgan, F. P. Scott: Results of obs'ns made with 9″ T.C. 1935–1945	Publ. USNO (2) **15** pt. 5 Washington 1948	1940 1935 1945	5446	 **510**
N. W. Zimmermann: Cat. of 2957 bright stars with dec's of −10° to +90°	Pulkovo Publ. (2), *61* 1948		2957	
J. Larink, A. Bohrmann, *H. Kox et al.:* Katalog von 3356 schwachen Sternen f.d. Äquin 1950, beob. an den Mk. d. Stw. Hbg.-Bgdf. v. Heidelberg	Hamburg- Bergedorf 1955	1950	3356	 **72***
A. I. Nefdyeva: Cat. of 2288 differentially observed stars of the KSZ	Kasan, Comm. Engelhardt Obs. No. 33 1963	1950	2288	

Table III-3 (*cont'd*)

H. A. Martinez: 6744 est. del Cat. Gen. de Boss compr. entre −47° y −82°	La Plata Pub. (Astr.) **30** 1959	1950 1943 1950	6744 0″.17 0″.17	 378
R. H. Stoy: Second Cape Cat. for 1950. Obs'ns of the Sun, Moon and Planets 1936–1959	London HMSO 1968	1950 1936 1959	6800 0″.43 0″.43	
H. A. Martinez: Cat. de 3710 est. galacticas australes	Ob. Astr. de la Plata, ser. Astron. **19** 1943	1950	3710	 377
E. G. Woolsey: Res. of obs'ns made with the rev. m.c. 1956–1961. Cat. of 3753 AGK3R stars.	Ottawa, Publ. Dom. Obs. **25,** 83 1966	1950	3753	
E. G. Woolsey: Res. of obs'ns made with the Ottawa rev. m.c. 1954– 1962. Cat. of 2665 st.	Ottawa, Publ. Dom. Obs. **25,** 228 1967	1950	2665	
Obs. de Paris Cat. de 11755 et. de la zone +17° a +25° et de mag's 9.5 a 10.5, destinees a servir de réf. pour la dét. des mouv. pr. des ét. du Cat. Phot. de Paris (photographic cat., see review by Dieckvoss, *Sterne* **28,** 73).	Paris CNRS 1950		11755	
Obs. de l'Univ. de Bordeaux Cat. mér. de 2024 ét. repères de la zone +11° a +18°	Paris, CNRS 1958	1950 1953 1953	2024	 301

Table III-3 (*cont'd*)

C. B. Watts, A. N. Adams: Cat. of 1536 st. fr. 1950.0 from Obs. in the yrs. 1936–1941. Red. w.o. p.m. to 1950.0	Pub. USNO (2) **16,** pt. 1, 255 1949	1950 1936 1941	1536	 **512**
C. B. Watts, F. P. Scott, A. N. Adams: Res. of obs'ns made w. the 6″ T.C. 1941–1949. Cat. of 5216 st. fr. 1950.0 Comp. of cat. $W2_{50}$ with GC and FK3.	Pubs. USNO **16** pt. 3, 447 1952	1950 1941 1949	5216	 **513**
A. N. Adams, S. M. Bestul, D. K. Scott: Res. of obs'ns made w.t. 6″ T.C. 1949–1956. Obs'ns of the sun, moon and planets, cat. of 5965 st. for 1950, corr'ns to FK4, GC and N30	Pub. USNO (2) **19,** pt. 1 1964	1950	5965	
J. Robertson: Cat. of 3539 zodiacal stars for the equi. 1950.0	Washington A.P. AENA **10** pt. 2 167 1940	1950	3539	 509

Chapter 4

COMPILATION CATALOGUES

4.1 The German Series of Fundamental Catalogues

4.1.1 THE ORIGIN AND IMPORTANCE OF FUNDAMENTAL CATALOGUES. The importance of systematic errors in astrometric work was only gradually realized. In 1797, the Astronomer Royal, Maskelyne, dismissed his assistant Kinnebrook because his meridian observations showed systematic deviations from those which he himself had made (Martin 1945). Only later was it realized that differences in the perception of transits are unavoidable between different observers. Thus, the personal equation was discovered, and with it came the realization that the judgment of the observer is one of the elements which decides the result of the observation.

Subsequently it became clear that there were many more sources of systematic errors than had originally been suspected. The result of this is that the positions in all catalogues, whether absolute or relative, are affected by more or less large systematic errors with respect to the ideal system which they are intended to represent.

Eventually, a situation like this must inevitably lead to confusion. Every conscientious observer probably feels that he has done everything in his power to keep systematic errors from creeping into his observations, and he might, therefore, feel that he is doing the correct thing by trusting his own fundamental observations above those of others. This practically unavoidable attitude, however, spells confusion for those who are interested in the best possible star positions (and proper motions) that could be compiled from the observed material available at any one time.

When the preparations for the observations of the zones of the AGK1 (see Section 5.1.2) were made, it was realized that the value of the planned catalogue would be greatly enhanced if all zones were on the same system, that is, referred to a set of reference stars which represented the best fundamental system that could be established at the time. To achieve this, A. Auwers compiled the Fundamental-Catalog (now termed the FC) of positions and proper motions. Later, as we shall see, when more material became available, it became feasible for him and subsequent investigators to derive better approximations to an ideal inertial system.

The fundamental catalogues can in principle be divided into two groups, as described below.

The only purpose of the catalogues in the first group is to define a fundamental system; they, therefore, list only the most extensively observed stars, with preference given to those that were observed absolutely, that is, fundamentally. Into this category belong the German fundamental catalogues: The Fundamental-Catalog (FC), the Neuer Fundamental-Katalog (NFK), the Dritter Fundamentalkatalog des Berliner Astronomischen Jahrbuchs (FK3) and the Fourth Fundamental Catalogue (FK4). Each of these catalogues, in turn, replaced its predecessor as a representation of the ideal inertial system.

The second group, besides defining the system by (mostly bright) extensively observed "fundamental" stars, includes a rather large number of fainter stars with fewer observations whose positions were never or seldom determined absolutely. The most important of the catalogues of this type is the General Catalogue (GC) by Benjamin Boss, which in turn replaced the Preliminary General Catalogue by Lewis Boss. Another modern catalogue of this type is the N30 catalogue. All of these will be discussed later in this chapter.

During the course of the years, quite a number of fundamental catalogues have been in use which today have only historical interest. In this connection, the fundamental systems of Newcomb (see Section 4.2.1) and that of Eichelberger (Section 4.2.2) must be mentioned. These catalogues were quite important at their time, but were quickly superseded by more thorough or more comprehensive discussions of the material which became available as time progressed.

As pointed out above in Sections 1.6.6 and 3.2.3.1, the establishment of a rigorous relationship between a true inertial system and a fundamental system is impossible because of the unavoidable accidental errors in the positions of those stars which, in their entirety, constitute the fundamental system. For the same reason, it is impossible to establish rigorously the systematic differences between two fundamental systems. It is, therefore, only possible to give estimators of these systematic differences, their standard deviations and their dependence on appropriate parameters, usually position on the sphere. Unfortunately, this rigorous statistical way of thinking has only recently been introduced into the type of astrometric research with which we are concerned here. In any case, we may assume in what follows, without in principle changing the discussion, that rigorous relationships do exist between any two fundamental systems and that they can be established from available material.

Consider now, at a certain epoch of orientation and place, two fundamental systems which are, of course, given by complete catalogues, each defined by a sufficiently large number of stars and of which a sufficient number is common to both. The star positions at this epoch will show certain systematic differences, which may be considered as corrections (with the appropriate algebraic sign) necessary to reduce the positions from one of the systems to the other.

Since the proper motions, as given in each of two systems, also show systematic differences, the systematic differences between the two systems will depend on the epoch at which they are compared, as discussed above in Section 3.2.3.3.

4.1.2 THE FUNDAMENTAL-CATALOG (FC). The purpose of the FC is clear from the complete title which, translated into English, reads "Fundamental Catalogue for the Zone observations . . . , issued at the request of the Zone Commission of the Astronomische Gesellschaft." The catalogue was issued in two separate parts (Auwers 1879 and 1883) for the northern (539 stars) and the southern (83 stars) hemisphere, respectively.

The characteristic of the German series of fundamental catalogues, which the next sections are going to deal with, is the rather heavy reliance of personal judgment in the determination of weights of the different catalogues incorporated. The position system of the FC depends actually on the Pulkovo 1865 catalogue. This catalogue consists of two series of observations, namely, the series of right ascension observations by Wagner, and the series of declination observations by Nyrén. There were actually two series of observations for both of the coordinates, namely, those of the Hauptsterne (principal stars, usually those brighter than fourth visual magnitude) and those for the "Zusatzsterne" (additional stars) between about the fourth and the sixth magnitude. On the northern sky, there are 336 Hauptsterne and 203 Zusatzsterne. The proper motions were determined without using the Pulkovo positions, but rather by comparing positions determined by Bradley around 1775 with those in the Greenwich 1861 catalogue. The northern part of the FC is therefore based on positions whose system is defined only by the two series of Pulkovo observations, and proper motions that are based on the comparison of only two series of Greenwich observations.

The published positions are the weighted means of positions that were reduced to the epoch 1875 by means of the proper motions described above, and to the system of the Pulkovo catalogues (which had been reduced to the epoch 1875 with the same proper motions) by corrections of the type given by formula (3.20a), which were derived by comparison with the Pulkovo positions reduced to 1875. In addition to the Pulkovo 1865 catalogues, which were responsbile for the position system, five more catalogues were used, namely (1) Pulkovo 1845, (GC:Pulk 45); (2) Greenwich 1872, (9-yr. catalogue, GC: Gch 72); (3) Cambridge (Mass.), "Aust" in Ristenpart's (1901) lists, (Harvard Obs. Annals **10**); (4) Leipzig 1866 and (5) Leiden 1870, after they had been reduced to the system of the Wagner and Nyrén catalogues. For details, in particular the weights of the catalogues and other data, reference must be made to Auwers' (1879) paper.

The northern FC was published in two "Abteilungen," that is, parts: Abteilung I: principal stars, and Abteilung II: additional stars. The catalogue columns give the following information:

Left-hand pages: serial number, (visual) magnitude, Flamsteed number and name of the star, Bradley No., right ascension epoch, declination epoch, RA (epoch and coordinate system 1875.0) to $0.^s001$, annual proper motion in RA (to $0.^s001$

or $0.^s0001$), the first three terms of the precession series (the third term only if it is significant) based on the precession constant of Struve.

Right-hand pages: serial number, declination (coordinate system and epoch 1875.0), annual proper motion in declination (to $0.''001$ or $0.''01$) and the first three coefficients of the precession series in declination. Then follow corrections to a preliminary catalogue of these stars originally published in the Vierteljahrsschrift der Astronomischen Gesellschaft and of merely historical value, even when the FC was published, and a column of remarks. The proper motions also refer to the coordinate system 1875.0. Sometimes, however, two values of the proper motion are given; in these cases, the upper one refers to the coordinate system 1865, the lower one to that of 1885.0.

In 1883, Auwers published in the same periodical the positions of 83 southern stars as a continuation of the FC to the southern hemisphere. The northern catalogue extends only to $-10°$ because the AGK1 zone program, for which it was to yield the reference stars, was at that time planned to cover the sky only as far south as $-2°$. The 83 additional stars cover the declination belt from $-10°$ to $-32°$. In addition to the catalogues which provided the material for the northern part of the catalogue, positions from eight more catalogues (compare Auwers [1883], page 2) were incorporated. The position and proper motion systems were determined in the same manner as for the northern part of the catalogue, although the declinations of Auwers' own fictitious "Greenwich 1861" catalogue (actually a combination of the two seven-year catalogues) were carefully discussed with particular attention paid to refraction and flexure effects.

The construction of the positions of these 83 stars is described in great detail. On pp. 27 and 29, Auwers (1883) communicates tables with the $\Delta\alpha_\alpha$, and so on, for the reduction of the individual catalogues to the system of the FC. Weights of the catalogues determined from residuals are found on p. 31. The rms. errors of a single observation (around 1865) come out to be about $0.^s035$ and $0.''60$ in RA and declination, respectively.

The list of the southern positions is arranged quite analogously to that of the northern ones. The "corrections to the preliminary catalogue" are, of course, absent and so are the third precession terms, while columns containing the weights of the positions in both coordinates were added. The secular variations of the entire FC were computed without considering the influence of the proper motions.

4.1.3 THE NEUER FUNDAMENTAL-KATALOG (NFK). J. Peters (1907) published the "Neuer Fundamentalkatalog des Berline Astronomischen Jahrbuchs nach den Grundlagen von A. Auwers für die Epochen 1875 and 1900" as No. 33 of the "Veröffentlichungen des Königlichen Astronomischen Rechen-Instituts zu Berlin," which set the pattern for the German fundamental catalogues that were to follow.

Since the appearance of the FC (which Auwers in his own publications always calls "A.G.C."), Auwers had been constantly working on the improvement of its systematic as well as its accidental accuracy, and likewise on its extension to the south pole. Several papers (Auwers 1889, 1897, 1898, 1904, 1905) are the result of this effort.

While the position system of the FC is based on only the Pulkovo 1865 observations, and the proper motion system on two series of Greenwich observations, the NFK system makes much more complete use of the available material of fundamental observations.

Besides the relative catalogues which were used for improvement of the accuracy of individual positions, but had, of course, no influence on the determination of the system, about fifty catalogues containing absolute positions were available for stars on the northern sky, and about fifteen for the stars south of $-30°$. Although Auwers' judgment still influenced the system by the weights assigned to the fundamental catalogues, the NFK essentially represents a genuine mean of the material which was then available. Proper motions and positions were computed by the method of least squares; however, they were not computed directly from the original positions but from normal points into which, on the average, about ten original positions had been combined. Auwers, incidentally, was completely aware of the fact that the combination of the original material into normal positions decreased the weight of the results, but only in this way could the amount of calculation be kept within reasonable limits in this pre-computer era.

The weights of the various catalogues were determined from the residuals after forming the fundamental catalogue (Auwers 1900) and are, therefore, reasonably free from the effects of arbitrary judgment. If one uses the above-mentioned weight tables for star catalogues by Auwers (1900), one should bear in mind that the weights given there refer only to those stars which were used in forming NFK, that is, the bright stars, and that the weights of the positions of the faint stars might be substantially smaller. In any case, these tables are now completely obsolete.

Tables of weights and tables for the reduction of catalogues to the FC system, and of the FC system to the NFK system are given in Auwers' various papers: tables for reduction to the FC system particularly in those of 1900 and 1903, and tables for reduction of the FC system to that of the NFK in that of 1904 (pp. 237–238).

Note that Auwers in his 1904 paper also corrects the FC system for magnitude equation, for which he adopts the value determined by Küstner (1902).

Peters' contribution, therefore, consisted chiefly in compiling, arranging and editing Auwers' work, and also in changing the entire material from Struve's value for the constant of precession to that of Newcomb, which had by then become internationally accepted. A few additions to and corrections of Auwers' material are given in the brief introduction to the NFK, which mainly contains an account of the exact meaning of the "precession terms" $d\alpha/dt$, and so on, that are listed in the NFK.

The catalogue contains on the left-hand pages, serial number and name of the star, its visual magnitude (in parentheses for variable stars), and the average epoch of the incorporated observations. The position, and so forth, of every star is given for epoch and coordinate system both 1875 and 1900; one printed below the other: the right ascension, $d\alpha/dt$, $d^2\alpha/dt^2$, μ_α (referred to Newcomb's constant of precession), $d\mu_\alpha/dt$ (for 100 years).

The right-hand pages repeat the serial number, give declination proper motion μ_δ and its derivative, $d\mu_\delta/dt$, the number in the FC, the number in Auwers (1905), the number in Auwers (1889), and the number in Newcomb's (1899) fundamental catalogue. Note that the $d\alpha/dt$, and so forth, are the time derivatives not only as concerns precession, but also as concerns the influence of proper motion. The stars in the polar caps are given in appendices. Altogether, the positions and centennial proper motions of 925 stars are given in the NFK.

The publication also contains, as appendices to the catalogue, a discussion of stars with peculiarities, mostly double stars, and a comparison of the NFK positions with those in Newcomb's fundamental catalogue (Section 4.2.1). The central epochs pertaining to a position (that is, the weighted mean of the epochs of the catalogues used for the determination of its position) are not listed in the NFK, but can be found in papers by Auwers (1904, 1905). They are generally in the neighborhood of 1877.

4.1.4 THE FK3.

4.1.4.1 THE DRITTER FUNDAMENTALKATALOG DES BERLINER ASTRONOMISCHEN JAHRBUCHS. The immediate reason for the correction and improvement of the NFK was the resolution of the Astronomischen Gesellschaft to repeat their AGK1 (see Section 5.1.2) and establish the AGK2 (see Section 5.1.3). The catalogue was published in two parts (Kopff 1937, 1938). Details of the work were published in the Astronomische Nachrichten as progress was made; bibliographies are given on p. 100 (right ascensions) and p. 101 (declinations) of Kopff's (1937) paper. The contents of this paper, which is the first part of the FK3 and which contains the revision of NFK for only those stars which are in the NFK (without the Zusatzsterne), is as follows: Introduction (A. Kopff), new catalogue of the Auwers, that is, NFK, stars (with explanations, Appendix I: Components of the double stars, Appendix II: Auxiliary tables for the reduction of star catalogues to the FK3 system, Appendix III: The catalogue comparisons made at the Astronomisches Rechen-Institut (A. Kopff), Appendix IV: The right ascension systems of the FK3 (A. Kahrstedt), Appendix V: The declination system of the FK3 (K. Heinemann).

The revision was again carried out in two steps: The improvement of the individual positions within the system of the NFK, and the revision of the NFK system itself.

The first part of these reductions was carried out in pretty much the same way as for the revision of the FC, with the exception that the fundamental catalogues which had in the meantime become available were not reduced to the NFK system, so that the positions of the NFK were actually, even then, systematically improved, although the equinox was still left unaltered. The results of this preliminary revision of the NFK were published by Kopff in two papers (1928, 1931) which defined the *preliminary working systems* A_v and A_vS, respectively; their proper motion systems were identical with that of the NFK. Details concerning the reduction process, in particular the list of incorporated catalogues, are found in these papers.

The essential details of the revisions of the NFK system (and the formation of the FK3 system) are given, as indicated above, in Appendices IV and V by A. Kahrstedt and K. Heinemann, respectively. Altogether 77 catalogues (partly absolute, partly relative) were available for the revision of the NFK.

In particular, the *right ascensions* were treated in the following way: Deviations of the incorporated catalogues from the NFK were established allowing, in effect, for corrections of the form $\Delta(\alpha, \delta) + \Delta\alpha_m$, that is, without constraining the corrections to the form $\Delta\alpha_\alpha + \Delta\alpha_\delta$. The magnitude equations $\Delta\alpha_m$, however, were constrained to be linear in the magnitudes. As it turned out, only three of the 21 absolute catalogues that were available for the revision of the NFK system required a correction for magnitude equation. The correction of the system of the NFK positions to that of the FK3 was obtained by taking weighted means of the $\Delta\alpha$ values for the 21 catalogues which provided absolute right ascensions. The corrections were postulated (except for the polar caps) in the "canonic form" (3.20a) with a linear magnitude equation. However, this was done independently in five zones (not containing the polar caps), so that for all practical purposes a more general correction in the form $\Delta\alpha = \Delta\alpha(\alpha, \delta) + \Delta\alpha_m$ of the NFK position system to yield that of the FK3 was established. The position of the equinox was determined from observations of the Sun made at Pulkovo and at Washington, and from observations of the Sun and the planets at Greenwich and at the Cape of Good Hope. For this purpose, only observations made with self-registering impersonal micrometers were utilized. This explains the difference of $0.^s012$ from an almost simultaneous investigation by Morgan (1932), who utilized all observations, no matter what recording was employed for them.

The *systematic corrections to the right ascension proper motions* were computed in the form $\Delta\mu_\alpha + \Delta\mu_\delta$. Altogether, thirty-six catalogues were used which had been observed at 9 different instruments. Only two of them date from before 1845, and eleven of them had *not* been used for the establishment of the NFK. The omission of the catalogues before 1845 is one of the characteristic features of the FK3. The argument in support of this procedure is that the earlier catalogues are allegedly affected by systematic errors whose average is significantly different from zero.

Kahrstedt argues that the proper motions which are most free from systematic errors are those determined from series of observations at the same instrument. He, therefore, determined the $\Delta\mu_\delta$ in principle by postulating that the $\Delta\alpha_\delta$ of the NFK at the average epoch of the FK3 are caused by erroneous proper motions only, as one may safely assume that the NFK position system is essentially free from systematic errors at its average epoch, namely 1877. This procedure resulted in the fact that the $\Delta\mu_\delta$ depend virtually only on the Pulkovo series for the northern, and the second Cape series for the southern sky. Kahrstedt quotes reasons to justify this procedure, but only the future will show whether such an extensive exercise of judgment in favor of taking unweighted means has produced the best results which the available material was capable of yielding. The determination of the $\Delta\mu_\alpha$ went along analogous lines in three zones: equatorial, higher northern and higher southern zone. All the series of observations were reduced to the mean of those series which confirmed Kahrstedt's assumption that the systematic errors of the NFK outside its mean epoch were produced only by the systematic errors of its right ascension proper motion system. A magnitude equation in the right ascension proper motion system of the NFK could not be found and was, therefore, not applied.

K. Heinemann signs as responsible for the establishment of the *declination system* of the FK3. Essentially, the principles employed in the reduction of the right ascension material guided the establishment of the declination system, except that no magnitude equation was provided for. Twenty-one catalogues could be utilized for the derivation of the $\Delta\delta_\delta$, while only seventeen catalogues with absolute declination observations contributed to the determination of the $\Delta\delta_\alpha$. As for the right ascensions, sets of $\Delta\delta_\alpha$ were determined for more or less wide declination zones, so that the corrections can in effect be regarded to be in the form $\Delta\delta(\alpha, \delta)$.

Difficulties arose in the determination of the $\Delta\delta_\delta$. The procedure applied previously by Kopff (1928) and by Boss (GC.I), see Section 4.2.4, of correcting the refraction constant by matching pairs of south and north hemisphere catalogues led occasionally to such impossible values for the constant of refraction, that this approach was abandoned. Also, a separate determination of the $\Delta\delta$ from vertical circle series on the one hand and transit circle observations, on the other, gave results which differed at the equator by $0.''5$, with differences between determinations from individual catalogues up to $2.''$ (equator point). This led to the decision to utilize only those series of observations for the determination of the $\Delta\delta_\delta$ which were made at an instrument at which observations of the Sun and/or Mercury, Venus, and the Moon had also been made during the period in which the catalogue was observed. From these, the equator point of a catalogue can be corrected as discussed above in Section 2.2.5.4. This procedure was critized by Oort (1943) who is of the opinion that the mean of all available catalogues would have given just as accurate a declination system as that determined only from the corrections obtained from observations of the Sun and the major planets, and that a combination of the two alternatives might have produced an even better declination system.

Oort also criticizes the assumption that absolute proper motions are best obtained from long series of observations at the same instruments. He points out that the difference in the personal equations of the different observers alone, even when they observe at the same instrument, are sufficient to generate systematic errors in the proper motions.

If one tries to follow the many steps that led to the establishment of the FK3 from the NFK, one cannot help but gain the impression that corrections were piled upon corrections, sometimes with not very convincing justification. One may well ask whether a simpler and more straightforward procedure would not have produced a system whose average deviation from a self-consistent system would not have been any larger than that of the FK3.

This is confirmed by considering those $\Delta\delta_\delta$ and $\Delta\mu_\delta'$ corrections to the FK3 system which Oort (1943), Tab. 2, p. 420 derives from considering all absolute catalogues that were available for computing the FK3, and not only those whose equator point was determined by means of observations of the Sun and some planets. Table IV-1 compares Oort's suggested corrections with those actually found in the computation of the FK4 (Fricke, Kopff et al. [1963], p. 131). One can see that especially on the southern hemisphere, the procedure suggested by Oort would have led to a system which is closer to that of the FK4, than the one actually arrived at by the very selective procedure used.

Table IV-1 **Acutal and Predicted Corrections to the FK3 System for 1950**

δ	$\Delta\delta_\delta$ (*Oort*)	$\Delta\delta_\delta$ (*FK4*)	$\Delta\mu_\delta'$ (*Oort*)	$\Delta\mu_\delta'$ (*FK4*)
+44°	+0.″14	+0.″02	+0.″0038/yr.	+0.″0007/yr.
+14	+4	0	+12	−8
+4	+3	−3	+16	−9
−6	+11	−3	+26	−8
−16	+13	+3	+20	−2
−26	−43	−9	−60	−2
−36	−32	−16	−36	−20
−46	−29	−16	−42	−21
−56	−25	−19	−39	−29
−66	−17	−12	−30	−20
−76	−7	−12	−26	−24

The *system of proper motions in declination* was treated analogously. Fifty catalogues were employed in the determination of the $\Delta\mu_\delta'$ and thirty-eight for finding the $\Delta\mu_\alpha'$. Their epochs ranged between 1846, and 1926 and 1922, respectively. The omission of catalogues before 1845 is again an important feature. The results also confirmed in general the assumption that the NFK is practically unaffected by systematic errors at its main epoch.

The arrangement of the FK3 catalogue proper is essentially the same as that of the NFK. In addition to spectral data, the FK3 gives the central epochs in α and δ (on the average around 1900), the rms. errors of α, δ, and the centennial μ_α and μ_δ. The numbers in the PGC and in Eichelberger's catalogue are also given.

The second part of the FK3 (Kopff 1938) contains the *positions and proper motions of the Zusatzsterne* (additional stars). These were computed anew from the catalogues by a linear adjustment after these catalogues had been reduced to the FK3 system. Thus, only the observations of the Auwers stars had any influence on the establishment of the system of the FK3. The Zusatzsterne were added in order to provide a more uniform star density; that of the NFK was noticeably lower in the equatorial regions than towards the poles. The list of the Zusatzsterne was first given in the Berliner Astronomisches Jahrbuch for 1936 (p. A53); their selection is explained by Kopff and Nowacki (1934).

Table IV-2, adapted from Zverev, shows the average rms. errors of the FK3 positions within the system at the average epoch 1900, as well as the rms. errors of the proper motion components (within the system).

Table IV-2 Random rms. Errors of FK3 Coordinates (ϵ_α and ϵ_δ) and Centennial Proper Motions (ϵ_μ and ϵ_μ') within the System

Declination range From	To	$\epsilon_\alpha \cos\delta$ *(1900)*	ϵ_δ	$\epsilon_\mu \cos\delta$ *(1900)*	ϵ_μ'
+90°	+75°	$0^s.001$	0."02	$0^s.007$	0."12
+75	+35	2	3	12	16
+35	−35	2	3	10	14
−35	−75	4	6	30	37
−75	−90	2	3	13	22

From Table IV-2 it is evident that the accuracy of positions as well as of proper motions is considerably lower on the southern hemisphere. This, incidentally, applied to the systematic as well as the random (that is, within the system) accuracy.

Besides those at the end of the first volume of the GC, tables for the reduction of the GC system to that of the FK3 have been published by Kopff (1939) for various epochs. They are based on the tables previously established by Kopff (1939a), which give the differences between the systems of the FK3 and the GC at the epoch 1950 and the systematic differences of their proper motion system. The latter tables are actually more useful than the former for automatic computations. Tables, in the form of position corrections at various epochs, for the reduction of the PGC and the NFK to the FK3 system are given in Appendix II of Kopff (1937). More extensive tables, containing the relationship between Eichelberger's (1925) catalogue and the FK3 are given by H. Nowacki (1935) in the form of systematic position differences for the epoch 1925 and proper motion system differences.

4.1.4.2 THE "SUPPLEMENT-KATALOG DES FK3 (FK3 SUPP)." Geodesists as well as fundamental astronomers felt the necessity for a larger number of fundamental stars than are available in the FK3. In order to satisfy these needs, at least in part, Kopff (1954) published, on pp. (1) through (44) of the appendix of the "Astronomisch-Geodätisches Jahrbuch 1954," the accurate positions of 1142 stars north of −10° and a list of approximate positions of 845 stars south of −10°. The English translation of the title is "supplemental catalogue to the FK3."

The accurate positions in the FK3 Supp were not computed independently from the existing catalogues but are the N30 (Section 4.2.4) positions to which systematic reductions to the FK3 system have been added. The positions for those FK3 Supp stars which are not listed in the N30 were derived from GC positions that had been improved by inclusion of positions in new catalogues, and reduced to the system of the N30 by F. P. Scott of the U.S. Naval Observatory in Washington, D.C. The approximate positions were intended to serve merely as an observing list. Most stars range from 4^m to 7^m.

The various columns in the FK3 Supp give GC number, N30 number and magnitude, then α (to $0.^s001$) for epoch and equinox 1950, $d\alpha/dT$ and $\frac{1}{2}d^2\alpha/dT^2$ (with T in units of Julian centuries); these quantities reduce simultaneously to different epochs to place as well as orientation. The centennial proper motion in right ascension, μ, is given to $0.^s01$. The next group of columns contains analogous data in declination (to $0.''01$ and $0.''1$, respectively). The last column (Bemerkungen, that is, Remarks) gives common names, parallaxes over $0.''02$, numbers in the double star catalogues ADS and SDS, and information on variability.

4.1.5 THE FOURTH FUNDAMENTAL CATALOGUE (FK4)

4.1.5.1 HISTORY AND RATIONALE. Considering that the mean epoch of the FK3 positions is in the neighborhood of 1900, and also considering that catalogues with epochs before around 1845 were not included for the construction of its system, it might heuristically be argued that the systematic inaccuracy of the FK3 at the epoch 1955—which is as far removed from 1900, as 1900 is from 1845—is about as large as that of the catalogues at 1845. The accuracy should, however, be higher since the proper motions, whose errors would contribute the overwhelming part of the errors of the system at an epoch so far removed from the mean epoch, were not determined from the oldest catalogues at the extreme epochs alone but also from catalogues at intermediate epochs. On the other hand, the best measure for the errors of a system is the corrections which were later applied to the system due to the accumulation of new material. Judging the systematic errors of the FK3 system is easy because corrections, which establish the FK4 system, have already been derived and will be discussed later. These may also be compared with the estimate of the systematic errors in the FK3 by Oort (1943, 1943a); see also Table IV-1.

It will be recalled that the immediate reason for the establishment of the FC and the FK3 were the AGK1 and the AGK2 projects. In like manner, a resolution to establish the AGK3 was one of the reasons for the revision of the FK3. Another reason, of course, was the fact that fundamental systems must be revised from time to time in order to work in whatever new material has accumulated.

4.1.5.2 THE FK3R. The first step in the computation of the Fourth Fundamental Catalogue, or FK4 (Fricke, Kopff et al. 1963), was the establishment of individually improved positions of the FK3 stars by reducing newly available (absolute and relative) catalogues to the FK3 system. These catalogues were then used to improve the individual positions and proper motions within the system of the FK3. The catalogue of positions obtained in this manner is called the FK3R (Astronomisches Rechen-Institut, 1960). As with the establishment of the FK3, only a selection of the entire collection of absolute positions available was used to establish the improved system.

The details of the construction of the FK3R were described in a paper by Kopff, Nowacki and Strobel (1964). The basic material consisted of those catalogues which had become available since the end of the work on the FK3 and which had not been incorporated in the FK3. As a first step, the systematic corrections of these catalogues to the FK3 system and their weights had to be established. The results of this effort were later published by Nowacki and Strobel (1968). First the $\Delta\alpha$ and $\Delta\delta$ (in the sense Cat-FK3) were formed. (In the polar caps between $\pm 60°$ declination and the poles, the $\Delta\alpha \cos\delta$ were used.) The averages of the $\Delta\alpha$ and $\Delta\delta$ were taken in 2° wide declination belts and from these, triadic means were formed. Plots of these (somewhat smoothed) furnished the $\Delta\alpha_\delta$ and $\Delta\delta_\delta$. Zenith discontinuities were allowed for, whenever they were suspected.

The $\Delta\alpha_\alpha$ and $\Delta\delta_\alpha$ were computed from the $\Delta\alpha$–$\Delta\alpha_\delta$ and $\Delta\delta$–$\Delta\delta_\delta$. They were averaged separately within declination zones 10° wide, for each hour of right ascension, from which, again, triadic means were taken and graphed. When the inspection of the graphs justified it, some of the 10° declination zones were combined into wider strips. The $\Delta\alpha_\alpha$ and $\Delta\delta_\alpha$ were read off and tabulated from smoothed graphs of the triadic means of the above-defined differences against right ascension. Magnitude effects had to be taken into account for only a few catalogues that had been observed without magnitude compensating devices. The differences $\Delta\alpha - \Delta\alpha_\alpha - \Delta\alpha_\delta = v_\alpha$ and $\Delta\delta - \Delta\delta_\alpha - \Delta\delta_\delta = v_\delta$ formed the raw material for finding the individual corrections, within the FK3 system, to the positions and proper motions in the FK3. From the data in the FK3, the normal equations that led to the positions and proper motions in the FK3 were reconstituted after the material had been collected in normal positions. The terms in these equations consisted of weighted product sums. These product sums, generated by the new positions, were then augmented by the proper additions, and the equations solved anew. The new solutions now incorporate all material including that already used for the compilation of the FK3. Identical results would have been obtained had the method described in Section 3.2.3 been applied.

The weights for the catalogued positions in this analysis had been computed in the manner described in Section 3.2.4.1.

4.1.5.3 THE FK4 RIGHT ASCENSION SYSTEM OF POSITIONS AND PROPER MOTIONS. A full report on the FK4 system of right ascensions and proper motions in right ascension was given by Gliese (1963). The fundamental procedure, as it had been established by Auwers, was not changed; that is, the system of proper motions was established by considering series of observations on the same instruments, namely: The Cape Reversible Transit Circle, the 6-inch and 9-inch Transit Circles of the U.S. Naval Observatory at Washington, the large Transit Instrument at Pulkovo and the Airy Transit Circle at Greenwich. Gliese recognizes that restricting the establishment of the proper motion systems to these series will by no means remove all problems. The assumption had been that all positions carefully observed on the same instrument would have essentially the same systematic errors thereby assuring that the proper motions, derived from these positions, would not have any appreciable systematic errors. Almost a century of experience has demonstrated, however, that this is simply not so. One must, in particular, realize that the same instrument, with different observers, will produce different systematic errors. In particular, the declination dependent systematic errors in the right ascensions (that is, the $\Delta\alpha_\delta$) are known to vary from catalogue to catalogue, even when they have been observed on the same instrument.

Altogether, series based on 11 catalogues were utilized for the establishment of the system of right ascension proper motions. They represent 5 series of observations with an average epoch range of 38 years and a total epoch range from 1897 to 1956.

The system of right ascensions was established from 45 absolute catalogues observed on 14 different instruments with epochs between 1918 and 1956. This shows that all absolute catalogues before 1918 were excluded from determining the system of right ascensions. The reason for this exclusion, which Gliese (following his predecessors) gives, is that the systems of the older absolute catalogues differ considerably from each other and are therefore suspect of containing systematic errors. Their inclusion, it is alleged, would have spoiled the system rather than improved it. If the catalogues were properly weighted according to their systematic accuracy, the effect of the inclusion of the absolute catalogues with epochs between 1900 and 1918 would have been mostly to produce a mean epoch of the right ascension system of 1924. The mean epoch of the right ascension system as it was derived is about 1934, varying slightly according to the declination. In Gliese's opinion, the accuracy of the system would hardly have been any better as a consequence of the inclusion of the older catalogues.

The positions in the catalogues used for the derivation of the right ascension system (or, rather corrections to the FK3 right ascension system) refer to different epochs. The corrections to the FK3R system thus had to be analyzed in various regions of the sky according to the formula (3.19), $\Delta\alpha(t) = \Delta\alpha(t_0) + (t - t_0)\Delta\mu$, where t_0 represents the mean epoch of the right ascension system. $\Delta\alpha(t_0)$ and $\Delta\mu$ were found by least squares. However, only the $\Delta\alpha(t_0)$, the corrections to the right ascension system, were used. $\Delta\mu$, the corrections to the system of right ascension proper motions, were discarded since, as mentioned above, those ultimately used were determined from considering only series of observations on the same instruments. Since the $\Delta\mu$ determined from series were judged to be more reliable, the question may be raised as to why the values for the $\Delta\mu$ were determined together with the $\Delta\alpha(t_0)$ from a least squares solution. Would, instead, the substitution of the latter in the formula above not have led to more reliable $\Delta\alpha(t_0)$; albeit larger residuals? Some will feel that personal judgment in the assignment of weights and the exclusion of material was used excessively in the compilation of the right ascension systems of positions and proper motions of the FK4. This is the very procedure, used also for the compilation of the FK3, which Oort (1943) has criticized. From the systematic differences of the various absolute catalogues against the FK4 (Nowacki and Strobel 1968), one could, following Oort (1943a), find the system which would have resulted from the inclusion of all absolute catalogues. Only after reliable corrections to the FK4 have become available through observations in the future, will it become evident which is superior: the very selective procedure actually applied toward the construction of the FK4 system, or one that would have made broader use of the available material.

In using the term "absolute catalogue," especially with regard to right ascension, it will be well to remember Section 2.2.6 and particularly Section 2.2.6.3, where it was pointed out that observations can be partially absolute, that is, free of systematic errors that depend on δ but containing systematic errors that depend on α, and vice versa (α_δ and α_α catalogues, respectively).

Also, practically all "absolute right ascensions" are only absolute right ascension differences, and the locations of the equinox, that is, the zero point of the right ascension system, must be determined by a special effort (see Section 2.2.6.1 and 2.2.6.4). The equinox of FK3 was adopted for the FK4 without change.

As stated above, the system of the FK4 was established by revising that of the FK3R. This was obtained from the FK3 by reducing those catalogues that had be-become available after the completion of the FK3 to the system of the FK3, and then using the newly available positions of the FK3 stars in them to improve their positions within the FK3 system.

When finding the corrections to the FK3 system, Gliese adhered in essence to the "canonic" form: equation (3-20a), without allowing for the Δm, in various regions of the sky. He thus, in fact, allowed the corrections to be of the form $\Delta\alpha(\alpha, \delta)$. Of the forty-five catalogues used for the improvement of right ascension system, twenty-five were $\alpha_\alpha/\alpha_\delta$ catalogues, seven were α_δ catalogues, and only two were α_α catalogues. Gliese also paid careful attention to the above-mentioned zenith discontinuity (see Section 3.2.3.4) in the corrections.

The magnitude equations $\Delta\alpha_m$ and $\Delta\mu_m$ in the reductions from FK3R to FK4 were treated summarily, that is, not investigated for each catalogue individually. Morgan (1949) had shown earlier that there is a slight magnitude equation in the FK3 system. Gliese (1963) adopted Morgan's suggestion that the N30 is free of magnitude equation in α and in μ, and calculated from this assumption the corrections (FK4–FK3R) which he communicates in Tables 26 and 27 of page 45 of his paper.

From the discordances of the system of the catalogues used to establish the FK4 systems of α and μ, one may estimate the rms. errors of the system in right ascension and declination.

Table IV-3 is an adaptation of Gliese's Table 31 on page 51 of his above-mentioned paper. In the table below δ is the declination of the center of the zone to which the data refer, n_I the number of instruments and n_c that of catalogues used for establishing the system. $\epsilon_\alpha(t_\alpha)$ is the rms. error of the α system at its mean epoch—that is, t_α = 1935—and ϵ_μ that of the μ system. Colons indicate uncertain values, double colons very undertain values. The bars separating the ϵ_α and ϵ_μ values referring to the regions above and below $-25°$ are a reminder that the system south of $-30°$ is essentially determined by the Cape Reversible Transit Circle alone, so that the ϵ_α and ϵ_μ quoted for $-30°$ and south thereof only measure the internal agreement of the systems obtained on the Cape RTC.

Table IV-3 **Rms. Errors of the α and μ Systems of the FK4**

δ	n_I	n_c	$\epsilon_\alpha(t_\alpha)$ cos δ	ϵ_μ cos δ
> +80°	11	22	$0\overset{s}{.}0010$:	$0\overset{s}{.}0010$/cent.
+75	11	19–22	10	8
+70	11	19	15	8
+60	10	17	25	9
+50	11	20	25	10
+40	10–11	20–22	30	10
+30	10–11	21–22	30	8
+20	13	27–28	20	8½
+10	13	28–29	10	6
0	10–11	25–27	10	6
−10	7–8	19–21	15	7
−20	6	16–17	20	12
−30	2	4	50	13
−40	2	4	50	17
−50	2	4	45	17
−60	2	5	30	23
−70	2	6	30	22
−75	2	6	90	24
−81	2	7	40::	60::
< −83	2	9	?	?

Inspection of the very instructive Table IV-3 also illustrates the deplorable scarcity of absolute position determinations south of $-30°$, and the diminished realiability of the FK4 system in this region as a result thereof. Expeditions with Transit Circles have already been sent to locations on the southern hemisphere (Australia, Argentina, Chile) and will hopefully strengthen the fundamental system south of $-30°$.

4.1.5.4 THE FK4 DECLINATION SYSTEM OF POSITIONS AND PROPER MOTIONS. The establishment of the FK4 declination systems followed the same pattern used for the FK3 (Kopff, Nowacki and Strobel 1964). The basic idea is due to Newcomb, who applied it first for the construction of the system N2.

First, the "equator point," that is, the systematic correction $\Delta\delta_\delta$ at the equator, and its (linear) change with time were determined. 41 catalogues which contained positions whose epochs range from 1846 to 1956 were available, including 29 that had already been used for the same purpose during the establishment of the FK3 system. Simultaneously with the star positions in these catalogues, positions of the Sun, the Moon, planets, and in one case, minor planets had been observed. The theory of the motions of these bodies then allows one to determine the systematic correction of the catalogue declinations at the equator, or more accurately, along the ecliptic. (See Section 2.2.5.4.) These data were analyzed in a least squares adjustment. The weights alloted to the individual catalogues in this analysis depended on the number of FK3 stars they contained, the accuracy of the catalogue in general (corresponding to the weight it was given in the individual corrections) and "an assessment of the reputation of the catalogue." Whether this is the best weighting system that could have been used is an open question. Since the weights were assigned for the purpose of establishing a fundamental system, it might have been better to determine them from an investigation of the individual catalogues' systematic differences against their average unweighted system.

Throughout, the system of the FK3 was used as a first approximation. The solution finally adopted for the $\Delta\delta_\delta(\delta = 0)$ was, for the proper motions system, based on all 41 catalogues. In the solution for the position system the 16 catalogues with average epochs before 1900 were excluded because, with one exception, their equator points had been determined from observations of the Sun only. Thirteen new catalogues had become available for this purpose after the work on the FK3 was finished.

The $\Delta\delta_\delta$ of the individual catalogues against the ideal system were then derived in the following manner. It was assumed that, on the average, the $\Delta\delta_\delta$ of a catalogue varied linearly with the declination between the equator and the north celestial pole and remained constant between the equator and the southern horizon for catalogues observed at northern observatories, and vice versa for catalogues observed at observatories in the southern hemisphere. The averages of the thus corrected catalogue declinations, that is, those obtained by the removal of the $\Delta\delta_\delta$ which were postulated to vary in the just described way, were then regarded as the corrections to the declination system of the FK3, to establish the declination system of the FK4. The corrections $\Delta\mu'_\delta$ to the proper motion system were obtained by a least squares solution of the group averages of "reduced" declinations of catalogues, spanning approximately ten years. This process is analogous to the formation of "normal points." As stated above, the $\Delta\mu'_\delta$ system was derived from the solutions involving all available catalogues, that is, those containing solar and planetary observations; the $\Delta\delta_\delta$ system, however, was derived from a solution involving only the catalogues with epochs after 1900. The very small systematic differences between the systems of the FK3 (which was the one to which the corrections were actually applied) and the FK3R were also taken into account.

The corrections $\Delta\delta_\alpha$ and $\Delta\mu'_\alpha$ from the system of the FK3 to that of the FK4 were derived from the residual differences after removal of the $\Delta\delta_\delta$ and $\Delta\mu'_\delta$. These were obtained in the zones $+90°$ to $+60°$, $+60°$ to $+40°$, $+40°$ to $+0°$, $0°$ to $-30°$ and $-30°$ to $-90°$ from all catalogues contributing to the δ_α system of the FK4. Since these do not necessarily involve observations of the bodies of the planetary system, there are more catalogues available for this purpose than for the establishment of the δ_δ system. As for the establishment of the δ_δ and μ'_δ systems, least squares solutions were made in every zone for every hour of right ascension from normal points obtained by forming weighted means of positions from catalogues within suitable ranges of epochs. *The μ'_α system was derived from all available catalogues; the δ_α system only from those with epochs after* 1900 (with a weighted average epoch of 1925).

The size of the corrections from FK3 to FK4 is only a fraction of that of the corrections of NFK to the FK3. This shows that the FK3, all in all, already realized a fairly good fundamental system. These corrections are compared in Figs. 3 and 4, respectively, on pp. 17 and 18 of Kopff, Nowacki and Strobel's (1964) paper. The estimation of the uncertainty of the δ and μ' systems of the FK4 is more complicated than of the α and μ systems, since the dependence of the system on α and δ, respectively, were judged separately. The error $\epsilon_\delta(t)$ is given by

$$\epsilon_\delta^2 = \epsilon_{\delta\delta}^2(T_\delta) + (t - T_\delta)^2\epsilon_{\mu'\delta}^2 + \epsilon_{\delta\alpha}^2(T_\alpha) + (t - T_\alpha)^2\epsilon_{\mu'\alpha}^2 \qquad (4.1)$$

where t, T_δ and T_α are (in units of Julian centuries) the epoch at hand, and the mean epochs of the δ_δ and δ_α systems, respectively. These position systems were derived from the catalogues after 1900. The T_δ and T_α and the standard deviations $\epsilon_{\delta\delta}(T_\delta)$ and $\epsilon_{\delta\alpha}(T_\alpha)$, however, are those obtained from the solutions which included all catalogues, hence T_δ and T_α are in the neighborhood of 19.15. Tables IV-4 and IV-5 which give the data necessary for the evaluation of formula (4.1), were adapted from Tables 4a, 4b, 4c, 4d and 5a, 5b, 5c, 5d, 5e of the above quoted paper by Kopff, Nowacki and Strobel (1964), in which more detailed data may be found.

From Tables IV-4 and IV-5, it is apparent again that the accuracy of the system decreases considerably south of, say $-25°$, that is, at declinations that are not accessible to most of the instruments on the northern hemisphere. Therefore, the establishment of instruments in the southern hemisphere which are capable of making absolute position determinations is one of the most urgent requirements of fundamental astronomy. Fortunately, several observations in the northern hemisphere, for instance the U.S. Naval Observatory, Pulkovo and Bergedorf, have set up transit circles in the southern hemisphere. Strictly absolute observations, however, are made at this time (1970) only by the Pulkovo group.

It was mentioned above that the corrections to the FK3 system (at the mean epoch of the FK4) indicate that the FK3 was already a good system. According to W. Gliese, the right ascension corrections of the system at mean epoch of the FK4 remain small ($\Delta\alpha \leqq 0.^s006$) in the zone around the equator ($-15°$ to $+35°$). The maximum of the right ascension corrections on the northern sky occurs at (9^h, $+75°$) with a value of $+0.^s055$ (and $+0.^s145$ for the correction to the centennial proper motions). In the southern hemisphere, where the values are considerably less accurate, the maximum right ascension correction of $-0.^s046$ is found at (12^h, $-75°$), while the proper motion system requires a correction of as much as $-0.^s22$ per century at (12^h, $-80°$). One must, however, consider that the values at $-80°$ are particularly uncertain because the system of the FK3 shows a discontinuity there.

Table IV-4 **Error Data for the δ_δ and μ_δ' Systems**

δ	T_δ	$\epsilon_{\delta\delta}(T_\delta)$	$\epsilon_{\mu'\delta}$	δ	T_δ	$\epsilon_{\delta\delta}(T_\delta)$	$\epsilon_{\mu'\delta}$
+85	19.165	0″.011	0″.05	−5°	19.150	0″06	0″02
+75	19.165	9	4	−15	19.150	18	7
+65	19.165	10	4	−25	19.150	17	7
+55	19.165	10	4	−35	19.140	22	8
+45	19.165	6	3	−45	19.140	31	10
+35	19.165	10	4	−55	19.140	30	10
+25	19.125	13	5	−65	19.140	15	5
+15	19.125	10	4	−75	19.140	36	12
+5	19.125	8	3	−85	19.140	35	15

Table IV-5

Error Data for the $\delta\alpha$ and $\mu'\alpha$ Systems

Declination region	T_α	$\epsilon_{\delta\alpha}$ (unit: 0″.001) 0^h	3^h	6^h	9^h	12^h	15^h	18^h	21^h	$\epsilon_{\mu'\alpha}$ (unit: 0″.01/100 yrs) 0^h	3^h	6^h	9^h	12^h	15^h	18^h	21^h
+90° to 60°	19.11	29	10	13	9	11	11	13	10	12	4	5	4	4	5	5	4
+60 to 40	19.16	10	7	17	7	14	9	12	15	4	3	7	3	6	4	5	6
+40 to 0	19.14	14	9	9	6	9	8	10	10	6	4	3	2	3	3	4	4
0 to −30	19.15	8	9	8	9	15	17	11	12	3	4	3	4	6	7	5	5
−30 to −90	19.14	13	19	12	20	15	11	18	11	4	6	4	6	5	4	6	4

As was shown above in Section 4.1.4, the corrections generally bear out Oort's (1943a) estimate of the errors in the FK3, which he made long before work on the FK4 was even started. Oort criticizes the procedure of establishing the declination system of the FK3 only from those catalogues whose errors have been determined with the aid of planetary and solar observations. He points out that better values for the FK3 system could have been obtained by using all available catalogues, and not making the severe demands and sometimes arbitrary restrictions.

4.1.5.5 DESCRIPTION OF THE FK4. The Fourth Fundamental Catalogue (FK4) is at this time internationally accepted as the best available fundamental system. It resulted from the revision of the Third Fundamental Catalogue (FK3) carried out under the supervision of W. Fricke and (the late) A. Kopff (1963), in collaboration with W. Gliese, F. Gondolatsch, T. Lederle, H. Nowacki, W. Strobel and P. Stumpff, and was published as Volume 10 of the "Veröffentlichungen des Astronomischen Rechen-Instituts Heidelberg" in 1963, Verlag G. Braun (Publisher) in Karlsruhe. It represents one of the finest team efforts of positional astronomers. The introduction, which gives a brief description of the processes employed for the construction of the catalogue, occupies pages 1–8. The table at the head of page 8 could be used in lieu of our Table IV-3, IV-4 and IV-5 for the computation of rough errors of the system of the FK4 at various epochs, if very accurate information is not essential.

The catalogue proper "Catalogue of 1535 fundamental stars for the equinox and epoch 1950.0 and 1975.0" which occupies pp. 9 through 127 gives on the left-hand pages the data pertaining to the right ascensions, and on the right-hand pages those pertaining to declination. The left-hand page columns contain:

1: No. The FK3 number. Since the stars are arranged in the order of increasing right ascension, and since the Auwers-Sterne and Zusatzsterne were mixed in the FK4 (1950), these numbers are no longer in sequential order.
2: Name. The name of the star (after Joh. Bayer) when available.
3: Mag. and 4: Sp. The visual magnitudes and spectral types from the Henry Draper Catalogue.

Columns 4 to 8 carry two items of information for every star: on the upper line, the information for epoch and equinox 1950.0, and on the lower, for 1975.0.

4: α, the mean right ascension for epoch *and* coordinate system 1950.0 (or 1975.0) in hours, minutes and seconds of time, to $0.^s001$.
5: $d\alpha/dT$, the first derivatives of the mean right ascension with respect to time (T being reckoned in tropical centuries) as a consequence of proper motion and precession, to $0.^s001$.
6: $\frac{1}{2}d^2\alpha/dT^2$, half the second time derivatives of the catalogued mean right ascension (T in tropical centuries) due to precession and proper motion, to $0.^s001$.
7: μ, the centennial proper motion in right ascension to $0.^s001$, based on Newcomb's value of the constant of precession.
8: $d\mu/dT$, the time derivative (T in tropical centuries) of the centennial proper motion to $0.^s001$.
9: Ep(α), the central epoch (minus 1900) in tropical years of the observations from which, *within the system*, the right ascensions were computed.

10: $m(\alpha)$, the standard deviation (rms. error) of the right ascension at the central epoch (in units of $0.^s001$).

11: $m(\mu)$, the rms. error (in units of $0.^s001$) of the centennial proper motion.

On the right-hand pages, the columns give the following information:

1. No., as on left. Columns 2: δ, 3: $d\delta/dT$, 4: $\frac{1}{2}d^2\delta/dT^2$, 5: μ', 6: $d\mu'/dT$, 7: $Ep(\delta)$, 8: $m(\delta)$, 9: $m(\mu')$ carry the information (2 through 6 to $0.''01$) which corresponds precisely to that for right ascension in columns 4 through 11 of the left-hand pages, except that $m(\delta)$ and $m(\mu')$ are in units of $0.''01$. Finally, columns 10 and 11 of the right-hand pages give the numbers of the stars in the GC (see Section 4.2.4) and the N30 catalogue (see Section 4.2.5).

The precise meaning of the first and second derivatives listed in the catalogue, and their use, are explained on pp. 9 and 10. There, it is also shown how a reduction accurate to the third order from one of the catalogued epochs and equinoxes to another pair of equal epochs of place and orientation can be performed. Unfortunately, these data are not of much use when the catalogued positions are to be referred to one epoch of place, but a different one of orientation (that is, equator and equinox). In this case, it will be necessary to employ the methods previously described in Section 3.2.1.

The pattern of the catalogue changes somewhat for the circumpolar stars ($|\delta| > 81°$; pp. 112–123), where the star positions, proper motions and their derivatives are given for the epochs of place and orientation of 1950, 1955, 1960, 1965, 1970 and 1975. Pages 125–127 give tables likely to be useful when one works with data concerning binaries in the fundamental catalogue. On pages 128 and 129 the so-called foreshortening effect is discussed which depends, in first approximation, on the product of radial velocity, parallax and proper motion. There are 31 stars known in the FK4 for which this effect significantly influences the derivatives of their proper motions, and thus the first and second derivatives of their coordinates. It is explained that the influence of the foreshortening effect was included in the tabulated values of these quantities. (Note, however, that the recently published formulas for the rigorous reduction of proper motion components, radial velocities and distances [Eichhorn and Rust 1970] make a special term for the foreshortening effect superfluous.)

On page 130 is an explanation of the Tables for the Systematic Differences FK4-FK3, and detailed instructions for their use; the tables themselves are given on pp. 131–134. The differences of the position systems hold for the epoch 1950.0, and for the proper motion system they are, of course, time independent. The corrections follow the canonical form equation (3.20a) except that the $\Delta\alpha$ and $\Delta\mu$ for $|\delta| > 60°$ are multiplied by $\cos\delta$. The δ-dependent corrections are given for every degree, the α dependent corrections in double entry tables for every hour of right ascension and every tenth degree of declination, except that they are given for $\pm 85°$ instead of 90°.

The systematic differences FK4-GC and FK4-N30 were given by Brosche,

Nowacki and Strobel (1964); the FK4-GC were also given in Brosche's (1966) paper. These publications provide graphs as well.

The FK4 ends on pp. 135–144 with a list of all catalogues from which positions were incorporated in the FK4 either for the purpose of establishing the system or deriving individual corrections to the positions published in the FK4. Of the first group of catalogues (numbers 1–156), those with numbers 1–77 had already been used for the compilation of the FK3. The numbers (in the first column) assigned are arbitrary. The second column shows the code for this catalogue in the GFH or in Kopff's list (see Sections 3.3.2 and 3.3.4). Column 3 shows the observatory at which the catalogue originated, columns 4 and 5 give the epoch of orientation and the mean epoch of place, respectively, and an "i" in the sixth column indicates that the positions in the catalogue were used for the individual corrections of the positions within the FK4 system. The seventh column finally shows identification, that is, the title, source, and when applicable, the name of the observer or principal reducer. The catalogues are primarily arranged in the order of increasing epochs of orientation, and those which are oriented to the same coordinate system are arranged alphabetically according to the observatory at which they were observed. The earliest average epoch of place is 1817, the latest 1953.

This table is followed by one which is similarly arranged but in which the catalogues are numbered from (1) through (85). All the positions in the catalogues in this table were used for the individual improvement of the positions within the FK4 system. Mean epochs of place are not given. The epochs of orientation range from 1745 through 1905.

4.1.5.6 THE PRELIMINARY SUPPLEMENT TO THE FK4 (FK4SUP). This "Catalogue of 1987 supplementary stars to FK4 for the equinox and epoch 1950.0" was compiled at the Astronomischer Rechen-Institut (1963) under the direction of W. Fricke, with the assistance of P. Brosche, W. Gliese, T. Lederle, H. von Lüde, H. Nowacki, W. Strobel and P. Stumpff.

The stars are those selected by Kopff for the FK3Supp (see Section 4.1.9.2) who had taken their positions and proper motions primarily from the N30 (and when not available in the N30 from the GC) and reduced them to the FK3 system. The FK4sup lists these positions and proper motions reduced to the FK4 system by means of the reduction tables in the FK4. The approximate positions in the southern sky were taken from the N30 or, when not listed there, from the GC. In the latter case, their positions had been frequently improved by F. P. Scott of the USNO. The accuracy of the star positions in the FK4sup is therefore that of the positions of these stars in the N30 or the (improved) GC, respectively, except that the uncertainties of the reduction from the N30 to the FK4 are added. According to Fricke, "The main purpose of the FK4sup is to stimulate further observation within absolute and differential programs of the stars listed therein, so that an improved fundamental system may include the supplementary stars as new fundamental stars."

The FK4sup supplies the following data: serial numbers starting with 2001 and increasing with the 1950 right ascension, except for the stars in the polar caps ($|\delta| > 85°$) which are treated in the appendices. Star names, which can be found in the FK3Supp, have been omitted.

The second and third columns (Mag. and Sp.) give visual magnitudes and spectral types, essentially from the Henry Draper catalogue. Next, follow the right ascensions and centennial proper motions (to $0.^s001$), declinations and centennial proper motions (to $0.''01$) for epochs of place and orientations 1950.0, as close as possible within the FK4 system. Further columns indicate the stars' numbers in the GC and the N30 catalogue, and a remark column supplies primarily information concerning binaries.

Central epochs as well as standard deviations of positions and proper motions, which are so important for the improvement of the catalogue by the incorporation of new data, were not given, but are identical with those quoted in the N30 catalogue for the same stars, if, as is true for the majority of them, their positions are also listed in the N30 catalogue. These data are not available in the printed literature for those stars that occur only in the GC.

4.1.5.7 THE APPARENT PLACES OF FUNDAMENTAL STARS (APFS). For most purposes of practical astronomy, it is not the mean places (as published in the catalogues) but the topocentric apparent places (see Section 1.6.4) that are needed. Since these would depend on the observer's location with respect to the Earth, they could not be tabulated in a volume intended for global use. Also, the difference between true and mean positions (at any epoch of orientation) is determined by the influence of nutation. Since there are terms in the expression for nutation that have periods of two weeks and less, it is clear that a tabulation of true (or apparent) star positions at intervals of ten days would not allow one to correctly interpolate the influence of the short period nutation terms.

In order to satisfy the various needs of the observers who require apparent positions at odd epochs, the International Astronomical Union has, since 1941, sponsored the annual publication of a volume entitled "Apparent Places of Fundamental Stars" (APFS). From 1941 through 1959, the APFS were published at the Royal Greenwich Observatory. Since 1960, The Astronomisches Rechen-Institut at Heidelberg has been in charge of the production of these volumes. The APFS volumes give the geocentric apparent positions of 1535 fundamental stars at 10-day intervals (for each day for the stars near the poles), at the instants of their upper transit at Greenwich. The apparent places appearing in the APFS are based on their heliocentric mean positions in the FK3 before 1963, and in the FK4 thereafter. In deriving the apparent places, the long-period terms of nutation and, when its influence is significant, the annual parallax are taken into account. For reasons just explained, the short periodic terms of nutation are left out. The constants used until 1967 were those adopted at the "Conférence Internationale des Étoiles Fondamentales" held at Paris in 1896. Starting with the 1968 volume, the newly adopted value for the constant of aberration ($20.''496$) has been used (compare Fricke et al. 1965). Starting with the 1960 volume, the effects of annual aberration are no longer calculated from a closed formula, but by numerical differentiation with respect to time of the coordinates of the center of the Earth, referred to the barycenter of the solar system as zero point.

Woolard's (1953) formulas are used for the computation of the long periodic terms in the expression for the nutation. All terms whose amplitude exceeds 0.″0002 are included. The reduction from heliocentric to geocentric positions is performed for those 721 stars whose annual parallax exceeds 0.″010 in Jenkins' (1952) catalogue.

In addition to the ephemerides of the stars, the APFS also contains several auxiliary tables among which are the Besselian day numbers, data to allow the computation of the short period terms of the nutation, aids for the conversion of time from sidereal to mean, and so forth, coefficients in the various formulas used for interpolation, and finally, means for the computation of the diurnal component of stellar aberration. (See Section 1.6.4.)

4.1.6 CRITIQUE OF THE FK4. Estimates of the actual errors of the FK4 system, and even suggestions for corrections to it, were first given in a paper by Guinot (1961), two years before the FK4 appeared in print. He bases his investigation mainly on his own catalogue, which was observed by means of a Danjon-type prism astrolabe (see Section 2.2.4). Therein, he also compares his results with those derived from other catalogues obtained with a prism astrolabe at Neuchâtel, and with a photoelectric transit instrument operated by N. N. Pavlov at Pulkovo.

For the precision of the FK4 system at the equator (for the epoch 1957) he suggests rms. errors of 0.″006 in right ascension and 0.″010 in declination, in fairly good agreement with the values suggested by the Heidelberg Rechen-Institut and reproduced in this chapter. If one may generalize Guinot's confirmation, it would appear that the nominal Heidelberg values for the errors are fairly correct for the whole sky. It is interesting to note that, judging from their internal agreement, a combination of the catalogues used by Guinot defines a better fundamental system in right ascension (with a mean error at the equator of only 0.ˢ004 at the epoch of about 1957) than the FK4, while Guinot's definition of the declination system (m.e. 0.″10) is as good as that of the FK4. This is the result of only about five years work. One must still consider, though, that the astrolabe observations are not really absolute.

Guinot's values do not contain, of course, the uncertainty of the equinox, which cannot be determined by a prism astrolabe; also, his discussion applies only to the northern hemisphere.

The situation on the southern hemisphere is unfavorable because the fundamental system south of about −30° is principally defined by the Cape transit circle. Any systematic errors in the positions obtained with this instrument have strongly influenced the fundamental system on the southern polar cap (from −30° down). One can only hope that the systematic errors will not be more than, say,

0.″30. Values of such errors, and a better definition of the fundamental system in these regions of the sky, must necessarily await the establishment of more observing stations in the southern hemisphere.

Since the early critique by Guinot, several more series of absolute observations have become available which allow one, even at this time, to suggest rather reliable corrections to the system of the FK4. These were subjected to a fairly comprehensive discussion by Gliese (1970) who gives graphs of the systematic differences between the means of the positions obtained from recent observations (that is, those which were not incorporated into the FK4) and the FK4. Their mean epoch is about 1957. Graphs for $\Delta\delta$ and $\Delta\alpha_\alpha \cos\delta$ are given for various declinations down to $-20°$. From this, one can see that maximum errors of the position systems of the FK4 appear to be somewhat smaller than 0.″1. This means that the systematic errors in the proper motion systems are about three to five times that large, if the system of the FK4 at its epoch is regarded as error-free, considering that the epoch of the α system in the FK4 is about 1935 and that of the δ system is about 1925. Systematic errors of this magnitude are a serious impediment to the reliable determination of Oort's Constants A and B of galactic shearing and galactic rotation, since these constants are of the same order of magnitude as the average systematic errors of the proper motion system.

In the southern hemisphere, the systematic errors are even larger, see, for instance, the papers by Zverev (1968), and Anguita et al. (1968). $\Delta\delta_\delta$ of the FK4 between $-20°$ and $+90°$ were graphed by Zverev (1968) on the basis of work in the U.S.S.R., while Gliese graphs the $\Delta\alpha_\delta \cos\delta$ on the basis of most of the available material. It is interesting to note that, at least north of $-20°$, the $\Delta\alpha_\delta \cos\delta$ (for the epoch 1958.5) show that the system of the N30 is in better agreement with the modern observations than that of the FK4. South of $-30°$, the various instruments yield somewhat discordant systems, (α_δ), and the apparent superiority of the N30 systems begins to disappear.

Generally, the μ_δ system appears to be much more accurately determined than the μ_α system.

In a paper presented to the International Astronomical Union at the Prague General Assembly, Fricke and Gliese (1968) discuss the desiderata for the FK5, that is, the improved and revised version of the FK4. Their paper was written with the realization in mind that data processing is no longer the drudgery it was when the previous fundamental catalogues were compiled, and, also, that observing techniques have become quite a bit more efficient, through automation, than they were a few decades ago. Thus these authors recommend an increase in the number of fundamental stars to about 5000 and emphasize that uniformity of distribution should no longer be an all-important consideration; especially since there is a concentration of especially interesting objects in the galactic plane. One of the most important requirements for a fundamental system is its freedom from magnitude effects, especially in the proper motions. One of the many reasons for this is that secular parallaxes are derived from the systematic trends in the proper motions, and an undetected magnitude effect could lead to wrong astrophysical conclusions of far-reaching consequences. It is virtually only since the widespread use of impersonal observing methods that fuller confidence has developed in the magnitude independence of absolute—and relative—position observations.

In this connection, Fricke (1970) pointed out an intriguing possibility: The accuracy of the absolute determination of positions of discrete radio sources corresponded then to a rms. error per coordinate of about one second of arc. At this time (1973), the accuracy obtainable for absolute positions of radio sources corresponds to a rms. error of less than $0.''1$. The pioneer in this field is C. Wade of the National Radio Astronomy Observatory. It is safe to say that the potential of radio astronomical techniques for fundamental astrometry are now just beginning to be tapped, and we can predict very exciting developments during the near future. The visible objects corresponding to these radio sources are frequently fainter than 18^m. Murray, Tucker and Clements (1969) have established optical positions on the FK4 system of a number of these with rms. errors of the coordinates of about $0.''25$, while taking careful precautions to avoid the introduction of a magnitude equation. Thus, a link is established between very faint and very bright objects. Providing the accuracy of radio positions continues to improve, there might be, sometime in the future, a mixed fundamental system encompassing objects from the brightest stars down to some fainter than 18^m, where the tie-in between the bright objects measured at a transit circle and the faint ones, whose positions depend on radio measurements, would be accomplished by methods similar to those used by Murray et al. (1969) mentioned above. It must be pointed out, however, that the determination of the equinox by radio observations will, at best, be an extremely difficult task, since it requires the observation of bodies in the solar system whose radio radiation is at this time not suitable for the accurate determination of their positions by radio methods.

According to the above-mentioned paper by Fricke and Gliese (1968), the methods to be used for the computation of the next fundamental catalogue (FK5) should be carefully scrutinized and, when necessary, the traditional methods should be changed. The traditional German School of constructing fundamental catalogues was always very selective in the utilization of material, and it appears that Fricke and Gliese now believe that some of this ought to be changed. In particular, they mention the new method for the comparison of catalogues developed by Brosche (1966), and express the opinion that the establishment of the proper motion system in right ascension should be accomplished by an adjustment from the positions in all absolute catalogues rather than from a series of observations on the same instrument as has been the practice in the FK series of the past.

4.2 The American Series of Fundamental Catalogues

4.2.1 NEWCOMB'S "CATALOGUE OF FUNDAMENTAL STARS." The astrometric conference at Paris in May of 1896 requested Newcomb to complete his work, begun in 1895, toward the establishment of a fundamental catalogue to serve as the basic star list of the American Ephemeris, and possibly also of other national ephemerides.[1]

Newcomb's catalogue appeared in 1899 and contains the mean positions for epochs of place and orientation 1875 and 1900 of 1257 stars. He also gives a finding list of 1597 stars for which he proposes to derive fundamental positions; there were, however, not enough observations to include all of them as fundamental stars in his final catalogue.

Altogether, positions from 44 catalogues whose epochs of place range from 1750 to 1895 were incorporated into the finally published positions, although not all of these were used for the purpose of establishing the system.

The *right ascension system of the catalogue* was established as follows:

The equinox was to be the same as that of the N_1, as defined by the right ascensions of Newcomb's (1880) previous "Catalogue of 1098 Clock and Zodiacal Stars." For the model of the $\Delta\alpha_\alpha$, Newcomb chose a constant plus a simple sine wave since, in his opinion, this was the only physically justifiable form, and he clearly states that he does not feel that any purely interpolatory or empirical correction formulas should be admitted.

For the formation of the system, Newcomb felt that the average of the systematic errors $\Delta\alpha_\alpha$ would constitute as good a system as one formed by the application of sine wave corrections prior to the taking of the mean. Thus, except for the eight Greenwich catalogues to which constant terms and sine wave corrections were applied, all other catalogues were reduced by the addition of constant terms only (equinox corrections). Newcomb was confirmed in this conclusion by not finding any systematic $\Delta\alpha_\alpha$ difference between N_1 and the FC that could be considered significant.

The $\Delta\alpha_\delta$ to be utilized for the formation of the fundamental system were determined as corrections to the $\Delta\alpha_\delta$ system of the FC. Auwers' $\Delta\alpha_\delta$ corrections, derived for the purpose of establishing the FC system, were primarily based on the system of Pulkovo observation only. Newcomb felt that a less selective process for establishing the system would have been preferable. He therefore regarded, in essence, the mean of the α_δ systems of all catalogues utilized to be the correct one and thus derived, from catalogues ranging in epoch of place roughly from 1750 to 1890, the corrections to the FC system (which he calls the A.G.C., as does Auwers). The corrections were derived to the position as well as the proper motion system of the FC (they were in the form $x + Ty$, with T being the epoch of place). Values were determined for every full 5°. Very little material was available for establishing the system south of $-30°$.

Although Newcomb devotes a considerable amount of space to the discussion of magnitude equation—its cause, and suggestions for its elimination—he did not

1. The fact that this task was assigned to Newcomb and not to Auwers was lamented by Ristenpart (1901, pp. 465–466) who felt that Auwers' FC should have been made the foundation for the apparent star places in the various national ephemerides.

apply any correction $\Delta\alpha_m$ to the positions used for forming the system of the right ascensions of his catalogue.

The *most significant* contribution of this catalogue is probably the *method devised for the establishment of the δ_δ system,* one of the most difficult problems of fundamental astronomy as every knowledgeable investigator knows. There is no need to repeat its description here because it was already fully described in the discussion of the formation of the $\Delta\delta_\delta$ system of the FK4 (Section 4.1.5.4; also 2.2.5.4). The underlying principle is the dynamical fact that, on the average, the orbits of the bodies in the solar system must be in planes, and the time averages of their declination must be zero in the long run. This is only a rough statement of the principle. Its actual application must, of course, be based on theories of the motions of the bodies utilized, and Newcomb was one of the pioneers in establishing these theories. The latest available results from applying this method is due to Pierce (1970). Since the major planets and the Sun are restricted to the zodiac, the establishment of the δ_δ system at declinations closer to the pole must be left to depend on the accurate determination of the instrument latitudes by observation of the transits of circumpolar stars in upper and lower culminations (see Section 2.2.5.3). The actually adopted δ_δ system in the intermediary declinations will then be one that varies linearly between that established at the equator[2] from the Sun and the planets, and at the poles from culminations. As a matter of convenience, Newcomb assumed that the equatorial correction to the declinations observed in the northern hemisphere remained constant toward the southern horizon, and vice versa for observations made in the southern hemisphere.

These were the only corrections to the system of the declinations which Newcomb applied in his catalogue. After a discussion of corrections of the form $\Delta\delta_\alpha$, caused by the variations of latitude of which he was well aware, he concluded that they would, on the average, cancel and decided not to apply any for the formation of his system. Besides, they had just then begun to be systematically observed. Likewise, no $\Delta\delta_m$ were applied, either in the formation of the system, or in the formation of the individual position and proper motions within the system which we are about to discuss.

It should be pointed out that Newcomb regarded Bradley's observations (reduced by Auwers) as very valuable, because of their age, for the determination of the proper motion system and the individual proper motions, and made special efforts to remove systematic errors from them. Frequently, he disagrees with Auwers on the way they should have been reduced and makes his own additional corrections.

Furthermore, it might be worth noting that the initially assumed system of declinations, to which Newcomb derived corrections, was established by L. Boss (1877).

---

2. Newcomb actually used the declination of $+5°$ which was the average of all observed declinations, since most observations were made at observatories on the Earth's northern hemisphere from which the planets are more likely to be observed in northern declinations.

The *final positions and proper motions* were formed by least squares solutions from the available positions in the catalogues included, after weights had been assigned, and corrections to the adopted system, formed as described above in this section, had been applied. These systematic corrections to the various catalogues are given on pp. 227–237, the weights on pp. 238–241.

The *catalogue* itself, which occupies pp. 250–369, gives data as follows. On the left-hand pages, the first column contains the serial number. These numbers have gaps in them, since fundamental data were not derived for all stars in the proposed list (pp. 91–122) of fundamental stars. The second column gives the stars' names, and the third, their visual magnitudes on the Harvard system. The fourth column carries, for every star, the numbers 1875 and 1900; one below the other, in order to indicate that for the pairs of data in columns 5–10, which concern the individual stars, the upper line refers to 1875, the lower to 1900.

These columns contain the stars' mean right ascensions in hours, minutes and seconds of time to $0.^s001$ for epoch of place and orientation as in column 4, then their centennial and secular variations, that is, the first and second derivatives (referred to 100 years as time unit) of the right ascension as a consequence of precession and proper motion, and next, under the heading of "Precession," the first derivatives of the right ascensions due to precession only and the centennial proper motions. Obviously, the sums of the numbers in columns 8 and 9 must equal the corresponding number in column 6. The precessions were computed from Newcomb's own values for the constants of precession, explained by him in the first part of the same volume which contains this catalogue (Newcomb 1899a). The (centennial) derivatives of the proper motions in column 10 (there called "Sec. Var. of Proper Motion") are given for both epochs, only if they are different for the two epochs. The last two columns (11 and 12) on the left-hand pages contain the differences between the FC right ascensions and centennial variations, respectively, for those stars common to both catalogues. All quantities pertaining to the right ascensions are given to $0.^s001$.

The data on the right-hand pages pertain mainly to declination. Columns 1 and 2 are repetitions of columns 1 and 4 of the left-hand pages, and columns 3–8 give the mean declinations, their total centennial first and second-time derivatives ("Centennial Variations" and "Secular Variation", respectively), the precessions, proper motion and their (centennial) time derivatives ("Sec. Var. of Proper Motion") to $0.''01$, completely analogous to the corresponding right ascension data. For the declinations, not only are the corrections FC minus Newcomb given for declinations and centennial variations, but also the corrections Boss minus Newcomb, when there were stars in common with L. Boss' (1877) original system. The final two columns contain the stars' numbers in Bradley's catalogue, and notes, respectively.

The catalogue is followed by tables occupying pp. 370–382 and containing mean epochs (central epochs, see Section 3.2.2), weights and probable errors of the published mean positions. The unit weights refer to rms. errors of $0.^s060$ and $0.''53$ in right ascension and declination, respectively. The data in these tables could be used for the improvement of the positions according to the method indicated in Section 3.2.2. The "Notes" in the catalogue are explained on pp. 383 and 384. The paper closes with an appendix giving mainly formulas and tables for the transformation of mean star positions from one epoch (of place and/or orientation) to another.

It would be interesting to compare the FC as well as the Newcomb system to the best available modern system, that of the FK4, in order to see which of these in its own time was closest to the truth—the very meticulously established FC system with all its selectiveness in the catalogues which it used for the definition of the system proper, or Newcomb's, which essentially was established by taking the mean of all these systems which were then available and regarded as absolute in one way or other.

4.2.2 EICHELBERGER'S STANDARD STARS. While the German Series of Fundamental Catalogues represents a more or less uniform and continuous sequence, its American counterpart does not. The only real continuity in the American Series is between the PGC (Section 4.2.3) and the GC (Section 4.2.4) which is the reason why Eichelberger's Catalogue is treated in this monograph before the PGC, although the latter antedates the former.

Eichelberger (1925) published his catalogue to satisfy the need of the American Ephemeris and Nautical Almanac (AENA) for accurate star positions, since those in Newcomb's catalogue had become inaccurate. For this he employed a procedure that was later applied by Morgan for the construction of the N30 catalogue (see Section 4.2.5): he formed normal positions at a recent epoch from newly available absolute catalogues and calculated the proper motions from comparison with the positions in an earlier fundamental catalogue at an epoch equal to the average central epoch of the stars contained therein.

In the case of Eichelberger's catalogue, there were four recent epoch absolute catalogues, of which two had been observed at the Royal Observatory at the Cape of Good Hope, and two at the U.S. Naval Observatory at Washington; one each at the nine-inch and the six-inch transit circles. (In addition to this, the positions of 32 southern circumpolar stars which had not been recorded in any of these catalogues were taken from the Cape Annals **11** pt. 5.) The Cape catalogues were the First and the Second Cape Fundamental Catalogues for the Epoch 1900; those at Washington were contained in the Publications of the U.S. Naval Observatory **9**, pt. 1 (9-inch) and **11**, (6-inch). The average of the epochs is about 1910.

The recent normal right ascensions were the weighted averages of the right ascensions in the four aforementioned catalogues, after they had been reduced to the average of the equinoxes of the four catalogues. No further corrections were applied to the right ascensions before their means were taken.

The recent normal declinations are the result of a more complicated process. At Washington, corrections to the constant of refraction had been computed by the method described in Section 2.2.5.3, while at the Cape, the refraction was computed from the Pulkovo Tables and used without any further correction. Corrections to the Cape refraction constant, and consequently the geographic latitude, were obtained by comparing the declinations in the Cape catalogues with those in the Washington catalogues, following the principles described below in Section 4.2.3. Furthermore, observations of the Sun, Mercury, and Venus at the Cape, and of the Sun, the Moon and major Planets at Washington provided corrections to the equator point following Newcomb's method described in Section 2.4.1. Eichelberger made a considerable effort to decide upon the analytic form which that part of the interpolated $\Delta\delta_\delta$, which is a consequence of the equator point correction, should take between equator and zenith. After quite extended trials, consisting mainly of investigating declination differences between the northern and southern catalogues, he decided in favor of a correction proportional to $\cos\delta$ for the Second Cape Catalogue and the two Washington catalogues, and one linear in the sine of the zenith distance for the Washington six-inch Catalogue. (For this purpose, zenith distances were reckoned positive on the southern, and negative in the northern part of the meridian.)

No right ascension dependent corrections were applied to the declinations before formation of the definitive positions which were the weighted means of the individual determinations. The weights of the individual determinations depended on the number of observations in the usual form—compare equation (3.2.4)—except that observations below the pole received reduced weights. The closer to the horizon an observation was made, the lower a weight was assigned to it.

The proper motions were determined by regarding as correct the positions in the PGC at its average central epoch 1865, except that the right ascensions had to be corrected by $-0.^{s}014$ because of the correction of the equinox.

One of the tables in Eichelberger's publication occupies pp. 28–51 and gives a comparison with Newcomb's catalogue. It also lists the average epochs of the recent normal positions, even residuals of all individual determinations (when existing). Another one (pp. 52–71) contains the differences against the PGC, again giving epochs and residuals of individual determinations. A third table (pp. 72–76), preceding the catalogue proper, gives a list of the various names assigned to the stars used in the national ephemerides.

The catalogue proper (pp. 78–127) contains right ascension information on the left, and declination information on the right-hand pages. Given are serial numbers, names, visual magnitudes (Harvard system), parallaxes (specially made available by F. Schlesinger) and spectra (from the HD catalogue). There is also a column of notes, which are, by and large, self-explanatory. The positions, as well as their first, second and third time derivatives are given for epoch of place and orientation 1925, the proper motions and their secular variations were computed with the use of Newcomb's values for the precessional quantities. Finally, the individual differences of the positions and proper motions against those in the PGC, the NFK and Newcomb's catalogue are available as $\Delta_B, \ldots, \Delta\alpha_A, \ldots$ and $\Delta\alpha_N \ldots$, respectively.

Appendix I (pp. 130–138) contains the ephemerides (at five-year intervals, between 1925 and 1950, inclusively) of the stars with $|\delta| > 83°$.

Appendix II (pp. 139–148) contains "corrections to obtain positions of components of double stars," and "secular variation of proper motion due to foreshortening." The latter were, incidentally, incorporated into the data in the catalogue when the accuracy aimed at required it.

Appendix III (pp. 149–155) gives tables for the systematic corrections of the position and proper motion systems of PGC, NFK and Newcomb's catalogue to those of Eichelberger's catalogue.

In Appendix IV, the corrections are determined which this catalogue produces in Newcomb's precession, motion of the equinox and coordinates of the solar apex.

The volume closes with a list of errata in Newcomb's Fundamental Catalogue.

4.2.3 THE PRELIMINARY GENERAL CATALOGUE (PGC). This catalogue is a precursor to the GC, described in the next section, and was completed by Lewis Boss (1910), who also initiated the work on the GC. The primary impetus for the construction of the PGC, as well as for the GC, was the desire to determine as many accurate proper motions of stars as the available material permitted. Boss' ultimate aim, realized only a quarter of a century after the completion of the PGC, was what eventually became the GC. He had previously, in a series of papers (Boss 1903, 1903a, 1903b, 1903c, 1903d, 1903e), constructed a fundamental system consisting of 627 stars distributed over the entire sky. This series was later published under the title "Catalogue of 627 Principal Stars." The methods used in its construction were fully and accurately described in the just mentioned series of Boss' papers, and are essentially the same as those used for the compilation of the fundamental system of the GC. Since none of these procedures were specifically used in the compilation of the PGC only, and since the principle underlying each and every one of these methods is fully described in various other places of this book, a detailed description of them is omitted, especially since the PGC is only of historical interest now.

The PGC, which was (except for magnitude equation) intended to be on the above described system, was constructed in order to make at least some of the motions available to the astronomical community while the GC was being constructed.[3] It was to contain all naked-eye stars and a number of fainter ones, namely, those whose positions had been determined in early catalogues, especially Bradley's. Boss was of the opinion that these old epoch positions were particularly valuable since they would facilitate the determinations of accurate proper motions of these stars. Note that exactly the opposite view was held by the German compilers of fundamental catalogues, notably Kopff (1954a), who states that the superiority of the FK3 over the NFK was mainly due to the leaving out of the FK3 the catalogues with epochs before 1845. In his view, those catalogues are affected

3. It must be pointed out that neither the PGC nor the GC are Fundamental Catalogues in the strict sense of the word, due to the fact that only few of the stellar positions and motions listed in them are derived from absolute observations. Only the "Catalogue of 627 Principal Stars" may properly be termed a Fundamental Catalogue.

with serious systematic errors that cannot be removed and, thus, have introduced serious systematic errors in the proper motions of the NFK.

Another point of disagreement between L. Boss and his German counterparts was the role to be played by those stars of which absolute positions had never been observed. Boss felt confident that the systematic errors of the positions recorded in the individual catalogues could be determined well enough by comparison with the 627 Principal Stars. He therefore felt that the positions of *all* stars, after correction for systematic errors, could be regarded as primary representatives of the system. Such views were expressly disavowed by the German Fundamental Catalogue compilers. The introductions to the Zusatzsterne of the FK3 (Kopff 1938), the FK3Supp (Section 4.1.4.2) and the FK4sup (Section 4.1.5.6) pointedly state that the stars in these lists must not be regarded as primary representatives of the system to which they were reduced.

Altogether, the data from eighty independent catalogues were used for the construction of the PGC. Their average epochs ranged from 1755 (Bradley) to about 1900. Besides these catalogues, some (then) unpublished observations of star positions, notably between declinations $-21°$ and $-37°$, specially made at the Dudley Observatory at Albany, New York, were also incorporated in the final catalogue positions.

Newcomb's (1872) equinox was adopted for the catalogue without change. The method of comparing the declinations listed in contemporaneous northern and southern hemisphere catalogues, which was used for the constructions of Eichelberger's Catalogue (see Section 4.2.2) and the GC (see Section 4.2.4), appears to have been applied for the first time for the purpose of constructing the PGC. The aim of this process is the improvement of the constants of refraction that had been adopted in the reduction of the individual catalogues.

The right ascensions of the "627 Standard Stars" were corrected by $-0.^{s}0069$ (m–3.5) where the magnitudes are on the "historic" scale in which the logarithm of the intensity ratio corresponding to an interval of one magnitude is 0.36 rather than 0.4, as in the presently used (Pogson) scale.

The pages of the catalogue contain primarily the right ascensions and declinations, referred to the epoch of place and orientation 1900, of 6188 stars. The right ascension data fill the left-hand pages; the declination materials, the right-hand pages. Other data contained in the body of the catalogue are serial numbers (No.), names (Designation), visual magnitudes on the natural scale (Mag.), central epochs (Epoch.), annual and secular variations including the effects of the proper motions (An. Var. and Sec. Var.), the third term of precession ($3^{d}t$) corrected for proper motion effects only when these are significant, proper motions based on Newcomb's precession quantities and their centennial time derivatives (due to precession and proper motion; μ and 100 $\Delta\mu$), probable errors of the position at the central epoch, the centennial proper motion and the position at epoch 1910 (Prob. Errors αEp., 100μ, α10 and Prob. Errors δEp., 100μ', δ10, respectively). The right-hand pages also carry a column headed "Remarks." There, one will find frequently-used names of the stars that were not listed under "Designation," data referring to double stars, parallaxes when these are known, and referrals to the appendix when the star in question is in one of the appendices.

Appendix I contains ephemerides of the stars within eight degrees of the poles, from 1900 to 1925, because, as is well known, formulas based on truncated Taylor series are not sufficiently accurate to compute the effects of precession and proper motion on star positions for objects that lie within the polar caps.

Appendix II "contains the longer notes pertaining to individual stars in the Catalogue," such as irregular proper motions. There, the foreshortening effects on the proper motions of Groomb 1830 and 61 Cygni are discussed.

Appendix III contains tables for the reduction of the positions in the 80 catalogues incorporated into the PGC, to the system of the PGC, and their weights. The methods for the derivation of these reduction tables and weights were discussed in L. Boss' above-mentioned series of papers in Volume 23 of the Astronomical Journal—especially in (Boss 1903e).

The introduction to the PGC is well worth reading. In it, the author presents (among other things) the rationale for carrying out work on star positions, and the perceptive reader will notice the high degree of Boss' insight into one of the raisons d'être of positional astronomy.

In the Introduction, the proper motions are also analyzed for their effects on Newcomb's constants of precession and his motion of the equinox, and the results are given. It should be pointed out that the corrections derived by Boss would bring these quantities closer to the values now being suggested.

Another section presents a "Method of Correcting the Catalogue Positions and Motions by Means of Additional Observations." A later, more sophisticated, version of the procedure was presented by Eichhorn and Googe (1969), and was reported on in Section 3.2.2.

The PGC is no longer of value for the originally intended purpose, especially since it was superseded by the now-also-obsolete GC.

4.2.4 THE "GENERAL CATALOGUE" (GC). The "General Catalogue of 33342 Stars for the Epoch 1950" by Benjamin Boss (1937) and six collaborators is a five-volume work.[4] The first volume gives the introduction, a description of the

---

4. It is generally referred to as GC. Recently, some people, mainly ones whose primary background is not in astronomy, have taken on the habit of calling it "The Boss Catalogue." This may, however, lead to ambiguities. Newcomb, for instance, uses this term in his own Fundamental Catalogue to refer to the catalogue of declinations compiled by Lewis Boss (1877). There are, besides, at least three other important catalogues in whose computation either L. Boss or his son B. Boss were involved. This writer feels that there is no justification to change abbreviations that are well established, and well and uniquely understood, and which were originally introduced by specialists in the field, just for the benefit of those who were unaware of their existence and felt that they had to create their own abbreviations, or names, for things and concepts with whose existence they had only recently become familiar.

catalogue proper, and, in three appendices, ephemerides of polar stars, peculiarities of proper motion and systematic corrections in right ascension and declination for many catalogues. These were referred to and discussed previously in Section 3.2.3.6.

The catalogue proper is contained in Volumes II through V. The scope and purpose of the GC is stated in the introduction as follows:

"The General Catalogue contains the standard positions and proper motions of all stars brighter than the seventh magnitude, extending from the north to the south pole, and some thousands of additional fainter stars promising to yield reasonably accurate proper motions."

It was the result of a correction and extension of the PGC, which was discussed in the previous Section, 4.2.3. For the scope outlined in the introduction, two tasks had to be accomplished: (1) the improvement of the fundamental system, and, (2) the reduction of a large number of relative catalogues to this fundamental system, and the derivation of the positions and proper motions of the nonfundamentally observed stars within the fundamental system. This double purpose of the GC must be kept in mind. In contrast to other investigations of similar scope namely, FC, NFK, FK3 (with which it was practically contemporaneous), FK4, Newcomb and Eichelberger, its purpose was not *only* the establishment and definition of a fundamental system as close as possible to the physical system in terms of absolutely observed stars, but also to provide the positions and proper motions of a large number of only relatively observed stars on this system.

The plans for this huge enterprise were the result of the thinking of L. Boss, and took definite shape toward the end of the nineteenth century. Originally, it was intended to include about 20000 stars; however, as time went on, more and more stars were added, and eventually 33342 stars were selected. It was in 1912, when L. Boss died, that his son B. Boss took over the direction of the work. In addition to the available independent catalogues, additional meridian catalogues were observed at San Luis near Cordoba, Argentina, where a meridian observatory had been improvised, and at the Dudley Observatory at Albany, New York. The purpose of these two catalogues was to strengthen the proper motions of the stars in the GC by providing accurate modern positions. The San Luis catalogue gave the positions of 15333 stars, the one observed at Albany those of 20811 stars (see also Table III-3). During the period that work on the GC was progressing, the available material of meridian observations kept growing. Since indefinite waiting for catalogues still under observation and reduction would have made it impossible to ever complete the work, a deadline was established. Thus, some series of observations with excellent accuracy, which had become available during the course of the work, could not be incorporated into the catalogue.

The establishment of the GC System is described on pages 9 through 37 of GC.I and was accomplished through the correction of the system of the PGC (see Section 4.2.3), which is identical with that of Lewis Boss' "627 principal standard stars" (L. Boss 1903), which was established using 27 high-quality catalogues, nine of which could have been regarded as absolute.

The right ascension corrections were carried out in the usual way, that is, following formula (3.20a) which provides for $\Delta\alpha_\alpha$, $\Delta\alpha_\delta$, $\Delta\alpha_m$. In addition, a correction to the equinox was considered, for which a mean of Kahrstedt's (1931) and Morgan's (1932) determinations was adopted in the sense GC = PGC − $0.^s040$. Except for a magnitude equation in right ascension, the $\Delta\alpha_\alpha$ were postulated to be of the form

$$\Delta\alpha_\alpha = e + a \sin\alpha + b \cos\alpha + c \sin 2\alpha + d \cos 2\alpha \tag{4.2}$$

with $a, \ldots, e$ being constants to be determined by some form of adjustment, and α, of course, the right ascension. Altogether, 43 catalogues, with average epochs ranging from 1755 to 1925, and supposedly giving absolute right ascension differences (that is, catalogues absolute in $\Delta\alpha_\alpha$), were utilized for the establishment of the various $a, \ldots, e$. According to the epochs of the catalogues, the various a, b, c and d were subjected to least squares solutions of the form $a = a_0 + a't$, and so on, with a (etc.) being the "observed" value of the coefficient of $\sin\alpha$ for the catalogue with the mean epoch t, and a_0 and a' (and so forth) constants which would eventually determine the systematic differences between the systems at epoch $t = 0$ (in that case, a_0) and the system of proper motions described by a'. The results were judged to be not sufficiently well established, and, therefore, no corrections in the form given by equation (4.2) were finally applied to the PGC system to get the GC system. The advantages and disadvantages of applying these corrections and the methods of obtaining them are discussed at great length in GC.I, pp. 11–16.

One should remember here that the absolute determination of right ascension differences is not as trivial a matter as it might appear. It was pointed out in Sections 2.2.6.2 and 2.2.6.3 that the determination of right ascension differences from transit observations requires that the Earth's rotation be strictly in phase with the clock that is used to record the transits. It is now a well-known fact that this is not the case, although this had been assumed to be so until relatively recently. If one suspects, however, diurnal, annual, or generally nonperiodic irregularities in the motion of the Earth, one must take great care to make sure that these are not explained away by assuming that there are systematic errors in the observed right ascensions of a particular catalogue which can be represented in the form of equation (4.2). Rather sophisticated methods must be applied to find out whether waves of the form, as in equation (4.2), are actually errors in the observed right ascensions or due to nonuniformities in the Earth's rotation. Such methods were devised, for instance, by Cohn (1899, 1913). The replacement of pendulum clocks (which were subjected to tidal effects) by quartz and atomic clocks also improved the situation considerably.

Newcomb, when constructing his fundamental system (Section 4.2.1) took the uniformity of the Earth's rotation for granted, while B. Boss (1927) was very well aware of its variability and the precautionary measures that had to be taken so the phenomenon would not be absorbed in postulated but unreal α-dependent corrections to the right ascensions. In the article just referred to, B. Boss points out that the observed motion of the Moon in right ascension, which has by now become universally accepted as the tool for the establishment of ephemeris time (see Section 2.1.6), can provide a check whether right ascension dependent systematic errors of the right ascensions themselves are actually due to irregularities of the Earth's rotation. He made an analysis in which he correlated the $\Delta\alpha_\alpha$ of 100 catalogues (epochs 1815–1919) with the residuals of the observed lunar right ascensions against the theory of its motion, and correctly concluded that there are secular, as well as annually-periodic, nonuniformities in the Earth's rotation. This insight undoubtedly helped a great deal to make the α_α system of GC a much better one than it would have been otherwise.

The $\Delta\alpha_\delta$ and $\Delta\mu'_\delta$ were found from 95 catalogues with average epochs ranging from 1755 to 1926 (of which 47 were given most weight) that satisfied L. Boss as actually being absolute in α_δ. These corrections (to the PGC system) were not postulated in any particular analytical form, and are given as Table 9 (GC.I, p. 19).

Altogether 46 catalogues, with a range of epochs of orientation from 1875 to 1925, were used for the establishment of the α_m system. These catalogues were chosen because the observers used either objective screens to equalize the brightness of the stars as they appeared in the field of the instrument, or traveling (hand or motor driven) right ascension micrometer wires (see Sections 2.1.6 and 2.2.2.5) called "self-registering micrometers," or a combination of these to avoid the introduction of a magnitude equation during the observations. The α_m system was investigated in four declination zones, and the linearity of the magnitude equation was also checked. The conclusion was that the α_m system of the PGC was adopted for the GC.

It is understandable that for the purpose of investigating the α_m system of a proposed compilation catalogue, the star magnitudes in the catalogues to be incorporated were reduced to the same system—in this case, the Harvard system of magnitudes. Later, however, when the various magnitude equations of the catalogues used for the construction of the GC were tabulated (see Section 3.2.3.6), the arguments of the corrections (that is, the magnitudes of the stars) were still left on the Harvard system, so that the magnitudes as they appear in the independent catalogues must be either reduced to the Harvard system, or one will introduce an error in the zero point of the right ascensions (that is, the equinox adopted for the catalogue in question) and a residual magnitude equation. This poses no difficulty for those stars in the GC whose magnitudes are given there on the Harvard system; but in order to properly use the magnitude correction tables given in GC.I for the various independent catalogues, the magnitudes in them must first be reduced to the Harvard system. This, quite obviously, introduces an inconvenience.

The declination errors were modeled to be strictly of the form given by equation (3.20a), namely $\Delta\delta = \Delta\delta_\alpha + \Delta\delta_\delta$. The terms of the form $\Delta\delta_\alpha$ can, as we know, be produced by a variety of sources: instrument errors not properly accounted for, erroneous reduction constants, in particular, that of nutation, latitude variations and seasonal variations of the refraction. The corrections $\Delta\delta_\alpha$ and $\Delta\mu'_\alpha$ to the PGC system were investigated (forced to be of the form $a \sin \alpha + b \cos \alpha$) by analyzing the data in 53 appropriate catalogues whose average epochs range from 1816 to

1925; a dependence of the coefficients a and b on the declination was found (see also Raymond 1927), but the weight of these determinations zone by zone was considered not high enough to warrant their actual use. Therefore, no $\Delta\delta_\alpha$ correction was applied for converting the system of the PGC to that of the GC. In reducing individual catalogues, however, to the GC system, $\Delta\delta_\alpha$ were, of course, determined and applied.

The establishment of the δ_δ and μ'_δ systems of the GC was first drafted by Raymond (1926) in an extensive paper; the actually applied procedures were, however, written up by B. Boss himself and are found on pp. 27 through 37. Flexure, errors in the refraction, division errors of the circles, variations of the latitude and zenith discontinuities are the principal known sources of errors of the type $\Delta\delta_\delta$.

The establishment of the δ_δ and μ'_δ systems for the GC (as mentioned before, in the form of corrections to the corresponding systems of the PGC) were ultimately based on the assumption that the only serious and essentially uncontrollable errors in the δ_δ systems of the individual catalogues were produced by erroneous constants of refraction, and that all the other above-mentioned sources of errors of the $\Delta\delta_\delta$ type could (and would) be removed from absolute observations.

Both Rayomnd (1926) and Boss also pondered the establishment of the μ'_δ systems in essentially the same way the μ systems were established in the FK3 and the FK4, that is, by relying on the proper motions from homogeneous series on identical instruments. Raymond ascribes this basic idea to Kapteyn (1922). There were three instrumental series available for this purpose, namely, those observed at the Cape, at Pulkovo and at Greenwich. The catalogues constructed at these observatories were carefully scrutinized, re-reduced, when deemed necessary, by Raymond and later by Boss, and tentatively used for an independent establishment of the δ_δ and μ'_δ systems. These systems agree well enough with those determined from the "general solution" (to be described next), but Boss found that their weight was so inferior to that of the systems found from the general solution that he discarded them, and adopted the systems determined from the general solution only.

Boss also considered Newcomb's method of determining the equator point from the observations of the Sun and the major planets as worthless, although he admits the value of the asteroids for this purpose. Thus, his δ_δ and μ'_δ systems were determined as follows.

In Section 2.2.5.3 it was shown that observations of circumpolar stars in both culminations are in essence used for the determination of the latitude of the instrument and the constant of refraction. It is clear that erroneous values of the constant of refraction will not falsify the determination of the instrument-reading corresponding to the pole, but will increase their influence on absolute declinations determined by this method by an amount proportional to the polar distance.

Following Boss, let δ_c be the correct declination, and δ_d that determined with the use of an erroneous constant of refraction. Also, denote by k a function of $\Delta R/R$ of equation (2.1a). Assume, as in Section 2.2.5.3, that the total value of the refraction at a certain zenith distance (or declination) is given by $R\rho$, where ρ is regarded as known from some refraction theory. Since the influence of the error introduced into the absolute declinations by a wrong constant of refraction, that is $(1 + k)R$ instead of R, grows linearly with polar distance, the equation

$$\delta_c = \delta_d + k(\rho - \rho') \tag{4.3}$$

will be valid where ρ is the value of the "proportional" refraction at declination δ, and ρ' that at the pole. If we compare the declination of the same star which was determined at two observatories of different latitude, we have

$$\delta_{d1} + k_1(\rho_1 - \rho'_1) = \delta_{d2} + k_2(\rho_2 - \rho'_2), \tag{4.4}$$

since the "true" value of the declination δ_c must be the same, regardless of the observatory at which it was observed (provided the epoch of place is the same). We may now rewrite equation (4.4) as

$$k_2(\rho_2 - \rho'_2) - k_1(\rho_1 - \rho'_1) = \delta_{d1} - \delta_{d2}, \tag{4.4a}$$

and use it as an equation of condition in a least squares solution for the determination of k_1 and k_2. $\delta_{d1} - \delta_{d2}$ play the role of the observations, and the ρ and ρ' are available from refraction theory.

It is evident that the larger the latitude difference between the two observatories, the better conditioned a system of normal equation formed from condition equations (4.4a) will be for the determination of k_1 and k_2. Comparisons of the declinations of stars that were simultaneously determined at northern and at southern observatories, and published in otherwise carefully reduced catalogues, will thus produce the most accurate values of the k's that characterize the various catalogues.

Boss' "general solution" for the δ_δ and μ'_δ system, which was adopted for the GC, was carried out by determining the k's of 93 catalogues, assuming that after application of appropriate corrections, when necessary, their only errors of the form $\Delta\delta_\delta$ originated from the use of erroneous values of the constant of refraction. This is, of course, a hypothesis that is very hard to prove, and was severely criticized by Kopff (1949, 1954a). The various values of k were thus determined for the 93 above-mentioned catalogues by applying equations of the type (4.4a) to sufficiently contemporaneous pairs formed from 59 northern and 23 southern catalogues, whose average epochs ranged from 1821 to 1926.

It is worth noting that this method, when taken and interpreted at face value, produced some constants of refraction that were quite beyond the range of realistic possibilities.

The *individual positions and proper motions* in the GC were produced in the usual way by least squares solutions from positions individually published in 238 catalogues whose average epochs ranged from 1755 to about 1930.

The tables for the corrections of the systems of these catalogues to the GC system form most of GC.I, and were discussed above in Section 3.2.3.6.

A few observations with regard to the GC system and the way it was obtained may be worth recording.

As pointed out above, the systematic corrections to the catalogues contributing to the system were strictly in the "canonic" form (3.20a), that is, no allowance was made for, say, the $\Delta\alpha_\alpha$ to depend on declination. Although special investigations were made to justify this restriction, as explained in the introduction pages to the GC, it is uncertain, and as a matter of fact, even improbable, that the posutlated form of the corrections of the PGC to the GC system takes adequate care of all systematic differences, especially since no $\Delta\alpha_\delta$ terms were applied. This was pointed out by Kopff (1954a), who, in the same paper, also draws attention to the fact that the systematic differences between the N30 catalogue (Section 4.2.5) and the GC system are different for those stars which are, and those which are not in the FK3. Kopff inclines to the view that the GC system is not homogeneous, but differs for fundamental stars (which at the same time would roughly correspond to the FK3 stars) and the "field stars." Errors of this type may, according to Kopff's above-mentioned paper, also have crept into the N30 system, which, according to Kopff (1954a) is also not homogeneous. This means that the positions of certain groups of stars in the N30 (and in the GC) will require different corrections to the "true physical system."

A detailed description of the columns of the GC is readily available on pp. 44 through 53 of GC.I and, therefore, need not be repeated here.

Note, however, that the positions given in the GC, as those in all other fundamental catalogues, refer to epoch of orientation and place 1950.0; the "epochs" given in the appropriate columns are the central epochs necessary to compute the probable (or rms.) error of a GC position at a different epoch t from the simple equation (3.12). Note also that all errors quoted in the GC are probable errors.

The listed annual and secular variations are the total time derivatives of the mean positions, due to precession and proper motion. The annual variations were obtained from a least squares adjustment of the star positions referred to the individual unprecessed coordinate systems and original epochs, so that they are, strictly speaking, not based on any particular constant of precession. The listed proper motions were, however, obtained by subtracting from the listed annual variations their precessional values as computed with Newcomb's constant of precession, so that the GC proper motions are actually based on Newcomb's precession constant, which has also been used in the computation of the secular variations and the third terms of the precession.

Appendix I of GC.I contains the ephemerides of stars near the celestial poles where precession, computed with the three conventional terms, will not be sufficiently accurate over longer periods of time.

Appendix II treats peculiarities of proper motion, mainly those of binaries.

Appendix III contains the above discussed (Section 3.2.3.6) systematic corrections of about 250 catalogues to the GC system.

Again, remember that the argument of the corrections $\Delta\alpha_m$ are not the stars' magnitudes as listed in the catalogue in question, but the magnitudes on the Harvard system. If only BD magnitudes were available, these were reduced to the Harvard system by means of the tables found in the "Annals of the Harvard College

Observatory," Volume **72**. In order to correctly apply the $\Delta\alpha_m$ corrections to positions found in a certain catalogue, the relationship of the magnitudes listed in that catalogue to the Harvard system magnitudes must be known.

The accidental rms. errors of the positions given in the GC vary strongly from one star to the next. However, at the epoch they are on the average about 0.″15 in both coordinates, and rise to an average of at least 0.″70 in 1965 because of the uncertainties of the proper motions (Schlesinger and Barney 1939a). It is thus obvious that most of the stars in the GC do not satisfy the requirements for a catalogue of accurate star positions in 1965. We have previously discussed (Section 3.2.2) procedures that can be used to update the GC positions by combining them with newly available positions and thus decrease their accidental errors at recent epochs. Unless this is done, the GC is rapdily losing its value as a source of accurate positions of nonfundamental stars at contemporary epochs.

4.2.5 THE N30 CATALOGUE. The average central epoch of the positions in both the FK3 and the GC is around 1900. For the epoch 1950, the accuracy of the positions in both these widely used fundamental catalogues had considerably deteriorated due to the influence of the uncertainties of the individual proper motions. Also, untraceable and, therefore, uncorrectable systematic errors in the old catalogues had introduced noticeable systematic errors into the proper motion systems of both the GC and the FK3.

As soon as enough new catalogues became available (around 1945), it was felt that a revision of the fundamental system had a chance of producing a significant improvement. Work on the problem started at the U.S. Naval Observatory in 1947 and was finished in 1951, and published under the title "Catalog of 5268 Standard Stars, 1950.0, based on the Normal System N30" (Morgan 1952).

In evaluating the procedures used for the computation of the N30 catalogue, and their result, one must consider that there was a rather large amount of material to be worked on with a limited crew. The work was done in the precomputer age, thus rigor and maximum achievable weight had to be traded for labor-saving shortcuts. This becomes particularly understandable if one considers that H. R. Morgan was of advanced age when the work was begun; he had to trade maximum achievable accuracy for speed and efficiency if he expected to see the result of his efforts before the end of his active career.

The basic procedure thus adopted was as follows. From a number of catalogues, whose average epochs ranged from 1917 to 1949, normal positions of the stars were established whose average central epochs were around 1932 but which ranged in a few extreme cases from before 1920 to after 1940. These define the *system of positions in the N30 catalogue*. Of these catalogues about half had already been used for the construction of the GC. They were newly analyzed and combined with other material to find corrections to the GC system at 1900, the average central epoch of the mean positions published in the GC. At this epoch, the GC system was, in Morgan's judgment, less affected by systematic errors than at any other epoch, and the N30 proper motions were straightforwardly found by dividing the differences at the new central epochs and the corrected GC positions at epoch 1900 by the difference of the new central epoch minus 1900.

The details of these procedures were as follows:

Sixty catalogues, ranging in epochs from 1917 to 1949 were used for the establishment of the N30 catalogue positions. Thirty-three of the catalogues could be regarded as totally or partly absolute in one or both of the coordinates, namely, 26 in α_α, 29 in α_δ, 24 in δ_α, 27 in δ_δ, and all 33 in α_m and δ_m. Only these were used to establish the system of positions.

A *correction of the equinox* was attempted from observations of the Sun, Moon and the planets between 1750 and 1945, with the result that no significant change of the equinox of the GC could be found.

The *equator point* of the GC system was corrected by means of positions of the Sun, Moon and major planets (observed over the period from 1750 to 1949). As a result, the equator point of GC was corrected by $+0.''05$ at the epoch 1900, and by $+0.''10$ at 1930.

All the sixty catalogues used for the establishment of the position system were reduced to the equinox of GC and equator point of GC $+ 0.''10$. While the compilers of the FK3 and the FK4, and also Newcomb, let the corrections to the declinations vary linearly between the equator and the pole, and kept the equatorial value between equator and horizon, these corrections were applied in the work on the N30 catalogue in the form of a perhaps somewhat more realistic model, namely, $\Delta\delta = 0.7\ \Delta\delta_0[\sin(\phi - \delta) + \cos\phi]$, where ϕ is the observatory's latitude and $\Delta\delta_0$ the equatorial correction to the declinations for the catalogue in question.[5]

No further corrections were applied to the average systems of the 60 above-mentioned catalogues in the coordinate with respect to which they were absolute. After that, the deviations of the systems of the thus corrected absolute catalogues from the average systems were "quite small." From the system formed by these 60 and other previously available catalogues, some of which had epochs going back to 1845, corrections $\Delta\alpha_\alpha$, $\Delta\mu_\alpha$, $\Delta\delta_\alpha$ and $\Delta\mu'_\alpha$ to the GC system were computed in the form of waves with periods α and $\alpha/2$. Declination dependent corrections were formed only for the position system at epoch 1930 and not for the proper motion system, and only from those catalogues that were absolute in α_δ and/or δ_δ. Also corrections to the GC system of the forms $\Delta\alpha_m$, $\Delta\mu_m$, $\Delta\delta_m$ and $\Delta\mu'_m$ were derived, apparently, from a limited number of stars. This was the system to which the relative ones in the list of the 60 catalogues were reduced. The averages of the thus reduced positions of the chosen stars in all of these 60 catalogues form the position system of the N30 catalogue.

Note that for this purpose the positions in the N30 catalogue at the central epochs were obtained by taking the weighted means without solving for proper

5. This formula will render $\Delta\delta = 0$ for $\delta = 90°$, that is, at the pole, but make it equal to $\Delta\delta_0$ for $\delta = 0°$ rigorously only for $\phi = 45°$ (0.7 being really $1/\sqrt{2}$). Since the latitudes of all observatories are sufficiently close to 45°, this is of little relevance. A rigorous formula, which yields $\Delta\delta(90°) = 0$, would have been $\Delta\delta = \Delta\delta_0[\sin(\phi - \delta) + \cos\phi]/(\sin\phi + \cos\phi)$.

motion. Since the positions in the N30 refer to coordinate system and epoch 1950, they were brought to epoch 1950 by application of the *N30 proper motions*, which were derived in the following way.

Since the average epoch of the GC is around 1900, the proper motions were formed by comparing the N30 normal positions (at the epoch, on the average around 1932) with the GC 1900 positions, which had been corrected only for $\Delta\alpha_n$ and $\Delta\delta_n$ terms, essentially by means of the corrections whose formation was described above in this section.

For the 781 stars not in the GC, the proper motions were derived similarly by establishing normal positions from observations around 1900 and earlier, and comparing them with the second epoch normal positions (that is, around 1930).

Due to the variability in the number and quality of observed positions which contributed to the positions finally published in the catalogue, the rms. errors of the N30 proper motions and positions (within the system, of course) varies considerably from one star to the next. Table IV-6 shows the number of stars in the N30 epoch and the proper motions. The results hold for both coordinates.

Table IV-6 Rms. Errors of N30 Data

No. of stars	*rms.e. at N30 Epoch 1932*	*rms.e. of p.m.*
100	0″.039	0″.0012
200	48	16
200	48	19
300	61	25
500	71	36
1000	77	37
2200	104	59
700	134	68

Pages vii through xviii contain the detailed description of the formation of the material published in the catalogue. Pages xix and xx contain tables for the systematic differences of the position and proper motion system for N30-GC, and pp. xxi and xxii for N30-FK3.

The α-periodic terms $\Delta\alpha$, $\Delta\delta$, $\Delta\mu$ and $\Delta\mu'$ are given in a small table on p. xxiii. From inspection of this table it is immediately apparent that the systems of FK3 and N30 differ much less than those of GC and N30—a confirmation of the superiority of the FK3 over the GC. The systems of the FK3 and GC were derived by different workers at about the same time from the same material of fundamental observations, while additional material, accumulated over a nearly 20-year-period, was available for the establishment of the N30 and FK4 systems. Pages xxiii through xxvii analyze, in a simplified way, the influence of essentially the N30 proper motions on the correction to the constant of precession and the motion of the equinox.

A description of the contents of the columns of the catalogue proper (pp. 109–321) is given on p. xxviii and ends Morgan's introduction to the N30 catalogue, which should be carefully read by all users.

As in the other fundamental catalogues, the data pertaining to any one star run over two pages.

The left-hand pages give running serial numbers (No.), the numbers in BD (+89 to −22°), CoD (−23° to −51°) and CPD (−52° to −89°) as DM, and (Mag) the visual magnitudes on the Harvard system. The third through seventh columns give mean right ascensions for epoch of place and orientation 1950, and the total first, second, and third time derivatives, referred to a century as time unit. The R.A.'s are given to $0.^s001$, the derivatives to $0.^s01$. The eighth column contains the centennial proper motions in R.A. to $0.^s01$ and its centennial time derivative when the latter exceeds $0.^s005$. The ninth column gives the above described epoch minus 1900 in terms of years. Since the positions and proper motions were not obtained from least squares solutions in the usual way, these epochs are not central epochs in Newcomb's sense (Section 3.2.2), and cannot be used as such for improving the N30 positions by incorporating new material. The true central epochs could be found from taking the weighted mean of 1900 and the N30 "Epoch +1900." The weight for 1900 would be the weight computed from the GC position for 1900, and that for the N30 Epoch, the weight (Wt) of the N30 position at the Epoch "Ep." The unit of weight in both coordinates corresponds to a rms. error of $0.''43$. Columns eleven and twelve give, for every individual star, the quantities N30-GC for positions and (centennial) proper motions, respectively.

The columns on the right-hand pages give, as customary, the data pertaining to declination (to $0.''01$ and $0.''1$, respectively) completely analogous to those for R.A. on the left-hand pages. The only other differences in the corresponding columns are that on the right-hand pages, column 2 contains the GC number, column 3 the (Harvard) spectrum, and an added 13th column gives the number of the star in the FK3. Remarks, pertaining mostly to doubles, are given at the bottom of the pages.

The proper motions were, of course, computed using Newcomb's constant of precession.

The N30 catalogue has been criticized by A. Kopff (1954a), who was responsible for the establishment of the FK3, and also for setting the procedure for its revision which resulted in the FK4.

Kopff's main points are: Over 5000 stars is too large a number to enable one to make an accurate and careful reduction; besides, there are not that many stars whose positions have been observed fundamentally. He objects to the failure to apply declination dependent corrections to the 1900 GC positions, which have been used as "first epoch normal places," and thus predicts that the N30 catalogue proper motions will be falsified through the systematic errors of the GC positions.

He also notes, and it is in this writer's opinion the most important point, that the systematic differences N30-GC are different for those stars which are in the

FK3 and for those which are not. The explanation offered for this fact is, that the GC system is not homogeneous but different for genuine fundamental and field stars; this inhomogeneity will, of course, be propagated into the off-epoch N30 catalogue positions.

Concerning Kopff's remarks, we must consider that, as was pointed out at the beginning of this Section, the way in which positions and proper motions were obtained for the N30 catalogue was the only one that enabled the staff working on it to cope with the huge amount of numerical calculations. Besides, neither Morgan nor Kopff foresaw the emergence and easy accessibility of computers and other automatic devices at the instruments. Thus, many of Kopff's points have become invalid.

The N30 system may have been not only useful for bridging the gap between FK3 and FK4, but may in certain regions of the sky actually be superior to that of the FK4 according to the results of recent absolute observations; see Section 4.1.6.

Kopff's (1954a) paper also gives systematic differences N30-GC and N30-FK3. His values (understandably in the sense FK3-N30) are given in more detail, and are based on a more extensive investigation than Morgan's.

4.3 Nonfundamental Compilation Catalogues

4.3.1 THE SMITHSONIAN ASTROPHYSICAL OBSERVATORY "STAR CATALOG" (SAOC)

4.3.1.1 ORIGINAL PURPOSE AND HISTORY. This most numerous of all existing compilation catalogues, which we shall refer to in what follows as "SAOC," grew out of the needs of the space age. In the United States, a considerable amount of planning had preceded the launch of the first artificial satellites. They were to be tracked by various methods, one of which was photography. For this purpose, very fast cameras ("Super Schmidt") had been specially designed by James G. Baker and built by the Joseph Nunn Co. They became known as the "Baker-Nunn Cameras." Their effective focal length is about 600 millimeters, and they take photographs on film—the field size being about 5×30 cm^2. These cameras were designed to allow one to derive the positions of the artificial satellites with rms. errors of about 2″ in both coordinates, timed accurately to about one millisecond. This was the highest accuracy which the planners, at the time, thought could be utilized. In retrospect, it is astounding that they did not foresee that the accuracy requirements for photographic satellite positions would rise tremendously, such that, nowadays, positions with rms. errors of better than 0.″1 would be welcome if they could be made available. This brings to mind a dictum, allegedly due to the late Ejnar Hertzsprung: "We don't know what data the next generation of astronomers will want. But we do know that they will want them more accurately."

The Smithsonian Astrophysical Observatory (SAO), located in Cambridge, Massachusetts and closely associated with Harvard University, was given the responsibility for the optical tracking of the satellites. Even with the new cameras, it soon became clear that a homogeneous and sufficiently numerous source of reference stars needed for the reduction of the satellite positions was not readily available, especially in the southern hemisphere. There were, of course, catalogues galore, mostly the AG-type catalogues (Sections 5.1, 5.2 and 5.3) and the Astrographic Catalogues (Section 5.5), but the positions were valid for old epochs and were on a variety of systems that were quite frequently not very precisely defined.

The SAO thus undertook the compilation of a complete catalogue, with a density of no fewer than four stars per square degree in any part of the sky, in which the positions at epoch of place 1970 would have rms. errors of less than 1″, and which would be, by and large, on that fundamental system which was then generally in use, that is, the FK3.

The result of these efforts is the "Star Catalog" (Smithsonian Astrophysical Observatory, 1966), subtitled "Positions and Proper Motions of 258,997 Stars for the Epoch and Equinox of 1950."

It is now clear that the intended goals were met. Haramundanis (1967), in her discussion of the SAOC, points out that the typical rms. error of a coordinate at the epoch 1970 is 0.″5, one-half of the maximum acceptable figure, and Scott and Smith (1971) have shown that the deviations of the system represented by the SAOC from that of the FK4 only rarely exceeds 0.″2. Considering the various limitations under which this catalogue had to be computed, the whole enterprise must be termed a success. It might be the perfectionist's nightmare, but currently being the most numerous and most homogeneous source of mean positions and proper motions of stars down to about 9th magnitude, it is invaluable for the astronomical and geodetic communities at large, particularly for the nonspecialist users.

The work began in 1959 and was performed by a few members of the staff of the SAO, notably Katherine Haramundanis, Joan Sears and William Day, under the general supervision of George Veis. The procedures employed followed by and large the suggestions by Heinrich Eichhorn, who was a consultant on the procedures. The SOAC was published in 1966.

Financial limitations dictated that this would have to be a "no frills" project. If, therefore, one particular source (catalogue) met the accuracy requirement stated earlier above by itself, it was used alone for establishing the SAOC positions and proper motions, although the use of additional available material certainly would have produced positions and proper motions more accurate than those published. It should also be emphasized that the SAOC is strictly a compilation catalogue, that is, no positions were especially observed for the purpose of its construction. It was computed solely on the basis of available material.

4.3.1.2 DETAILS OF CONSTRUCTION. First, a list of sources had to be compiled from which star positions and proper motions could be extracted which satisfied the previously established accuracy requirements. If proper motions were not already available, sources for their computation had to be found. Secondly, the positions and proper motions in these sources had to be reduced to the system of the FK4. Thirdly, proper motions had to be computed when they were not available in any catalogue.

The sources were used following, partly, a hierarchy of preference, and, partly, the dictates of sky coverage. FK4 data were used whenever available.[6] In the absence of FK4 data, GC data were used when available.[7] Next in the order of preference were the Yale zones (see Section 5.2) and the Cape Photographic Catalogue (CPC) (see Section 5.3), since these are complete catalogues, that is, they also give proper motions. For the regions not covered by the foregoing, positions and proper motions had to be computed from the AGK1 (Section 5.1.2), the Melbourne Meridian Catalogues Me3 (Ellery and Baracchi 1917) and Me4 (Mt. Stromlo Observatory, 1959), the AGK2 (Section 5.1.3) and the Greenwich zone of the AC (see Section 5.5). Positions and proper motions in the zone −52° to −40° were taken from the "Cape Astrographic Zone" (Gill and Hough 1923) and Spencer Jones' and Jackson's (1936) work, respectively. Table IV-7 (taken from Eichhorn 1969) shows the sources for the nonFK4-nonGC star data in the SAOC. Those (about 15% of the total) CPC stars for which no proper motions are quoted in the CPC were omitted from the SAOC.

Table IV-7 Sources for SAOC Data

Declination		*Source*
From	*To*	
+90°	+85°	Yale
+85°	+80°	AGK2—Greenwich AC
+80°	+60°	AGK2–AGK1
+60°	+50°	Yale
+50°	+30°	AGK2–AGK1
+30°	−30°	Yale
−30°	−40°	CPC
−40°	−52°	Cape Astrographic Zone
−52°	−64°	CPC
−64°	−90°	Me4 and Me3

Me3 and Me4 contain proper motions. Those proper motions of Me3 stars, which

6. When the work on the SAOC was initiated, the FK4 was not yet available. Thus the FK3 was initially the all-overriding source for data. When in 1963, during the second phase of the construction of the SAOC, the FK4 became available, the FK3 data were removed and replaced by those in the FK4, and all other data were reduced to the FK4. Details are given in this Section.

7. This was probably a misjudgment since nonFK4 stars in the GC have, at 1970, positions of an accuracy inferior to that which would have been obtained had the GC been ignored completely.

were newly derived in Appendix I of Me4, were used in preference to those given originally in Me3. Stars for which these catalogues failed to give both μ and μ' were omitted from the SAOC as well. The fact that no systematic corrections of the Me3 and Me4 proper motions to the FK4 system—or even the GC system—were applied, has probably produced systematic errors in their SAOC off-central epoch positions (compare Scott and Smith 1971). Mönnichmeyer's corrected Bonn AGK1 zone (see Section 5.1.2) was, of course, used instead of the original one.

The corrections necessary to reduce all data to the FK4 system were performed in two steps. The FK4, which contains the tables for the differences between the FK4 and FK3 systems, became available only while the work on the SAOC was in process. All data were originally reduced to the FK3 system, and then by application of the systematic differences FK4–FK3 to that of the FK4. Thus, the final data were frequently obtained from those originally published by applying several sets of corrections. Considering this, the systematic differences FK4-SAOC found in the above-quoted paper by Scott and Smith (1971) are certainly within acceptable limits. These would undoubtedly have been smaller if the modern, revised and accurate reduction tables by Brosche, Nowacki and Strobel (1964) could have been used for the frequently necessary transformation of the data from the GC system to that of the FK4, instead of going from the system of the GC to that of the FK3 by means of Kopff's (1939a) tables and from this to the FK4 system by means of the tables in the FK4.

The reductions to the GC were performed by means of the tables in the GC.I (Section 3.2.3.6) for these catalogues for which they were there available. The Yale catalogue data were reduced to the GC system by means of Barney's (1951) tables. The corrections required to reduce the Greenwich AC zone to the GC system were assumed to be the same which perform this task for the system of its reference stars (taken from the Greenwich Second Nine Year Catalogue), with the addition of a magnitude equation given in the introduction to the Greenwich AC. Corrections to the Cape zone ($-40°$ to $-52°$) proper motions were available from the work of Williams (1947). The published data in the AGK2, CPC and the Yale zones $+85°$ to $+90°$, and $+50°$ to $+60°$ were assumed to be on the FK3 system. For the Yale zones, this is not true; see, for instance, Table V-2.

Precession operations were performed by rigorous formulas, essentially as described in Section 3.2.1.

4.3.1.3 Description and critique of the catalogue data. The SAOC was published in four volumes (parts). The data are arranged in zones, called "bands" in SAOC, of 10° width. Parts 1, 2, 3 and 4 contain the zones $+80°$ to $+30°$, $+20°$ to $+0°$, $-0°$ to $-20°$, and $-30°$ to $-80°$, respectively. Within the zones, the data are arranged in the order of increasing right ascension. The introduction occupies pp. xi through xxvii of part 1, where detailed descriptions of the data in the SAOC and of the method of their construction can be found. Needless to say, anyone who makes use of the investigations that involve accurate star positions and proper motions ought to make himself familiar with the facts presented in that introduction.

The catalogue manuscript itself was actually produced by photographing the pages off a TV screen!

There is little point in describing here in detail the data in all columns of the SAOC, much of which are either self-explanatory or whose accurate meaning may be found by consulting the "Introduction." Thus, data presented in the pages of the SAOC will be discussed below only if there is a particular reason for it.

All positions and proper motions are referred to the coordinate system (epoch or orientation) 1950.0. The mean positions are, however, given for the epoch (of place) 1950 (α_{1950}, δ_{1950}), and the "original epoch" (α_2, δ_2), respectively. These "original epochs" (ep.), referring to α_2 and δ_2, mean different things depending on the source of the data. For GC and FK4 stars, they are the true central epochs, but for the rest of the stars, they are the epochs of the positions in the more recent catalogue from which the proper motions were derived. For proper motions derived by comparing AGK2 and AGK1 positions, for instance, "ep." is the AGK2 epoch. This feature makes it necessary to hunt for the epochs of the positions, and their rms. errors in the first epoch catalogues used for the calculation of the proper motion, if one wants to find the real central epoch and the rms. errors of the coordinates at that epoch. This could be accomplished, after careful study of the introduction, from the data given under the heading "SOURCE-CAT." Remember that it is these data that enable one to improve with minimum effort the positions and proper motions by the incorporation of new positions. (Section 3.2.2). σ and σ' denote the rms. errors of α_2 and δ_2 at their "ep." respectively. The fact that σ is given in seconds of arc instead of seconds of time is rather anomalous, and further complicates the utilization, and restricts the utility, of these numbers. The same is true for σ_μ, which is given in seconds of arc on the great circle, while μ itself is given in seconds of time on the parallel, in the conventional way.

The quantity found under the heading σ_{1950} is intended to be an estimate of the standard deviation of the difference between the catalogued and the unknown "real" position of the star at epoch 1950, measured in seconds of arc on the great circle, and it is conceivable that this would be useful for some investigators. Unfortunately, the σ_{1950} were computed by means of a formula which is grossly wrong (equation (10) on p. xiv of the introduction), since it assumes the "ep." to be the genuine central epochs, and the rms. errors of the right ascensions (on the great circle) to always equal those of the declinations. This means that the σ_{1950} are really only crude (typically too large) approximations to the intended quantities, and one will do well to disregard them in critical investigations. (See also Eichhorn 1969, p. 233.)[8]

One of the features in the SAOC which, in this writer's opinion, is worthy of imitation, is the omission of the data usually given to facilitate the computation of precession. These are superfluous if an electronic computer is available.

---

8. Formulas (8) on p. xiii of the introduction contain a misprint; they should read $\sigma_\mu = (\sigma_2^2 + \sigma_1^2)^{1/2}/(t_2 - t_1)$ and analogously for (σ_μ'); the correct formulas were, however, used for the calculation of the rms. errors μ and μ'.

For those interested, copies of the SAOC on magnetic tape are available and may be purchased through the SAO.

4.3.2 THE AGK2A. When the Astronomische Gesellschaft decided to repeat the AGK1 (see Section 5.1.2) for the purpose of deriving the proper motions of the stars contained in it by constructing the AGK2 (see Section 5.1.3) by photographic means, it was decided to establish a catalogue of reference stars by independent observations. The only sufficiently numerous source of reference star positions for the reduction of the AGK2 plates would have been the GC which, however, appeared in print only about one decade after the plates for the AGK2 were taken. Besides, the GC positions at the average epoch of the AGK2 plates were not sufficiently accurate for the purpose intended.

The catalogue (Kopernikus-Institut 1943), which is the result of this effort to observe reference positions for the AGK2, appeared under the title "Katalog der Anhaltsterne für das Zonenunternehmen der Astronomischen Gesellschaft. Nach Beobachtungen an den Sternwarten Babelsberg, Bergedorf, Bonn, Breslau, Heidelberg, Leipzig und Pulkowo." No author is given although A. Kopff, who signed the preface and was the Director of the Rechen-Institut, states that the late J. Peters had been mainly responsible for the organization and direction of the work.

The AGK2A, as it is commonly known, gives the positions of 13747 stars mostly of magnitudes 8^m and 9^m rather uniformly covering the region between $-5°$ declination and the north celestial pole, a slight extension beyond the coverage of the AGK2 itself. The catalogue positions were to be reduced to a well-defined fundamental system as rigorously as possible, namely through frequent observations of "series" (see Section 2.2.7) at all participating transit circles, as well as through the time-honored procedures for making relative observations.

The work at the transit circles was distributed among the seven observations indicated in the title. At Berlin-Babelsberg, there were two transit circles, both covering the entire region without duplication, that is, the program stars observed on one instrument were not observed on the other, and vice versa. In order to provide the material for the combination of the two halves of the Babelsberg zone into a homogeneous catalogue, about 1200 common stars were observed on both transit circles. These stars were also observed at Bonn. The regions observed by the other observatories were essentially as follows—Hamburg-Bergedorf: all stars, Breslau: $+90°$ to $+60°$, Pulkovo: $+60°$ to $+45°$, Bonn: $+45°$ to $+20°$, Leipzig: $+20°$ to $+50°$ and Heidelberg: $+5°$ to $-5°$.

On the average, the program stars were observed six times and the "1200 common stars" ten times. The combination of all observations into the AGK2A and the reduction to the fundamental system of the FK3 (for the establishment of which the AGK2A was the immediate reason) were carried out at the Astronomisches Rechen-Institut at Berlin-Dahlem (then named "Kopernikus-Institut") from the data delivered to them by the individual participating observatories. The catalogue positions refer on the average to the epoch of place 1930. Their rms. errors $\epsilon_\alpha \cos\delta$ and ϵ_δ are given in Table IV-8.

Table IV-8 Rms. Errors of AGK2A Positions

Zone	*6 obs'ns.* $\epsilon_\alpha \cos\delta$	ϵ_δ	*10 obs'ns.* $\epsilon_\alpha \cos\delta$	ϵ_δ
$-5°$ to $+5°$	$0^s.008$	$0''.19$	$0^s.007$	$0''15$
$+5$ to $+20$	11	20	9	16
$+20$ to $+35$	9	22	7	16
$+35$ to $+45$	9	21	7	16
$+45$ to $+60$	7	20	6	15
$+60$ to $+90$	8	18	7	15

The main body of the catalogue contains in column 1 the serial number. An asterisk behind this refers to a remark (these are quite self-explanatory) at the bottom of the page. Column 2 contains the visual magnitude on the Harvard System obtained as the mean from independent estimates at the transit instruments. Columns 3 and 9 list the mean right ascensions (to $0.^s001$) and declinations (to $0.''01$), respectively, of the various stars for the epoch of orientation 1950, on the system of the FK3. In columns 4, 5, 6, and 10, 11, 12 the user finds the quantities necessary for the calculation of precession according to equations (3.1a). Note that these quantities do not take into account the changes of the stars' positions due to proper motion. Columns 7 and 13 contain for right ascensions and declinations, respectively, the excesses of the epochs of place beyond 1900 in units of tropical years (to $0.^y1$). The centennial proper motions μ_α in α (column 8) and μ_δ in δ (column 14) make no claim to high accuracy, and were mostly determined by comparison with the AGK1.[9] The numbers in the BD are found in the last column (15) where an asterisk indicates one of the "1200 common stars" whose catalogued positions are based on 10 rather than 6 individual observations.

There are three appendices. I: The ephemerides of stars at high declinations; II: the precessions from 1950 to 1875 (computed with Newcomb's constant of precession) to facilitate comparison with the AGK1, and III: the individual errors of the fundamental stars of the FK3. At this time the scientific value of these appendices has become insignificant.

4.3.3 THE "KATALOG VON 3356 SCHWACHEN STERNEN" (KSS). This catalogue, whose full title is "Katalog von 3356 schwachen Sternen für das Äquinoktium 1950, beobachtet an den Meridiankreisen der Sternwarten Hamburg-Bergedorf und Heidelberg-Königstuhl von J. Larink, A. Bohrmann, H. Kox, I. Groene-

9. Note that the purpose of the catalogue did not require particularly accurate proper motions, since the differences between the epochs of the plates and those of the positions in the AGK2A are usually only one or two years.

veld, H. Klauder" (1955) was observed with the purpose of establishing accurate positions of faint stars with magnitudes mostly between m_{pg} 8^m and $10\frac{1}{2}^m$ on the system of the FK3 such that there is no magnitude equation between the FK3 star positions and those in the present catalogue (KSS). The importance and significance of extending the fundamental system to faint stars was emphasized previously in Section 4.1.6. Every published position depends on four observations, each at the transit circles of Bergedorf and Heidelberg; the sky region covered is from $-5°$ declination to the north celestial pole. The reason for the observation at two different instruments was, of course, the desire to reduce the expectations of systematic errors. The observations were relative to the FK3 stars; the reductions to their system was effected by "series" (see Section 2.2.7) and by observing fundamental stars together with program stars. In order to reduce the magnitude equation between FK3 stars and program stars to the smallest possible amount, the fundamental stars were reduced to the average magnitude of the program stars through the use of objective screens. The introduction to the catalogue gives an accurate description of the observations, their reduction and combination into the catalogue. The average rms. errors of the individual Bergedorf observations are $0.^s019\sec\delta$ and $0.''39$, and of the Heidelberg observations $0.^s015$ and $0.''31$, in right ascension and declination, respectively, with very small differences between the accuracies of the positions of the brighter and fainter stars. The systematic errors between the Bergedorf and the Heidelberg observations were carefully investigated and turned out to be insignificant, so that the published positions are the straight means of the 4-each observations at the two participating transit circles. The internal rms. errors ϵ_α and ϵ_δ of a published position are $0.^s0059\sec\delta$ and $0.''143$, respectively, for the "bright" stars (average $8.^m6$); and $0.^s0067$ and $0.''152$, respectively, for the faint stars (average $9.^m3$).

The main body of the catalogue contains serial numbers (column 1), BD numbers (column 2), photographic magnitudes from the AGK3 and spectra determined by A. N. Vyssotsky in columns 3 and 4. Mean right ascensions and declinations for the epoch of place (in column 15), and epoch of orientation 1950 on the system of the FK3 are given in columns 5 and 10 respectively, while columns 6, 7 and 11, 12 contain the first two terms for the calculation of the precession following generally the model of equation (3.1a).

The centennial proper motions were derived by comparison with the AGK2 positions (μ_α and μ_δ; columns 8 and 13, respectively), and Küstner's (1900) catalogue (Table III-3, μ'_α and μ'_δ; columns 9 and 14 respectively), after the latter was reduced to the system of the FK3. In spite of the reduction, the μ and μ' show systematic differences against each other and are not very accurate, since the epoch difference against the AGK2 is on the average only 19 years, and the larger epoch difference of 50 years between the KSS and Küstner's catalogue did not result in a significantly higher accuracy for the μ' because of the larger rms. errors of Küstner's positions. The centennial proper motions are thus not very accurate, and quoted to $0.^s1$ in right ascension and $1''$ in declination. Column 15 finally contains the excess of the epoch over 1900 in tropical years (to $0.^y1$). Their average is about 1949. The observations were made between 1947 and 1953. Most average epochs, however, are in 1948 or 1949. An asterisk after the epoch indicates a note at the bottom of the page; those notes are quite self-explanatory.

The individual corrections (derived separately from the Bergedorf and the Heidelberg observations) to those 872 FK3 stars that were used as fundamental stars are a useful by-product of the main catalogue; these precede the main body of the catalogue, and were utilized for the improvement of the individual positions of the fundamental stars in the construction of the FK4.

4.3.4 THE AGK3R. This catalogue, which is at the time of this writing (January 1971) not yet available in print, is the catalogue of the reference stars for the AGK3 (see Section 5.1.4), in the same sense as the AGK2A was the source of reference star positions for the AGK2. It provides about 25 reference stars per $5° \times 5°$ plate, as compared to a reference star density of 15 stars per $5° \times 5°$ for the AGK2, which was considered to have been insufficient. The plans for the repetition of the AGK2 go back to 1953. While the AGK2A was observed (with the exception of Pulkovo) on transit circles of German observatories and reduced in Germany, the AGK3R was thoroughly international in scope. The coordination of the work on the AGK3R was in the hands of a committee appointed in 1955 by the Commission 8 (Positional Astronomy) of the International Astronomical Union (IAU) at its general assembly in Dublin. The observations started in 1956 and were, in essence, completed in 1963. The reductions were in the hands of F. P. Scott (the chairman of the committee) at the U.S. Naval Observatory, who has regularly reported on the progress of the work. The definitive report to the IAU (Scott 1967) was delivered at its general assembly in Prague. An abbreviated version of this report was published in the following year (Scott 1968).

Positions of 21,499 stars between $-5°$ declination and the north celestial pole for the mean epoch 1958 were observed on 11 transit circles at 10 observatories in six countries. The average magnitude of the program stars was close to $8.^m4$, with extremes of about 6^m and $9.^m5$. The list of program stars is a combination of two star lists of nearly equal size, one compiled by F. P. Scott and the other a Catalogue of Faint Stars (KSZ) compiled at the Sternberg and Pulkovo observatories (Zverev 1951, 1951a).[10]

10. The KSZ is actually not a catalogue in the sense in which we have been using this term, but a search list. One of its primary purposes was to stimulate the observation of accurate positions of stars between 8th and 9th magnitude, and the later spectral types. These would be used in the Soviet program of finding quasi-inertial proper motions of the stars by referring them to the system of background galaxies. This list forms the basis of most astrometric programs in the Soviet Union and some other countries. Zverev also selected a list (FKSZ) of 931 stars to form a fundamental system of faint stars, and gave a search list (Zverev 1951b) for those 645 FKSZ stars that have declinations between $-30°$ and the north celestial pole. The KSZ and FKSZ contain serial number, BD number, magnitude, spectral type, α_{1950} (approximate), annual precession in α, δ_{1950} (approximate), annual precession in δ, and the source. Much of the fundamental astrometric work in the Soviet Union and some other countries concerns the stars in the FKSZ. The entire FKSZ was published by the Pulkovo Observatory in 1956.

The Scott stars are on the average somewhat brighter and considerably bluer than those of the KSZ, although there is some overlap between the two lists. The Scott stars were selected specifically for the reduction of 5° × 5° plates exposed with an instrument equipped with an objective prism which would produce a 3.5 magnitude difference between the central image and the first order spectrum of a star. In the KSZ list, preference was given to fainter and redder stars which best suited the Russian program for relating proper motions to the galaxies. If C denotes 3782 stars common to both lists, S the 10,031 Scott stars and Z the 7692 (Zverev) (that is, KSZ) stars, the commitment of the various observatories toward observing the AGK3R stars is shown in Table IV-10. Column 3 in this table indicates the number of AGK3R stars on the program of the observatory. Typically, each star was observed twice at each observatory, except that 2972 stars of the Bordeaux Program, 3551 of the Paris program and 1169 of the Strasbourg program were observed four times. The positions of the C and the S stars in the catalogue depend on ten observations, those of the Z stars (except of the bright Z stars, $m \geq 8.8$, in the zone +5° to +40°) on only eight.

According to the plans adapted for the programs, all individual observations, reduced by use of nightly parameters based on the FK3R, were to be sent to the U.S. Naval Observatory for reduction to the FK4 on a nightly basis prior to their use in the formation of a catalogue of final positions of the reference stars. The data eventually sent to the U.S. Naval Observatory conformed to the above plan with the exception of the results from Heidelberg, Strasbourg and Paris. The two former observatories, having fully discussed their observations, sent mean results instead: Heidelberg on the system of the FK3R and Strasbourg on a system very close to the FK4 which resulted from the use of a synthesis method during the final reductions. (Ann. de l'Observatoire de Strasbourg, VI, pp. 39–37.) In this method a weight was derived for each night's work which led to a marked improvement in the results. Although Paris sent individual observations, the reduction to the FK4 was carried out at that observatory.

The reduction of the remaining observations to the system of the FK4 and the compilation of the catalogue of final positions were carried out under the direction of F. P. Scott.

Since the data transmitted to the U.S. Naval Observatory were not homogeneous, it was necessary to use mean results for all observatories in the formation of the catalogue. The final catalogue was accomplished by means of successive approximations in which the variances of the systematic differences (observatory minus mean of all observatories) in 5° zones were used to derive weights for the next approximation. The process was started by giving the same weight to a single observation made at all observatories. This permitted the evaluation of a set of variances to be used in assigning weights for the next approximations. In all, three approximations were made. The variances at that point indicated that but little improvement could be expected by carrying the process any further. It should be mentioned that SAOC proper motions were used in computing the systematic differences, but not in the formation of the means. Table IV-9 shows the average rms. error in each coordinate after each approximation.

Table IV-9
Average rms. Errors in α and δ at the End of Each Iteration

	RA	*Decl.*
1st	$\pm 0^s.0053$	$\pm 0.''133$
2nd	.0051	.123
3rd	.0050	.116

Table IV-10 **Observations of AGK3R Stars at the Participating Observatories**

Observatory	*from*	*to*	*Total number of stars*	*Re-marks*
Bergedorf	$+15°$	$+90°$	12124	1
Bordeaux	$-5°$	$+20°$	4565	2
Greenwich	$-5°$	$+90°$	17329	3
Heidelberg	$-5°$	$+15°$	4324	3
Nicolaiev	$-5°$	$+25°$	9994	
Ottawa	$-5°$	$+90°$	3754	4
Paris	$+20°$	$+50°$	5332	5
Pulkovo	$+25°$	$+90°$	11511	
Strasbourg	$+50°$	$+70°$	2777	5
USNO	$-5°$	$+90°$	13002	6

Key to the remarks:

1: Z stars also from $-5°$ to $+15°$
2: no C stars from $+15°$ to $+20°$
3: no Z stars
4: C stars only
5: No C stars
6: Zone divided between the 6″ T.C. (40%) and 7″ T.C. (60%)

Enterprises of this type, which are currently under way (for instance the SRS program, see Section 4.4.1) may profit from a remark which Scott made in his above quoted paper (Scott 1967, p. 6). "The writer is now convinced that the final catalog would have been just as accurate and that considerable time would have been saved, if each observatory had derived its own mean positions before sending them to Washington. At any rate, it would not have required three years to put the voluminous data on punch cards, verify the punching, correct thousands of typographical and punching errors, and resolve blunders by correspondence which the observers could have disposed of in an instant had they been discussing their own work."

Undoubtedly, the task of compiling the catalogue was not made any easier by the virtually complete lack of guidelines (except for the list of program stars) for the observations and preliminary reductions. This is one of the principal differences between the work on the AGK2A and the AGK3R. The work on the former followed the same rather strict and inflexible guidelines that were binding on all participating observatories. Whereas strict guidelines might have insured greater homogeneity of the result, the variations between the methods used for the work on the AGK3R at the various observatories may have resulted in a better systematic accuracy, since uniform methods could have generated undetectable systematic errors that would have been of the same nature at all instruments. Uniformity is undoubtedly easier to enforce in an essentially national undertaking (such as the AGK2A), especially when backed by a strong and powerful authority, than in an international one. Only the future will decide about the advisability of following strict and uniform guidelines. This author feels that their application or non-application will not have very profound influences on the results, provided that all work is performed competently and conscientiously.

Comparisons of the final AGK3R positions with those in the FK4 were made with the use of a third catalogue as intermediary since there is virtually no overlap. Significant deviations of the AGK3R positions from the FK4 system were not found, and are probably in the order of only a few hundredths of a second of arc.[11] The rms. errors of the positions are, on the average, $\epsilon_\alpha \cos\delta = 0.^s0050$, $\epsilon_\delta = 0.''116$, with some slight variation depending on the declination zone and the magnitude. The positions of faint stars are predictably somewhat less accurate. More detailed data concerning the accidental errors were given above in Table IV-9.

The catalogue itself has not yet appeared in print, and therefore cannot be described here. It is, however, finished and arrangements may be made with the U.S. Naval Observatory in Washington, D.C. for the acquisition of copies of the catalogue data on magnetic tape. Work is also being done to derive accurate proper motions for all stars in the AGK3R, and publication may be deliberately delayed to await their completion.

4.3.5 THE PFKSZ. This catalogue (Polozhetsev and Zverev 1958) could prop-

---

11. In contemplating this result, it may be well to remember the discussion of the definition and meaning of the term "system" given above in Sections 1.6.1 and 1.6.6. There were, of course, significant systematic differences between the positions derived from the observations at the individual instrument, and Scott (1967, p. 10) writes: "In the absence of a definition of what is meant by relating a set of stars to the Fundamental System and, further, in the absence of prescribed observing procedures for accomplishing the 'relationship,' the appearance of systematic differences between observatories is not startling."

erly be termed "Preliminary General Catalogue of Fundamental Faint Stars" (Zverev 1960) and contains accurate mean positions of 587 Fundamental Faint Stars, selected from the FKSZ. It is not a fundamental catalogue, since the system of its positions is based on relative observations gathered in 14 catalogues with mean epochs 1943–1958 which were reduced to the FK3 system by the observers. These catalogues originate at the observatories at Bucharest, Golosseyevo near Kiev, Kazan (Engelhardt Observatory), Kharkov, Kiev, Moscow (Sternberg Institute), Odessa, Pulkovo, Tashkent and Breslau (now called Wrocław). The system of positions was obtained in two steps. First, a mean system was formed from all catalogues, and weights were assigned to the individual catalogues according to the variance of the systematic deviation of the catalogue system from the mean system. The final system adopted was the weighted mean. The mean epoch of the position system is about 1949.

The proper motions were intended to be on the FK3 system and were obtained by comparison of the PFKSZ positions with those in the AGK2 (see Section 5.1.3), the Yale Photographic Catalogues (see Section 5.2) and the GC after the positions in the latter two had been reduced to the FK3 system by means of the tables by Barney (1951) and Kopff (1939), respectively.

Zverev claims an accuracy corresponding to average rms. errors $\epsilon_\alpha \cos\delta = 0.^s0068$, $\epsilon_\delta = 0.''127$, $\epsilon_\mu \cos\delta = 0.^s029$, and $\epsilon_{\mu'} = 0.''650$ for the position and centennial proper motion components, respectively, of the PFKSZ.

Comparisons with the systems of the FK3, GC, N30 and the α1 catalogue of Pulkovo, for the epoch 1930, were made with the use of (because of virtually non-existing overlap) the AGK2 as an intermediary. These comparisons show that the reduction to the FK3 system was as successful as could have been expected, the systematic differences generally being of the order of a few hundredths of a second of arc.

As its title indicates, this catalogue is clearly intended not to be of permanent use, but rather an ad hoc solution for the pressing requirements of certain problems. Its compilers hope that its existence will suggest to observers to include the stars which it lists on their programs for the absolute determination of star positions.

4.4 Compilation Catalogues Now Being Constructed

4.4.1 THE SOUTHERN REFERENCE STAR PROGRAM (SRS). In various sections of this book it has been pointed out that the situation with regard to accurate star positions is much more unfavorable in the southern than in the northern hemisphere of the sky. This is true not only concerning the accuracy of the fundamental system, but also concerning the availability of relative star positions of high accuracy. Equivalent tasks were always achieved in the northern celestial hemisphere earlier than in the southern one. Thus, while the acquisition of accurate proper motions from photographic zone catalogues (AGK2 and AGK3) has already been accomplished (see Section 5.1.4), astronomers have just now, in essence, finished the first epoch coverage of the southern sky by photographic zone catalogues (Yale Catalogues, Cape Photographic Catalogue; see Sections 5.2 and 5.3).

It was clear that the second epoch coverage of the southern sky by photographic AG-type zone catalogues could only be a question of time,[12] and therefore the need for an adequate system of reference stars for the photographic work—analogous to the AGK3R—was obvious. Strict contemporaneousness of the reference star work and the exposure of the photographic plates is not necessary, if reasonably accurate proper motions for the reference stars are available. These considerations prompted the formulation of plans during the period from 1956 to 1960 for the organization of a Southern Refernce Star program (SRS) for the southern celestial hemisphere, analogous to the AGK3R for the northern one. The committee for the coordination of the necessary efforts was appointed at the general assembly of the International Astronomical Union in 1961 at Berkeley. The chairman was F. P. Scott of the U.S. Naval Observatory. The first observations toward this goal started at the Cape Observatory in 1961, and the completion of the observations for the SRS is expected in 1973.

Scott (1967) gave a description of the program; the latest available progress report was written by Scott and Schombert (1970). The SRS program covers the sky between declination $+5°$ and the south celestial pole by 20495 fairly uniformly distributed stars, 18796 of which lie between the equator and the south celestial pole. Their mean magnitude is about $8.^m5$, while the range of the magnitudes slightly exceeds $6.^m9$ to $9.^m2$. The most frequent spectral type is K. Between $+5°$ and the celestial equator, the SRS stars are identical with the AGK3R stars. Between the equator and $-30°$ declination, the stars were selected at the U.S. Naval Observatory by adding stars to the KSZ list selected according to the criteria used for choosing the AGK3R stars. In the region not covered by the KSZ, that is, south of $-30°$ to the south celestial pole, a list was selected at the Cape Observatory, following the criteria for the establishment of the KSZ, which was then augmented by brighter stars that satisfied the criteria for inclusion in the original U.S. Naval Observatory list for the AGK3R. The resulting density of sky coverage is thus about equal to that in the AGK3R.

The commitments of the various observatories to the program is illustrated in Table IV-11.

Every star will be observed at least six times. The accuracy of the positions obtained in the SRS program should be just as good as that of those of the AGK3R.

Although the question has not been entirely settled at this time (1971), it is quite likely that at least the U.S. Naval Observatory commitment to the SRS program will be referred to an improved fundamental system, which will be established from absolute observations made concurrently with the 7-inch transit circle while at El Leoncito in Argentina and with the 6-inch transit circle in Wash-

12. It is, as a matter of fact, now being actively prepared by astronomers at the U.S. Naval Observatory in Washington. Its execution may be indefinitely delayed because of lack of funds.

ington. In this way, it is hoped that the U.S. Naval Observatory SRS catalogue will be on a "smoother" (that is, more self-consistent) system than that represented by the FK4, so that it would later be easier to reduce it to an improved, generally adopted fundamental system, perhaps that of the FK5. (Scott and Schombert 1970).

Table IV-11 **Breakdown of the work on the SRS Program**

Observatory	*Zone* *from*	*to*	*No. of stars*	*Obs'ns. per star*
Abbadia	+5°	−15°	1560	4
Bordeaux	+5°	−15°	1560	4
Bucharest	+5°	−10°	1176	4
Nicolaiev	0	−20°	5984	2
San Fernando	−10°	−30°	3709	4
Tokyo	−10°	−30°	3560	4
USNO 6-in	+5°	−30°	{8706	2
			{1233	4
Cape	−30°	−90°	10082	4
Santiago-Pulkovo	−25°	−90°	11496	4
Bergedorf (Bickley)	+5°	−90°	20495	4
USNO 7-in	+5°	−20°	7683	2
(El Leoncito)	−20°	−90°	13503	4
San Juan	−40°	−90°	7190	2

4.4.2 DIECKVOSS' AGK2/3—AC GENERAL CATALOGUE. The general nonspecialist user of high-accuracy star positions and proper motions seldom has the knowledge—and time—to derive the positions and motions he needs, with the required accuracy. It is indeed an unfortunate state of affairs that only a small portion of the huge material gathered during more than two centuries has been more or less completely used for the calculation of star positions and proper motions, that is, principally for the compilation of the various fundamental catalogues. The number of stars covered by the latter is very small, compared to the number of stars for which accurate positions have been gathered. In particular, only very scant use has been made of the numerous material gathered in the various AG type zone catalogues and the Astrographic Catalogue (AC) which will be described below in Sections 5.1, 5.2, 5.3 and 5.5. Even though the Yale Zones, and the CPC, list proper motions that were derived mostly by comparison with earlier epoch meridian catalogues, and even though the AGK3 was constructed specifically with the purpose of deriving proper motions by comparison with the AGK2, no comprehensive uniform treatment has yet been attempted of the entire material, including the use of the AC positions which are particularly valuable because of both their early epoch and their (generally) high accuracy. In this connection it may be interesting to recall that the AGK2 was undertaken for the avowed purpose of deriving proper motions by comparison with the AGK1. This aim was later abandoned because of practical difficulties experienced in deriving reduction of the AGK1 to a homogeneous fundamental system because of the scarcity of stars in common with available fundamental systems. Such reductions could be accomplished only through the use of intermediary catalogues such as the GC which, for the precision required, would be inadequate. This led the German astronomers to temporarily write off the AGK1 and to concentrate on the AGK3. This decision had considerable merit. At any rate it stimulated interest in the production of the AGK3 which, with the AGK2, should produce proper motions to backdate modern positions to the epochs of the AGK1 and the AC. Such positions should enable one to derive systematic differences, including magnitude equation, for the reduction of AGK1 to the fundamental system. Likewise, they should also enable one to make a redetermination of the AC plate constants and a limited study of their magnitude errors and coma. If these expectations are fulfilled, even to a small degree, the intrinsic accuracies of both of these old catalogues may be brought into play for the improvement of proper motions of stars to the 10th magnitude and, in some cases, even those of fainter ones.

This author believes that the emergence of the electronic computer should trigger a reevaluation of the decisions taken years ago, when the drudgery of routine measurements and routine computations were among the most important elements in the construction of any catalogue. It will also be well to remember that all existing tables for the reduction of the positions in one catalogue to the system defined by those in another one were obtained from comparing the positions of only those stars common to both catalogues. Never, in the past, was any advantage taken of the fact that any one star (at a given epoch) can have only one correct set of coordinates and proper motion components. This restriction, applied to the nonfundamental stars, should yield a considerable increase in the accuracies of the systematic corrections required to reduce, simultaneously, the positions in a set of several catalogues to a given—fundamental or nonfundamental, but in any case well-defined—system. In other words, noncontradictory systematic corrections of various catalogues to a common system must also statistically dissipate the systematic differences between the position data of the stars common to any pair of catalogues of this set. This means that the "true" systematic corrections must not leave any systematic differences between the data in any combination of catalogues reduced to the same system.

An algorithm for the construction of a general catalogue, based on this principle, was outlined above in Section 3.2.5, but has never yet been applied in practice, although a completely analogous one is the basis of the plate overlap method (see Section 2.3.8.2) which has very decidedly increased the systematic accuracy of photographic astrometric positions.

The first plan for systematically combining data from at least more than two catalogues into a general catalogue was proposed by W. Dieckvoss at the Tampa Conference on Photographic Astrometric Technique in 1968 (Dieckvoss 1971). Two years later, he reported on the progress of the project at the International Astronomical Union Colloquium No. 7 on Proper Motions at Minneapolis (Dieckvoss 1970). Most of the AGK3 and the revised AGK2, both on the FK4 system, were available in 1970. The proper motions derived from these catalogues can be used to calculate the positions of the stars contained in them for the epochs of the various plates taken for the construction of the AC. They can thus be used as reference positions for the determination of reduction constants for the AC plates. With these, the rectangular coordinates measured on the AC plates can be converted to positions on the FK4 system. This was recently done systematically by Günther and Kox (1970) for the AC zones Greenwich, Vatican, Catania and Helsingfors which cover the area from the north celestial pole to +39°.

The accuracy (rms. errors) of the positions calculated in this manner can be estimated from the comparison of positions derived from overlapping plates. The positions of the AGK stars derived from the AC plates may then be used to improve the AGK2/3 positions and proper motions in the way described, for instance, in Section 3.2.2. This plan was actually carried out by Dieckvoss for the region in the sky indicated (+39° to +90°). It gives, for the individual stars, BD numbers, AGK2 magnitudes, positions for epochs of place at the central epoch, and epoch of orientation 1950,0, the central epoch (see Section 3.2.2) annual proper motion components (in α in seconds of time on the parallel, and seconds of arc on the great circle, in δ in seconds of arc), and the rms. errors of the position at the central epoch, and of the proper motion components. In Table IV-11, adapted from Dieckvoss' (1970) paper, typical values are given for the resulting central epochs, t_c, and the rms. errors of a position component $\epsilon(t_c)$ at the central epoch, and of a proper motion component (ϵ_μ), respectively, which result from the incorporation of the AC zones Helsingfors and Vatican that differ from each other, mainly by the different rms. errors of the positions extracted from them (0.″14 for Helsingfors and 0.″32 for Vatican).

Table IV-12 Characteristics for the AGK2/3–AC General Catalogue

AC Zone	t_c	$\epsilon(t_c)$	ϵ_μ
Hels (+43°)	1920	0.″09	0.″004
Vat (+56°)	1935	0.″10	0.″006

The extensions of this catalogue to the entire sky is no small task and must, in any case, wait for the completion of a second epoch photographic zone catalogue on the southern hemisphere. Table IV-12 shows, however, quite clearly that the data in this catalogue would be much more accurate than those in the SAOC and will indeed, when complete, render the SAOC obsolete.

This proposed AGK2/3–AC catalogue could also be used to find the systematic corrections of any other catalogue to the FK4 system. Therefore, the positions in this catalogue could be incorporated to further improve the accuracy of the catalogue data. In his above quoted paper, Dieckvoss reports on some experiments of this kind involving the Helsingfors-Gotha zone of the AGK1 (rms. error of a position $0.''70$; see also Table V-1) and the appropriate Yale zone (rms. error of a position $0.''26$, see also Table V-2). Due to the low accuracy of the AGK1 zone, the data referring to the $+56°$ zone in Table IV-12 will not change much: t_c to 1934, $\epsilon(t_c)$ to $0.''10$, and ϵ_μ to $0.''005$. The inclusion of the Yale zone will further change these values to 1932, $0.''09$ and $0.''005$, respectively.

Although this scheme is simple and straightforward, it suffers from the imperfection that the reduction to the FK4 system is accomplished only by the AGK2/3, a difficulty also pointed out by Dieckvoss. This could be avoided by a simultaneous reduction procedure (as outlined in Section 3.2.5), rather than the iterative one actually applied which follows the precepts set down in Section 3.2.2 where it is assumed that the positions in all catalogues to be incorporated into the general catalogue are already on the chosen system. It would, however, be possible to obtain data identical to those that would result from a simultaneous adjustment, by using for the determination of the systematic errors of an additional catalogue not only comparisons with the positions in the already existing general catalogue, but also comparisons with those in the fundamental system on which the final product is intended to be. One could probably, in principle, develop an appropriate algorithm for this without much difficulty.

4.4.3 PLANS FOR THE REVISION OF BOSS' GENERAL CATALOGUE.

There exist plans for the construction of what would essentially be a revision of the GC, at the Heidelberg Astronomisches Rechen-Institut. No details concerning the procedures to be applied toward this purpose have yet been discussed in the open literature.

Chapter 5

SYSTEMATIC ZONE CATALOGUES

In this chapter are discussed the catalogues that provide accurate positions of essentially all stars to a certain limiting magnitude in an extended region of the sky bounded by parallels of declination.

5.1 The Astronomische Gesellschaft Catalogues

5.1.1 INTRODUCTION. Although thousands of observations of the positions of stars had been made prior to 1865 there was, before that time, but little coordination of the work of the various observers to achieve a systematic coverage of all stars in the sky to a certain limiting magnitude. Each observer designed his program to satisfy his own desires. Of the many star catalogues produced before 1865, very few of them were confined to narrow enough declination belts to be called zone catalogues according to current terminology, and still fewer included all stars down to some defined limit of magnitude.

In August 1867, F. W. Argelander addressed the Astronomische Gesellschaft (then in the second year of its existence) on the subject of the establishment of catalogues which would contain the accurate positions of all stars down to magnitude $9.^{m}0$ (on the BD scale) and of a restricted number of fainter BD stars between declinations $+90°$ and $-2°$. Argelander was particularly well qualified to speak on the subject. It was he who had successfully completed the "Bonner Durchmusterung," which consists of a list of approximate star positions between the same declinations complete to magnitude about $9.^{m}5$, and a set of corresponding charts. Later, the BD was extended to 23° southern declination. Incidentally, the same ideas which eventually were realized by what is now known as the AGK1 (Section 5.1.2) had been expressed almost fifty years earlier by Bessel (1822), whose assistant Argelander had been at the time. Bessel himself had actually carried out his plans for a limited zone with very crude accuracy.

The work proposed by Argelander was to be executed according to the following guidelines:

(1) Since an enterprise of this kind could only be completed through the cooperation of many individuals and institutions, the greatest possible degree of homogeneity and uniformity in the working procedures should be maintained.
(2) The observations were to be differential with respect to a network of 500 to 600 fundamental stars to be observed at Pulkovo.
(3) Every star was to be observed at least twice.
(4) Utmost economy and efficiency were to be maintained in the observations.
(5) To avoid errors arising from observer fatigue, no individual observer was to work more than two hours at a time.

In retrospect, it appears that the strict adherence to these guidelines has at best only partially produced the intended effects. Almost the only advantage was the virtually identical format of the catalogues published on the basis of observations made at different observatories. The accuracies of the positions, however, vary considerably from one observatory to the other, and so do the systematic corrections that are required to reduce the different individual catalogues to the same system, which was intended to be that of the FC (see Section 4.1.2).

5.1.2 THE "CATALOG DER ASTRONOMISCHEN GESELLSCHAFT" (AGK1) AND ITS SOUTH AMERICAN EXTENSIONS. Argelander's plan for the observation of zones between declinations $+80°$ and $-2°$ was adopted and the work started almost immediately. This "Erste Abtheilung," which lists positions of the program stars in those zones for the coordinate system 1875 (1905 for one of the zones), was finished in the first decade of this century. In 1887 it was agreed that zones between $-2°$ and $-23°$ should be similarly observed. The positions of these stars, for the coordinate system 1900 appeared in the "Zweite Abteilung." The observation of stars between $-22°$ and $-82°$ was begun early in the present century at the Cordoba and La Plata Observatories. The guidelines for the AGK1, however, were not followed for these catalogues.

Table V-1 (p. 230) contains a list of the data pertaining to the AGK[1] catalogues and their counterparts in the southern hemisphere observed at the South American observatories. The following is a description of the features common to all these catalogues.

Every volume (except for Cordoba B and C) is, as usual, preceded by an introduction which provides information on the selection of the program stars, the instrument with which the observations were made, the mode of observation, an explanation of the steps taken in the reductions and an explanation of the columns of the body of the catalogue.

---

1. This is the name by which they are known at this time. Before the appearance of the AGK2, they were often called Astronomische Gesellschaft Catalogue, AG Catalogue, AGK or AGC.

The main bodies of the catalogues carry, as a rule, the following columns: serial number; magnitude (usually estimated during the observations); right ascension and declination oriented to 1875 (or 1900, 1925, 1935, 1950, respectively, for the southern hemisphere catalogues) for the epoch of place equal to the time of observation (that is, *not* reduced to 1875 or 1900 by the application of proper motion); the first two or three terms of the precession series in right ascension and declination, initially always computed with Struve's constants (annual and secular variations, third term), but in the last catalogues that have appeared, computed with the use of Newcomb's constant; the epoch to which the coordinates refer (occasionally α and δ refer to different epochs); the numbers of the zones in which the star was observed with comments (some of the catalogues also contain a list of the zones); and the BD or CoD number. Frequently, remarks are at the bottom of the pages. Positions of fundamental stars, as a rule, were taken from the fundamental catalogue (FC, see Section 4.1.2, Auwers 1879) and reprinted from there (reduced to the epoch of place 1875) in bold type. It would have been preferable to have given the positions of the fundamental stars as they resulted from the observations, because the format chosen restricts a direct comparison of these catalogues with the FK4 to the Zusatzsterne only. This is not enough to establish reliable systematic corrections.

Appendices give extensive remarks, too long to be printed at the bottom of the page, positions observed in addition to the regular program or with an instrument different from that with which the bulk of the observations was made, proper motions, comparisons with other catalogues (that is, a list of the differences between the positions for the same star from this and another catalogue), lists of errata, and so on.

It is quite remarkable that even now, almost one hundred years after its inception, the project is not yet quite finished, since only a part of the catalogue La Plata E has so far been published. Even as it stands, the AGK1 is a monument to the unselfish industry of the scores of astronomers who worked on it. It also illustrates very well the development of meridian astronomy around the turn of the century.

Nowadays, work of the same scope is no longer done with transit circles, but with the faster, more efficient and more accurate photographic methods described above in Section 2.3. A glance at the epochs of the La Plata zones of the AGK1 will show that all of the positions were observed after 1914, which is the epoch of the first photographic Yale catalogue. A similar examination will show that the positions in some zones (including Cordoba D) were observed around and after 1930, when the work on the photographic AGK2 and the Yale catalogues was at its peak.

One may wonder why the South American astronomers continued the work with transit circles in spite of the fact that the vastly superior photographic method was being used elsewhere. In 1915, an explanation might have been that the superiority of the photographic methods had not been established. (See, however, Schlesinger and Hudson 1914). This could scarcely have been a valid reason from

Table V-1

Zone From	*To*	*Name of Zone*	*Appeared*	*Author(s)*	*Ep. of Orient.*	*rms.* α	*rm[s.]* δ
+80°	+75°	Kasan	1898	D. Doubiago	1875.0	0″.45	0″.6
+75°	+70°	Berlin C	1910	L. Courvoisier	1905.0	.29	.3
+75°	+70°	Dorpat	1910		1905.0	.45	.7
+70°	+65°	Chistiania	1890	D. Fearnley and H. Geelmuyden	1875.0	.45	.8
+65°	+55°	Helsingfors-Gotha	1890	A. Krüger	1875.0	.86	[illegible]
+55°	+50°	Harvard	1892	W. A. Rogers	1875.0	.57	[illegible]
+50°	+40°	Bonn	1894	Deichmüller			
			1909	C. Mönnichmeyer	1875.0	1.23	[illegible]
+40°	+35°	Lund	1902	F. Engström and A. A. Psilander	1875.0	.60	[illegible]
+35°	+30°	Leiden	1902	J. H. Wilterdijk	1875.0	.51	[illegible]
+30°	+25°	Cambridge (E)	1897	A. Graham	1875.0	.70	[illegible]
+25°	+20°	Berlin B	1895	E. Becker	1875.0	.29	[illegible]
+20°	+15°	Berlin A	1896	A. Auwers	1875.0	.48	[illegible]
+15°	+10°	Leipzig I	1900	H. Bruns and B. Peters	1875.0	.77	[illegible]
+10°	+5°	Leipzig II	1899	H. Bruns and B. Peters	1875.0	.56	[illegible]
+5°	+1°	Albany	1890	Lewis Boss	1875.0	.56	
+1°	−2°	Nicolayew	1900	J. Kortazzi	1875.0	.66	[illegible]
−2°	−6°	Strassburg	1906	E. Becker	1900.0	.45	[illegible]
−6°	−10°	Wien-Ottakring	1904	L. de Ball	1900.0	.42	[illegible]
−10°	−14°	Harvard	1912	Arthur Searle	1900.0	.62	[illegible]
−14°	−18°	Washington	1908	A. N. Skinner	1900.0	.72	
−18°	−23°	Algiers	1924	C. Rambaud and F. Sy	1900.0	.50	
−22°	−27°	Cordoba A	1913	C. D. Perinne	1900.0	.63	
−27°	−32°	Cordoba B	1914		1900.0	.54	
−32°	−37°	Cordoba C	1925		1900.0	.57	
−37°	−47°	Cordoba D	1954	L. C. Guerin and J. Bobone	1950.0	.37	
−47°	−52°	La Plata F	1938	H. A. Martinez	1935.0	.50	
−52°	−57°	La Plata A	1919	P. T. Delavan	1925.0	.58	
−57°	−63°	La Plata B	1929	A. Aguilar and B. Dawson	1925.0	.36	
−62°	−66°	La Plata C	1924	H. A. Martinez	1925.0	.59	
−66°	−72°	La Plata D	1936	V. Manganiello	1925.0	.53	
−72°	−82°	La Plata E	1947	Numa Tapia	1925.0	.55	

The AGK1 and its South American Extensions

vg. No. of bs'ns per Star	Avg. Epoch	No. of Stars	Instrument	Method	Publication	Remarks
.4	1875	4281	Repsold 125–2000	eye-ear	AG	
.3	1905	3461	Pistor & Martin 189–2620	self-reg.	AG	1)
	1875				AG	1)
.9	1875	3949	Ertel 120–1525	eye-ear	AG	
.2	1876	14680	Reichenbach 8-inch	eye-ear	AG	2)
.1	1877	8627	Troughton & Simms 210–2860	reg. micr.	AG	
		18457	Pistor & Martin 116–1970	eye-ear	AG	3)
.6	1875					
.7	1881	11415	Repsold 157–2280	eye-ear	AG	
.2	1873	10209	Pistor & Martin 162–2600	reg. micr.	AG	
.0	1879	14441	Troughton & Simms 8 inch-9 foot	eye-ear	AG	
.7	1881	9308	Pistor & Martin 189–2620	reg. micr.	AG	
.5	1870	9789	Pistor & Martin 4 inch-5 foot	reg. micr.	AG	4)
.3	1880	9547	Pistor & Martin 6 inch	eye-ear	AG	
.3	1885	11875	Pistor & Martin 6 inch	eye-ear	AG	
.3	1880	8241	Olcott meridian circle (Pistor & Martin 200–3000)	reg. micr.	AG	
.3	1885	5954	Reichenbach 108–65 inch	reg. micr.	AG	
.9	1891	8204	Repsold	reg. micr.	AG	
.2	1895	8468	Repsold 123–1500	reg. micr.	AG	
.2	1892	8337	Troughton & Simms 210–2700	reg. micr.	AG	
.2	1895	8824	Pistor & Martin 227–3760	eye-ear	AG	
.3	1892	9997	Gautier 190–2300	self-reg.	Algiers Annals **8**	
.5	1895	15975	Repsold 125	reg. micr.	Cord. Res. **22**	5)
.7	1896	15200	Repsold 125	reg. micr.	Cord. Res. **23**	5)
.0	1897	12757	Repsold 125	reg. micr.	Cord. Res. **24**	5)
.7	1935	16610	Repsold 190–2250	reg. micr.	Cord. Res. **38**	6)
.7	1935	4828	Gautier w. Repsold micr.	self-reg.	La Plata Pub. **13**	
.4	1915	7412	Gautier w. Repsold micr.	self-reg.	La Plata Pub. **5**	
.5	1916	7792	Gautier w. Repsold micr.	self-reg.	La Plata Pub. **7**	
.7	1920	4412	Gautier w. Repsold micr.	self-reg.	La Plata Pub. **8**	
.0	1920	4513	Gautier w. Repsold micr.	self-reg.	La Plata Pub. **9**	
.7	1929	2486	Gautier w. Repsold micr.	self-reg.	La Plata Pub. **10**, 1	7)

1929 to 1935. Maybe the South American astronomers continued with transit circles because they had agreed to extend the AGK1 type survey to the south pole, and furthermore because they were not equipped to do photographic star catalogue work. One must still wonder why they did not follow the AGK1 guidelines, at least as far as the star list is concerned.

The first two columns in Table V-1 contain the approximate limits between which the catalogue contains stars. The third column shows the place of observation. If more than one zone were observed at any one observatory, they are distinguished by Roman numerals or capital letters according to established usage.

Column four shows the year in which the catalogue was published.

Column five gives the author (or authors) who signed as responsible for the catalogue in its final published form.

Column six gives the epoch of the coordinate system to which the catalogue positions are referred.

Columns seven and eight give the average rms. errors of a catalogue position, computed from the weights given either in the first volume of the GC or the Gyllenberg paper mentioned earlier, Section 3.2.3.6, or, where the catalogues were not contained in either of these lists (La Plata E and Cordoba D), from the errors quoted in the introduction to the catalogue.

It should be pointed out that the errors in columns seven and eight vary greatly from star to star, depending on how often it had been observed and, probably to some slight degree, on its magnitude. (See GC.I and Gyllenberg 1946.)

Column nine shows the average number of observations per star.

Column ten gives the average of the epochs of observation. The actual values for the individual stars often deviate from this figure by several years.

Column eleven records the number of stars in the main body of the catalogue. Frequently, the catalogues contain supplements listing the positions of stars which were observed only once or which were left out of the main catalogue for some other reason. These stars are not included in the numbers given in column eleven.

Column twelve lists the instrument used for the observations, giving the name of the manufacturer(s), and when available, the diameter of objective, and focal length. Unless indicated otherwise, the numbers are in millimeters.

Column thirteen shows the method of observation; either eye-ear, registration with a key and chronograph (reg. micr.) or with an impersonal (automatic) micrometer (self-reg.).

Column fourteen shows where the volume was published. The AGK1 proper, except the Algiers zone, was published by the Astronomische Gesellschaft.

Column fifteen refers to the explanatory remarks to the list:

(1) The zone from $+75°$ to $+70°$ was taken over by the Berlin Observatory after it had been almost completely observed and reduced, but not quite finished, at Dorpat. This explains why the epochs of place and of orientation (1905) of the positions in this Berlin C Catalogue deviate by thirty years from the rest of the catalogues in "Abtheilung I." A

history of the Dorpat observations is given in the introduction to Berlin C, in which it is pointed out that they are inhomogeneous and were often incorrectly reduced. The Dorpat catalogue, in its final form, is available only through the differences against the catalogue Berlin C in one of its appendices. Errors of the Dorpat positions could still classify it as a catalogue of average accuracy, in spite of the shortcomings pointed out by Courvoisier who signs as the author of Berlin C.

(2) All meridian observations of the Helsingfors-Gotha zone were made with the same instrument which was transported to Gotha after the observations at Helsingfors were finished. In dense regions of the Milky Way, some observations were made at a refractor with a micrometer. These are indicated in the catalogue by an R. One of the appendices contains a list of positions in the clusters h and χ Persei.

(3) The Bonn zone is the least accurate one of the AGK1. One of the reasons for this may be the fact that there were four observers who had different personal perception characteristics. The originally published catalogue contains the positions without homogenization. Later, C. Mönnichmeyer (1909) attempted to reduce all observations to the system of the observer H. Seeliger, and published a revised list of positions for which Boss found the low accuracy quoted in Table V-1. This figure makes it doubtful whether Mönnichmeyer's new reduction produced a substantial improvement. This remains a moot question, since the accuracy of the originally published catalogue was never investigated. Boss' figure (from the GC.I) for the rms. error of an AGK1 Bonn right ascension may be too high. A sample of seventy stars which was investigated in the course of some unpublished work by Eichhorn on the stellar association Cygnus VI gave an rms. error of about $0.''70$ for the Bonn right ascensions, but from such a small sample it is hard to draw a general conclusion.

(4) The zone Berlin A ($+20°$ to $+15°$) was observed by A. Auwers who was at that time the most respected influential positional astronomer in Europe, and one of the group of prime movers who got the AGK1 project under way. Although the rms. errors of his positions in this catalogue do not show this, he was an excellent observer as well. In this case he had, however, an inferior, rather short-focus instrument at his disposal.

The catalogue Berlin A is preceded by a very long and detailed introduction which is an excellent source for data concerning the history of the AGK1. Extensive comparisons with other catalogues, as well as proper motions for individual stars are also given there.

(5) The catalogue Cordoba A contains only a very short introduction, and B and C have no introduction at all. They also give neither the number of observations nor precession data. (The latter are also lacking in Cordoba D.) No author is indicated for Cordoba B and C, and it must be assumed that everything was analogous to Cordoba A. The copies of these catalogues in the Yale Observatory library have had entered, man-

ually, the number of observations of each star, and it is from these that we computed the figures quoted here.

(6) Reduction tables to the GC system are available neither in the GC nor given by Gyllenberg (1948). According to the introduction, the catalogue is on the system of the GC. Comparisons against the GC are given in the appendix. From this, reduction tables to the GC could be constructed.

(7) So far, only the first part of this catalogue is available. About one-third of the stars have been observed only once, or were excluded from publication for some other reason. Nothing is stated about the system of the catalogue, but an appendix gives a comparison with GC from which the reduction tables could be constructed.

The zone +90° to +80° was never observed, since Carrington's (1857) catalogue covers this in almost the same way that a regular AGK1 zone would have covered it. Besides, the Greenwich Second Nine Year Catalogue, and Courvoisier and Freundlich's catalogue (Freundlich 1916) give positions for all AG stars.

5.1.3 THE ZWEITER KATALOG DER ASTRONOMISCHEN GESELLSCHAFT (AGK2).

5.1.3.1 Introduction. For a long time the AGK1 provided the only source (apart from the Astrographic Catalogue) for positions of relatively faint stars. Due to the lack of proper motions or poorly determined proper motions, the accuracy of these positions deteriorates as time progresses. After Schlesinger and Hudson (1914) had demonstrated the suitability of short-focus wide-angle astrographs for photographic astrometric work, Richard Schorr and Fritz Cohn suggested in 1921 to the German Astronomische Gesellschaft that it sponsor a reobservation of the stars in the first part (Erste Abtheilung) of the AGK1 by means of short-focus wide-angle astrographs.

This proposal was received favorably and a study commission was appointed in the same year. Their report was delivered in 1924 (Vierteljahrsschrift der A.G., **59**, pp. 168 and 243.) The suggestions were: (1) to have all the positions in the final catalogue on the same fundamental system, namely, that which would result from a revision of the NFK. This requirement provided, incidentally, the immediate impetus for the formation of the FK3; (2) to establish a program and a plan for the observation of the necessary reference star positions at suitable transit circles. These were later published as the "Anhaltsterne" (AGK2A) and have been discussed above in Section 4.3.2; and (3) to expose the plates with astrographs of identical design at a small number of observatories and as contemporaneously as possible with the observation of the reference stars.

The *program list* was to contain stars of the Erste Abteilung of the AGK1, the "nonAG stars" in the Geschichte des Fixsternhimmels, and additional stars in areas with noticeably low star density. Altogether, a density of about 10 stars per square degree was aimed at. Originally, the Pulkovo Observatory was to take the plates covering the region between declinations +70° and +90°, the Bergedorf

Observatory that between $+70°$ and $+20°$ and the Bonn Observatory between $+20°$ and $-2°$. In 1937, however, the Pulkovo astronomers informed the commission that their part would be published by the Pulkovo Observatory (see Section 5.1.3.4). For this reason, the positions in the Pulkovo zone were duplicated by the Hamburg-Bergedorf Observatory, where, as if with remarkable foresight, the exposures had been extended to the pole when the photographs were obtained in 1929 and 1930.

The astrographs used for taking the plates have four lens Zeiss objectives of 110 mm effective aperture and a focal length of 2060 mm, so that one millimeter on the plates corresponds to 100″. The guiding telescope and the camera are mounted in the same tubes. The plates were squared on very carefully and special precautions were used to have every plate in the same position with respect to the objective, so that the coordinates of the tangential point on the plate could always be regarded as known and did not need to be determined from the adjustment to the references stars for every individual plate. In the re-reduction, to be described later (Section 5.1.4), the coordinates of the tangential points were solved for on every plate. The plates were arranged in such a way that a plate had the corners of four neighboring plates at its center, as in the Astrographic Catalogue (corner-in-center overlap pattern). Although the limiting magnitude was to be $9.^m0$, this must be regarded rather as the limit of completeness, as occasionally stars as faint as $12.^m0$ have been included. Altogether, the AGK2 records the mean positions of about 180,000 stars.

5.1.3.2 THE HAMBURG-BERGEDORF REGION ($+20°$ TO $+90°$). Most of the procedures followed in this region were also used in the Bonn and Pulkovo regions. The investigation of the *Hamburg AG Astrograph* was carried out by von der Heide (1937) and is also described in Vol. **1** of the AGK2.

The *radial distortion* of the field was carefully determined from ad hoc star photographs, by placing exposures of a pair of stars at a distance of about half-a-field width (that is, $2\frac{1}{2}°$) at various places on the plate. It was found that terms of the form Vx^3 and Vy^3 removed the effect of distortion from the measured coordinates. With x and y in millimeters, the V in the above terms was equal to $+0.''021 \pm 0.''003$ (rms. error). At a distance of 150 mm from the center, this leads to a distortion correction of $1.''52$. No account was taken of the possibility of decentering distortion.

The position of the *tangential point* on the plate, which is equivalent to the "*tilt*" of the plate, was regularly checked by the use of a collimating telescope (in the same way as in the Yale zones) and also by ad hoc exposures of displaced star pairs during the period in which the plates were exposed. Since all plates were taken almost perfectly centered on the meridian, the zenith distance alone practically determined the position of the camera since all exposures were made on the same side of the pier. As was explained in Section 2.3.3, the tilt of the plates is accounted for by terms $px^2 + qxy$ in x and $pxy + qy^2$ in y, where p and q are the coordinates of the tangential point with respect to the zero point of the coordinates. In the Hamburg zone, q showed a marked dependence on the zenith distance; however,

p and q were so small that the corresponding corrections were not applied except to plates taken during a specific interval of a few days.

The *measuring machines* had been specially designed and built in the machine shop of the Hamburg-Bergedorf Observatory. The construction is particularly sturdy and allows one to accurately rotate the plate carrier by 90°. There is only one lead screw and the integer millimeters of the measurements are displayed on a mechanical counter. In order to minimize the influence of vertical irregularities in the V-ways, the plates are located during measurement in the plane of the V (guiding) ways. The periodic and progressive errors of the lead screw were carefully investigated as were the errors of the V-ways, which for these machines are actually of cylindrical shape. The progressive errors of the screws (three machines altogether were built) were compensated by special mechanical devices. The errors produced by irregularities in the V-ways (cylinders) were under one micron. In this way, it was not necessary to correct the measures in any way for imperfections of the measuring machine. The temperature at the measuring machine was kept constant within about 1° C during the work. There were two pairs of crosshairs in the field of the eyepiece, separated by a half revolution of the lead screw. A star was measured by one setting each between both the pairs of parallel crosshairs. The rms. error of one such setting is 0.″14. A more detailed table is given in AGK2 Vol. **1**, p. E21, Table 17. Two measuring machines were kept busy at Bergedorf. The reversing prism was not used during the measurements for the catalogue. The plates were very carefully oriented, and measured in the manner indicated above, first in x, then rotated by 90° and measured in y. The reference stars were remeasured, with the plate still in the plate holder after the field stars had been measured, in order to check on changes of the situation of the plate in the machine. If changes were noticed, the whole plate was remeasured.

The AGK2A (Section 4.3.2) served as source for reference stars positions.

The second order terms of differential refraction as well as the distortion corrections were applied directly to the measured coordinates before the reduction. Six plate constants were determined on the assumption of an affine relationship of the type (2.14a) between the thus prereduced measured and the standard coordinates. Only for nine plates was it necessary to compute and apply tilt terms, that is, the coordinates p and q of the tangential point. *No* corrections for *magnitude equation, color magnification error or coma effect* were applied to the originally published coordinates.

The published spherical coordinates of the field stars were computed from the standard coordinates, after they had been computed from the measured ones with the reduction parameters determined through the adjustment. The rms. error, ϵ, of a catalogue position from the mean of two plates is given for both coordinates ($\alpha\cos\delta$) and δ by the formula:

$$\epsilon = 0.''145 + 0.''061(m_{pg} - 9.^{m}12)^2$$

where m_{pg} is the photographic magnitude of the star as given in the catalogue.

The volumes of the catalogue usually contain the positions in 5° wide belts, and are each arranged in declination zones of one degree width within which the stars are consecutively numbered. The second column gives the photographic magnitude as estimated on the plates themselves. The remaining columns give the mean spherical coordinates oriented to equator and equinox 1950 with the first two (from +90° to +70° the first three) terms of the precession power series, the number of images from which the catalogue position is derived, the epoch (minus 1900) and the BD number. A column with remarks ("Bemerkungen") contains information about ADS and BDS[2] numbers, names of variable stars, and so forth. In the "Bemerkungen," especially note that the positional data for stars with the remark "FK3-Ort" are simply copied from the FK3 and refer to the epoch 1950.0. Since these stars are generally bright, they often have large proper motions, and should, therefore, be excluded from investigations where AGK2 positions with a mean epoch of 1930 are required. Also, this feature makes it impossible to compare the catalogue directly with the FK3 (or the FK4). For the sake of uniformity, it would have been preferable to publish the positions of the FK3 stars which were derived from the plates, but perhaps the images of the brighter ones were too large to measure.

At the end of each one degree zone are the "Babelsberger Meridianbeobachtungen" (that is, positions observed at the Berlin-Babelsberg transit circle) which contain data in the same form as the main catalogues for stars which for some reason could not be measured on the plates—mainly close double stars.

For the stars within five degrees of the north celestial pole (1950.0), standard coordinates ($X = \cos\delta \cos\alpha \csc 1''$, $Y = \cos\delta \sin\alpha \csc 1''$) referred to this pole as tangential point are given together with the precession terms up to the third order.

After the publication of most of the AGK2 Hamburg-Bergedorf, special investigations showed that the positions are affected by systematic errors which had not been removed from the published positions. One of these is the *color magnitude equation* which is particularly serious for the Bonn positions. Diagrams for the removal of these errors are given on pages E3 (Hamburg-Bergedorf) and E4 (reduction from Bonn to Hamburg-Bergedorf) of Vol. **10** (Bergedorf-Bonn: +20° bis +25°) of the AGK2. Note that the signs in the lower diagram on p. E3 are in error and should be reversed for application. A color magnification error of the same nature as in most of the Yale zones was also discovered, but its effect on the published positions, being the mean of two positions, cannot normally exceed 0.″04. It is, however, systematic in areas of approximately $2\frac{1}{2}°$ by $2\frac{1}{2}°$, as one can easily see. Since the application of these corrections to the catalogued spherical coordinates requires one to know the plate coordinates of the contributing images, this error cannot be removed by the user. The problem has become somewhat academic, however, since the plate measurements for the AGK2 have been completely re-reduced with consideration given to all known systematic errors for the purpose of

---

2. Referring to the double star catalogues by Aitken and Burnham, respectively.

deriving the proper motions in the AGK3; see Section 5.1.4. *The printed volumes of the AGK2 have thus, in effect, become obsolete.*

The positions given in the zones +20° and +21°, and those in the southern half of zone +23° were obtained by combining the information from one Bergedorf and one Bonn plate each. The Bonn plates were reduced to the Bergedorf system by applying the corrections derived from the diagrams on p. E4 in Vol. **10** of the Bergedorf AGK2. The column "Bemerkungen" informs about the sources contributing to the positions.

Vol. **1** (+70° to +90°) of the Hamburg-Bergedorf zone of the AGK2 contains a detailed introduction to that zone, a description of the instruments, and a discussion of the various steps that led to the published positions.

5.1.3.3 THE BONN REGION ($-2\frac{1}{2}$° TO +20°). In general, the work followed the same pattern as at Hamburg-Bergedorf. A camera identical to that at Hamburg was mounted to the Bonn double astrograph, whose long focal length (about five meters) made it possible to guide the exposures very accurately.

The first tests of the objective system revealed a strong misalignment (decentering) of the components and the optics were, therefore, returned to the Zeiss works for realignment. Apparently, this was not very successful, since the adjustment residuals on the plates taken with this camera still reveal unmistakable evidence of uncompensated decentering distortion. Thus, the positions derived from the Bonn plates are affected by unremoved systematic errors which depend on the positions of the images on the plates. The scant data on p. (6) of Vol. **11** may make it possible to reconstruct these. The coordinates of the tangential point with respect to the geometric center of the plate were checked twice a week with a collimator device and found to be sufficiently constant. A description of the instruments and procedures used for the construction of the Bonn Region of the AGK2 was given by Kohlschütter (1957) in Vol. **11** of the Bonn AGK2, and also distributed as a separate reprint.

At Hamburg, the plates were exposed 10^m and 3^m throughout; the Bonn exposures generally lasted between 10^m and 20^m, sometimes even as long as 30^m.

The plate coordinates were measured on two machines. One, identical to the machines at Hamburg, was used to measure the northern half of the zone. An investigation of the errors of this machine proved them to be negligible. The coordinates for the southern half of the zone were measured on a machine manufactured by the Askania Werke following the same principles that guided the design of the Hamburg machines, but which was much sturdier than the Hamburg instruments. Its lead screw showed a progressive error which could not easily be compensated for—as it was, for instance, at Yale and at Hamburg—by an error-compensation curve on which an arm originating from the nut moves. The measurements made with the Askania machine were, therefore, corrected for this progressive screw error whose amplitude was 5 microns.

The image coordinates were measured by two settings each, separately in both coordinates. Between the settings, a dove prism, which was attached to the eyepiece, was turned by 90° so that the equivalent of a measurement of the plates in direct and reverse positions was achieved. The rms. error of a measurement based on two settings was 0.″08. This is an average value of all four measurers.

Since the adjustment model for the representation of the differences between the standard coordinates and the measured coordinates of the reference stars was chosen in the form of an affine transformation (equations [2.14a]) and thus involves six linear plate constants, the positions of the tangential point on the plate had to be carefully established.

Significant nonlinear terms in the relationship between measured plate coordinates and standard coordinates were generated only by refraction except, of course, by the previously mentioned decentering distortion which was not accounted for in the reduction model. Differential aberration produces no significant nonlinear effect on plates of this format ($5° \times 5°$) and, although there was evidence for cubic distortion of nonrotational symmetry (that is, with a different coefficient in x and y), no corrections were applied because most of its effect was absorbed by allowing for different scale coefficients in x and y.

It is interesting to note that the correction for the progressive errors of the screw were applied only after the least squares adjustment, and that they could have been derived much more accurately from the investigation of the adjustment residuals than by the method which was actually used. This is possible because the plates of neighboring declination belts were measured alternately in direct and in reverse positions, that is, those centered at 15° direct, those centered at $12\frac{1}{2}°$ and $17\frac{1}{2}°$ reverse, and so on. Corrections for the physiological magnitude equations of the measurers were small enough to be safely neglected.

The final adjustment involved the use of punched cards and Hollerith machines.

Because of the aforementioned residual misalignment of the objective components, the effect of systematic errors caused by (decentering) distortion remained in the measurements. They are tabulated on p. (26) of Vol. **11**, and could be removed by the user from the published coordinates only if the coordinates of the images on the plates were known.

As emphasized in Vol. **10** of the Hamburg-Bergedorf catalogue, the Bonn positions are strongly affected by a color equation which apparently does not depend on the position of the image on the plate, but does depend on the magnitude. This error was determined by comparison with the AGK2 Hamburg, the Yale zones, and the AGK2A. All these comparisons gave practically identical results, thus indicating that it is indeed the Bonn AGK2 which is responsible for the color equation. Corrections for this color effect were not applied to the published positions, but tables for their applications are given on p. (38) of Vol. **11** of the AGK2. However, A. Kohlschütter, who wrote the introduction to the Bonn AGK2, emphasized the preliminary character of these tables and promised better ones. This promise was realized by W. Dieckvoss who performed a comprehensive re-reduction of the AGK2 for the purpose of computing proper motions by comparison of the positions in the AGK2 with those in the AGK3, see Section 5.1.4.

Page 35 gives tables for the reduction of the positions in certain Yale zones to the FK3 system at 1930.0.

The accuracy of a catalogue position as the mean of measurements on two plates corresponds to rms. errors of 0.″155 in $\alpha\cos\delta$ and 0.″185 in δ, respectively.

The catalogues are arranged in a way similar to that in the Hamburg-Bergedorf section, except that the number of images contributing to a position is given under "Bemerkungen" only if it is different from 2, and in that the spectra are also given under "Bemerkungen."

5.1.3.4 THE PULKOVO REGION (+70° TO +90°). As mentioned above, the zone from +70° to +90° had originally been assigned to the Pulkovo Observatory near Leningrad. Although the Bergedorf Observatory also published positions in the Pulkovo zone, the Pulkovo astronomers went through with their plan and published their catalogue (Belyavsky 1947).

The procedures were practically identical to those employed at Bergedorf and at Bonn, except that the field size was only $4\frac{1}{2}$ degrees square, so that the centers of the plate belts were only 2° 15′ apart in declination.

The plates were measured in one position on an Askania long-screw measuring machine whose progressive screw errors were applied. Four settings were made in both coordinates on each star, two each in two different positions of the dove reversing prism.

The reductions were performed in a fashion almost identical to that at Hamburg and at Bonn. The rms. error of a catalogue position, however, amounted to 0.″28 in $\alpha\cos\delta$ and 0.″24 in δ, which is considerably higher than the corresponding numbers for the German AGK2 zones. The introduction to the Pulkovo catalogues gives no indication as to what might have been the reason for this. Belyavsky, however, reports that the lens gave satisfactory images only after repeated adjustments which were partly—and, apparently, unsuccessfully—performed at the Zeiss plant at Jena after the objective was returned there, and partly at Pulkovo. In spite of considerable effort which involved even the grinding of a new worm gear, a periodic error with an amplitude of about 6″ remained in the camera telescope drive.

It is possible that these circumstances are, at least in part, responsible for the low accuracy of this Pulkovo catalogue; it is, however, also possible that systematic errors of the same nature as in the Hamburg and the Bonn zones, which were not investigated at Pulkovo and might be considerably larger, have caused the inferior quality of the Pulkovo positions.

The catalogue gives serial number, photographic magnitudes, spectra (where known), positions and precession terms (up to the third order) for 1950, the epoch, BD number and remarks.

5.1.4 THE AGK3. Although the planned derivation of proper motions from comparison with the positions in the AGK1 was initially one of the main reasons for undertaking the AGK2, no proper motions were given in the AGK2. The reason for this is that the astronomers who determined the policy for carrying out the project felt that the AGK1 (which would *have* to be used as first epoch material) is neither sufficiently homogeneous, nor are its systematic errors known with sufficient accuracy to make the computation of proper motions worthwhile—especially since in some regions the proper motions had already been computed and published in the Yale zones. This was already pointed out in Section 4.4.2.

In the middle 1950's, the Director of the Hamburg-Bergedorf Observatory, O. Heckmann, initiated a plan for the repetition of the AGK2—to be called the AGK3—which is now finished and available on magnetic tape, and was printed in 1973. All the photographic work was carried out between 1959 and 1961 at Bergedorf with the same instrument that was used for exposing the AGK2 plates. The reference stars were observed contemporaneously at several transit circles. These observations were combined into the AGK3R (see Section 4.3.4).

After considerable counseling which included astrometrists from all over the world, it was resolved to carry out the original plan. The reference stars were to be observed on transit circles in several countries and collected at the U.S. Naval Observatory which was to compile the reference star catalogue under the direction of F. P. Scott. The plates were to be taken at Bergedorf contemporaneously with the meridian observations.

Most of the photographic and meridian work was done from 1959 to 1961. The plates were measured and reduced at Hamburg-Bergedorf. For the derivation of the AGK3 proper motions, the original measurements of the Bergedorf and Bonn zones of the AGK2, which were to serve as the first epoch material, were completely newly reduced. For this purpose, the stars of the AGK2A (see Section 4.3.2) served again as reference stars, after they had been reduced to the system of the FK4 by means of the tables in the FK4 (see Section 4.1.5).

Before the final reduction of the measured rectangular coordinates, they were corrected for all kinds of aberrations, so that the following model could be used for the conversion of measured $\{x, y\}$ to standard coordinates $\{\xi, \eta\}$ on all plates:

$$\xi = Ax + By + C + px^2 + qxy$$

$$\eta = Dx + Ey + F + pxy + qy^2. \tag{5.1}$$

This means that the coordinates of the tangential point were carried as unknowns in this re-reduction, while they had been regarded as known in the original reduction.

The *corrections applied to the measurements* were as follows:

(1) *Corrections for way and lead screw errors* of the measuring machines, which had originally been neglected, were eventually applied, but only to those finally derived original AGK2 positions at Bonn, which had been measured on the Askania measuring engine.

(2) Next, the measurements were corrected for the effects of *color magnitude equation.* This was done by means of a table which was constructed from the appropriately weighted contributions of three types of material:

(a) The plates for the AGK2 had all been taken with the camera on the same side of the pier, except for five which had been taken at lower culmination, so that a comparison of the positions on these plates with those obtained from plates taken in upper culmination yielded material for the determination of the color magnitude equation, that is, essentially, correction terms of the form $\Delta x(m, c)$ and $\Delta y(m, c)$.

(b) Further material for the determination of these correction terms was available through the comparison of the AGK2 and the Yale positions (Barney 1953 and 1954; see Section 5.2.6).

(c) Finally, from the residuals after a preliminary adjustment.

Furthermore, the measurements had to be corrected for a *color magnification error*, which had also been neglected in the original reductions. This was assumed to be proportional to the distance from, and in the direction to, the plate center. The material for its determination came from the investigation of systematic differences between positions derived from a selection of overlapping plates, and again from residuals after the preliminary adjustment. Corrections for radial distortion were applied to the Bonn as well as to the Bergedorf measurements. The measurements obtained at Bonn required a correction for *tangential* (*decentering*) *distortion*, which was derived from the residuals of the original adjustment and represented as a complete third order polynomial in both coordinates, similar to the zone corrections that were applied to the Cape zone measurements. Neither in Bonn nor in the Hamburg instruments was there any detectable coma effect. Finally, the quadratic terms caused by differential refraction were also removed.

The new reductions were carried out iteratively in the sense that the reference stars in the AGK2A were reduced exactly to the epoch of the plates, first by means of proper motions obtained by comparison with the AGK3R and other available sources, and then further improved by inclusion of the positions obtained from the AGK2-3 measurements themselves.

Through the new reduction, all known sources of systematic errors were removed from the AGK2 positions which were now available on the system of the FK4. The new reduction also succeeded in bringing the rms. error of a catalogue position (which typically depends on the measurements on two plates) down to $0.''15$. Details are available in the printed version of the AGK3.

The plates for the AGK3 were all taken with the AG astrograph at Hamburg-Bergedorf that was used for the photography of the Bergedorf zone of the AGK2. The plate centers were the same as for the AGK2; only this time, the plates were taken alternately east and west of the meridian (all in upper culmination). A coarse objective grating with a grating constant of about $3.^m5$ was used with these exposures

The measurements (all made at Hamburg) were subjected to a pre-reduction in a rather similar way to those on the AGK2 plates.

This means that the nonlinear effects of differential refraction (and for some plates, also aberration) were removed as were the way and lead screw errors of the measuring machine. The determination of the color magnitude equation for these plates was stronger than that for the re-reduced AGK2 positions because of the east-west of the pier arrangement of overlapping plates. The same color magnification corrections were applied that had been used in the re-reduction of the AGK2. An attempt to find a coma effect from the residuals of the preliminary adjustments failed; apparently it is not large enough to influence the positions significantly. Finally, the measurements were corrected for the effects of radial distortion, as were those for the establishment of the newly reduced AGK2.

The reduction model for the AGK3 was again that given by equation (5.1), where x and y are the measurements pre-reduced as just described.

The AGK3R served as the source of reference stars, and the reductions were again performed iteratively such that the reference stars were originally reduced to the epochs of the plates by means of proper motions which had been obtained by comparison with the AGK2A and other sources, and in later iterations by means of proper motions for the determination of which the AGK2 (newly reduced) and the AGK3 had also been used.

The plates used for the AGK3 project are not of the same excellent quality as those used for the AGK2. They have a higher background fog, and the images are not always as good. This caused the *rms. error of an AGK3 position* (typically dependent on measurements on two plates) to increase to about $0.''18$.

Up to the fall of 1973, the AGK3 existed only on magnetic tapes. Copies of these tapes are available to legitimate users through the Hamburg Observatory, and in the United States through the U.S. Naval Observatory. The AGK3 contains, besides the usual information (magnitudes, spectral types, BD numbers), the new positions whose epochs are around 1958, and the annual proper motions $\mu_\alpha \cos\delta$ and μ_δ (to $0.''001$) derived from direct comparison of the AGK3 positions with those in the newly reduced AGK2. The rms. errors of these proper motions are about $0.''009$/year. Since the epoch difference AGK3 minus AGK2 is also given, the newly reduced AGK2 positions on the FK4 system can be reconstructed from the AGK3 data; thus, the existing printed volume of the AGK2 have become obsolete. Note that the AGK3 proper motions (unlike the Yale proper motions; see Section 5.2) were in no way corrected by subjecting them to all kinds of conditions that restricted their systematic components, so that they are well suited to contribute to information concerning the improvement of the constant of precession, the motion of the equinox, secular parallaxes, and Oort's parameters of galactic shearing and galactic rotation (A and B, respectively), independent of any preconceived notions about galactic kinematics with which they have been made to conform.

A version of the AGK2-3, based on the measurements identical with those just described, was constructed (at this time only between $+40°$ and the north celestial pole) with the use of the overlap method (see Section 2.3.8.2) by P. Lacroute at Strasbourg, who has been provided with the raw material (that is, the measurements) by W. Dieckvoss since 1966. Apparently, the use of the overlap method at Strasbourg has produced, from the identical measurements, positions which in some regions deviate considerably (up to several tenths of a second) from those obtained at Hamburg. It is, at this time, too early to say which set of positions and proper motions is the better one. Should it be the set obtained at Strasbourg (and there are indications that this may well be so) it would be a good illustration of the superiority of the plate overlap method.

Tapes of the Strasbourg version of the AGK2-3 may be available to legitimate users upon application to the Strasbourg Observatory, at cost.

5.2 The Yale Catalogues

5.2.1 INTRODUCTION AND GENERAL DESCRIPTION. The pioneer of modern techniques in photographic astrometry, Frank Schlesinger, began in 1913 a series of experiments with the aim to use wide-angle photographic telescopes of moderate focal length for the establishment of a catalogue of star positions with essentially the same scope as the AGK1 described in Section 5.1.2.

To a man of Schlesinger's foresight the advantage of the photographic over the visual method of using transit circles was obvious. Not only can the accuracy of photographic positions be made much higher than that of relative meridian positions, but also the labor of obtaining them is considerably less.

Unlike the photographic catalogues produced by the observatories at Hamburg-Bergedorf and the Cape of Good Hope, the Yale catalogues are rather nonuniform and various groups of zones must be considered separately. Although this is definitely an inconvenience to the user, the value of the Yale Catalogues is unquestioned. Their nonuniformity can be explained by the fact that the Yale Catalogues are the oldest of this type. Better and more efficient techniques were applied as soon as they became available. Altogether five objective lens systems were used, and during the almost four decades that the work on the Yale catalogues has been progressing, the procedures were adapted to automatic processing whenever possible.

The introductions to the Yale Catalogues are singularly well suited to give the user an insight into various problems not only on photographic astrometric technique, but on astrometry in general. Due to the long time interval over which they were written, one may also learn much about the history of catalogue type photographic astrometry from studying them. They are a "must" on the reading list of every serious astrometrist.

Except for the polar caps, the plates were taken in "belts" which provided double coverage of the sky. All plates in one belt were centered at the same declination, with the right ascensions of the centers spaced in such a way that the vertical center line of a plate almost coincided with one of the vertical edges of one of its neighbors, and so that the lower edges of the plates in any one belt overlapped slightly with the upper edges of the plates in the belt immediately below (edge-in-center-line overlap pattern).

In what follows, volume numbers will refer to "Transactions of the Astronomical Observatory of Yale University" (Y. T.) unless stated otherwise.

After Schlesinger's death in 1942, Ida Barney continued the direction of the Yale catalogues. Since her retirement in 1959 Dorrit Hoffleit has been directing the work on these catalogues south of $-30°$. The Yale Catalogues were Dr. Barney's life's work, and are an impressive testimony to her industry, patience and good judgment.

Y. T., **23** (Barney 1951) is a "supplementary volume" to the Yale Zone Catalogues $-30°$ to $+30°$. When one uses any of these zones, attention should be paid to the following:

Section G of Y.T., **23** gives systematic differences between the positions and proper motions in the Yale catalogues and those in the GC, at the epoch of the Yale plates, for every hour and every degree for these zones published until then (up to Y.T., **22** which appeared in 1951). A pair of tables is assigned to each Yale zone. These tables are more accurate than those in the introduction to the individual zones as described below, and should be given preference in critical work. Pierce (1970) also gives corrections to the positions in the Yale catalogues for declinations between $-30°$ and $+30°$ from the analysis of Minor Planet Observations (see also Section 2.2.5.4). These corrections, however, do not refer to a commonly used fundamental system such as that of the FK4, but to "the equator and equinox defined by the geocentric solar ephemeris used to determine the computed positions of the minor planets: in this (that is, Pierce's) analysis, the equator and equinox are defined by Vol. XIV of the *Astronomical Papers of the American Ephemeris.*" If Pierce's corrections are to serve for the reduction of the Yale zone positions to another system, the relationship between this system and that defined in A.P.A.E., **14** must be known. In the sense of what was said in Section 2.2.5.4, Pierce's corrections may hopefully be construed as reducing the positions in the affected Yale zones to a system which is more self-consistent and closer to a true physical system than that of the FK4.

The second part of section G of Y.T., **23** gives tables with the systematic differences between the field stars in the regions where adjoining plate belts overlap. The fact that considerable systematic differences between the positions in these overlapping regions remain, which exceed $0.''5$ in one instance (Table 23, p. 99 in Y.T., **23**) in spite of an attempted reduction to the GC system, demonstrates that there can be systematic differences between the field stars and the reference stars on a photographic plate. Eichhorn (1960) has proposed a method for the reduction of photographic positions and proper motions which will not allow noticeable systematic differences between positions obtained from overlapping plates to remain (see also Section 2.3.8.2).

Section H of Y.T., **23** contains lists of misprints in Y.T., **9** through Y.T., **22.**

The table of $\Delta\delta_\delta$ for the belt from $+20°$ to $+30°$ should be replaced by the one given in Y.T., **23**, p. (10).

The proper motions of the fundamental stars which were not observed in the zones of the AGK1 but taken from the fundamental catalogues, and for which proper motions were therefore neither computed nor publ'shed in the original Yale catalogues, are given for the regions between $-30°$ and $+30°$ in Section E of Y. T., **23**.

The following table, adapted from Heckmann and Dieckvoss (1958) gives a survey over the Yale catalogues:

Table V-2

The Zones of the Yale Catalogu

Zone From	*To*	*Plate Epoch*	*Coord. System*	*Obj. System*	*Format of Field Measured*	*Meridian Circle Used for Ref. Stars*	*Reference System*	*rms.*	*Y.T. Vol*	*Re-mar*
+90°	+85°	1951	1950	5	10° × 10°	Washington 6-inch	6	0″.18	26/I	
+60°	+55°I	1916	1875	1	5 × 5	Leiden	1	0″.22	7	
+55°	+50°I	1916	1875	1	5 × 5	Lick	1	0″.24	4	
+60°	+55°II	1947	1950	4	10 × 11	Washington 6-inch	6	0″.13	27	
+55°	+50°II	1947	1950	4	10 × 11	Washington 6-inch	6	0″.13	26/II	
+30°	+25°	1928	1875	2	14 × 10	Lick	2	0″.20	9	
+25°	+20°	1928	1875	2	14 × 10	Lick	2	0″.20	10	
+30°	+25°r		1950				3		24	
+25°	+20°r		1950				3		25	
+20°	+15°	1940	1950	4	10 × 11	Greenwich	4	0″.17	18	
+15°	+10°	1940	1950	4	10 × 11		4	0″.17	19	
+10°	+9°	1940	1950	4	10 × 11	Greenwich	4	0″.17	22/II	
+9°	+5°	1937	1950	3	10 × 11	Greenwich, Cape	4	0″.17	22/I	
+5°	+1°	1937	1950	3	10 × 11	Greenwich, Cape	4	0″.17	20	
+1°	−2°II	1937	1950	3	10 × 11	Greenwich, Cape	4	0″.17	21	
+2°	+1°I	1914	1875	1	6 × 4	Lick	1	0″.24	3	
+1°	−2°	1914	1875	1	6 × 4	Lick	1	0″.23	5	
−2°	−6°	1933	1950	3	10 × 8	Cape	5	0″.16	17	
−6°	−10°	1933	1950	3	10 × 8	Cape	5	0″.16	16	
−10°	−14°	1933	1950	3	10 × 10	Washington 9-inch	2	0″.16	11	
−14°	−18°	1933	1950	3	10 × 10	Washington 9-inch	2	0″.16	12/I	
−18°	−20°	1933	1950	3	10 × 10	Washington 9-inch	2	0″.16	12/II	
−20°	−22°	1933	1950	3	10 × 10	Cape	5	0″.16	13/I	
−22°	−27°	1933	1950	3	10 × 10	Cape	5	0″.16	14	
−27°	−30°	1933	1950	3	10 × 10	Cape	5	0″.16	13/II	
−30°	−35°	1956	1950	5	10 × 10	Cape	3	0″.35	28	
−35°	−40°	1956	1950	5	10 × 10		3	0″.35	29	
−40°	−50°	1942	1950	3	10 × 10	Cape	7	0″.23	30	
−60°	−70°	1942	1950	3	10 × 10		3	0″.35		
−70°	−90°	1956	1950	5	11 × 11		3	0″.35	31	

Notes and remarks to Table V-2, listing the Yale catalogues.

NOTES:

In the column "objective system," 1 denotes the lens discussed in 5.2.2, 2 that in 5.2.3, 3 that in 5.2.4, 4 that in 5.2.5, and 5 that in 5.2.6.

In the column "reference system," 1 means the NFK (Section 4.1.3); 2 the Eichelberger system (Section 4.2.2); 3 the FK3 (Section 4.1.4.1); 4 that of the First Greenwich 1950 catalogue; 5 that of the Third Cape Catalogue 1925.0; 6 that of Publication of the U.S. Naval Observatory, 2nd Ser. Vol. **16**, Pt. 3; 7 indicates the system of the FK4. The rms. error refers to a position based on two plates, throughout.

REMARKS:

1. The old catalogues (used for the calculation of proper motions) were the AC Greenwich and the Hamburg, as well as the Pulkovo zones of the AGK2. The published positions are based on four plates; the quoted rms. error refers to a position based on two plates.

2. The positions published in these catalogues are not based on new plates, but the positions from the original zones were reduced to the FK3 and combined with those in the AGK2 (see Section 5.2.6). Proper motions were determined in case 2 by comparison with AGK1 Cambridge and the Greenwich 1910 catalogue; in case 3 the old catalogues were Berlin B and the AC Paris, reduced with the plate constants by Heckmann, Dieckvoss and Kox (1949, 1954).

3. New positions in Y. T., **23**.

4. New proper motions in Y. T., **23**.

5. Rms. errors estimated but possibly lower. Not yet published.

6. Preliminary catalogue in press. Accuracy refers to the mean of two plates. The actually published positions will depend on from one to twenty plates.

5.2.2 THE FIRST GROUP (Y. T., **3**, **4**, **5** and **7**). In the foreword to the first volume of the Yale zones to appear, Schlesinger (with Hudson, Jenkins and Barney 1926a) stated that they were eventually to cover the whole sky. This aim was never completely achieved. As Table V-2 indicates, the zones $+60°$ to $+85°$, $+30°$ to $+50°$ and $-60°$ to $-50°$ are lacking. The positions published in the volumes of this first group are definitely a pilot project. All zones covered were later repeated. The camera used has a relatively small usable field (about $6° \times 6°$), and the techniques used for the reduction of these three zones (the positions in Y. T., **3** and **5** were derived from the same plates) differ considerably from those applied for the reduction of the other zones, for which plate fields of $10° \times 10°$ were more typical.

The objective lens system designated by 1 was "an ordinary symmetrical doublet" whose four lens objective had an equivalent focal length of 1635 mm (scale: 1 mm = 126.″2), and an effective aperture of 72 mm. (The objective was stopped down from 100 mm in order to prevent vignetting.) The camera, which was clamped to the Thaw refractor of the Allegheny Observatory, had been made at the Allegheny Observatory machine shop, and had provisions for adjusting the objective lenses and squaring on the plates.

The plates were exposed for 25 minutes. The tangential point was determined experimentally with a collimating telescope. While this tangential point (which Schlesinger calls "the base") seemed relatively stable in the zones $-2°$ to $+2°$ and $+50°$ to $+55°$, there is evidence for its wandering in the zone $+55°$ to $+60°$. The one-lead-screw measuring machines were designed and manufactured to Schlesinger's specifications by Gaertner and Brashear, respectively, and allow one to measure a field of 18×18 cm². The plates were measured only in one position each in x and y, but neighboring plates in any one of the three belts were measured alternately in the direct and reverse positions, and referred to as "even" and "odd" plates, respectively. While thus on the "even" plates the measured coordinates in both x and y increase with the corresponding standard coordinates, the opposite is true for the "odd" plates. Since every "even" plate is overlapped by a pair of "odd" plates (and vice versa), the adjustment residuals may be tabulated against the screw (or V-way) readings, and then against the standard coordinates. In this way it becomes possible to separate, under certain conditions, the residual screw and V-way corrections from systematic errors in the positions that depend on the location within the field (and perhaps even other parameters), which include those caused by the aberrations of the lens. By the same token, the differences between the positions obtained from "odd" and "even" plates, respectively, may be tabulated against the screw readings, V-way readings or alternately the positions on the plate (that is, the standard coordinates) and other parameters such as magnitude and color equivalents. These differences also help determine the same quantities which the adjustment residuals of the reference stars determine. In these ways (that is, by considering adjustment residuals of reference stars and differences between the positions of the same stars obtained from overlapping plates), definitive screw and V-way corrections, deviations of the measured from the actual tangential point, the coefficients of cubic distortion and the coma effect (which Schlesinger calls "magnitude distortion"), a color term in the declinations of the equator zone (Y. T. **3** and **5**) and the coefficients of physically unexplained empirical terms XY^2 in x and Y^2 in y were found, although not always applied (see below).

The measuring machine was provided with a reversing prism (dove prism) at the eyepiece, which was reversed between two groups of two or three settings on one image, so that it approached the crosshairs from different directions. This was done to eliminate a possible psychological magnitude error of the measurer (the same procedure was followed at Bonn in the measurement of the AGK2 plates). All plate centers in any one zone were at the same declination, and the images of all stars whose positions were printed in the catalogue were measured on exactly two plates.

For measuring the zone $+55°$ to $+60°$ (Y. T., **7**) an important change was made in the measuring machine. The tube of the microscope was cut off just above the crosshairs, and the eyepiece was mounted on a separate bridge. This eliminates the danger of disturbing the alignment of the microscope during the work by bumping it with the nose, and so forth.

The periodic errors of the measuring screw were found to be negligibly small, and the corrections for the progressive errors of the lead screw and V-way errors were applied to the published positions.

The plate constants were determined by the six-constant method, that is, by assuming an affine relationship of the form (2.14a) between the standard and the measured coordinates. The *reference stars* for the equator zone and that from +50° to +55° were observed by Tucker, almost contemporaneously with the plates on the transit circle of the Lick Observatory; 602 for the equator zone providing about 10 comparison stars per plate, and 1070 for the zone from +50° to +55° providing for 23–24 comparison stars per plate. The 1033 reference stars for the zone + 55° to +60° were observed by Hins at the Leiden transit circle, providing an average of 29 comparison stars per plate in this zone. All comparison stars were about $8.^m5$, oriented to 1875 and supposed to be on the NFK system. The epoch of the Leiden stars is about 10 years later than that of the corresponding plates, so that for the reductions they were reduced to the epochs of the plates by means of proper motions that had been derived by comparing the Leiden positions with those in the AGK1. In the zones −2° to +2° and +50° to +55°, the comparison stars were restricted to the zones −2° to −1°, +1° to +2°, and +50° to +51°, +54° to +55°. Before the adjustment, the reference stars in the equator zone, which is at a large zenith distance at Pittsburgh, were corrected for differential refraction.

The additional corrections due to the progressive errors of the lead screw and those of the V-way were investigated as indicated above, and found to be mostly under one micron, but still applied. In all three zones, the various methods of comparing adjustment residuals and positions of the stars revealed a displacement of the tangential point ("tilt") from the experimentally determined one in right ascension, but only one in declination in the equator zone. While in the equatorial zone, and in that which covers +50° to +55°, this displacement appeared to remain essentially constant, it varied apparently significantly for some plates in the zone +55° to +60°. Corrections for this were applied to the right ascensions of the zone, and the declinations of the other two zones, with individually different values of the tilt coefficient for ten plates in the zone +55° to +60°.

Besides those for tilt, the following corrections were applied to the catalogued positions: in the equator zone, none. In the zone +50° to +55°, revised screw corrections. In the zone +55° to +60°, corrections for coma effect when it amounted to more than $\pm 0.''05$ in the published coordinate, which is the result of measurements of the plates.

In the equatorial zone, *proper motions* were not formed. In Y. T., **5**, differences, Yale minus AGK1 and the epoch intervals, are given instead, and Y. T., **3**, containing the zone +1° to +2°, contains no proper motion data at all. This was done because the accuracy of the Nicolayev AGK1 positions is definitely inferior to that of the Harvard ones, and also the epoch differences between Yale and Nicolayev are smaller.

The proper motions in the other two zones were determined from the raw proper motions obtained from comparison of the Yale positions with those in the AGK1. Under the assumption that the AGK1 positions are subjected to all kinds of systematic errors, in particular magnitude errors, the "crude" proper motions (that is, the raw differences Yale minus AGK1 divided by the epoch difference) especially the μ_α, were subjected to all kinds of adjustments before they were regarded definitive.

The stars were divided in groups of progressing magnitude and the average proper motions (taken over the right ascensions) were formed for these various magnitude groups. These came out rather significantly different for the different groups. Schlesinger postulates that there is no intrinsic reason why the right ascension averages of the μ_α of stars of varying magnitudes should be different; as a matter of fact, he postulates further that the right ascension averages of the μ_α should be zero for all magnitude groups.

The latter would be true only if the motion of the equinox (see Section 2.4.3.4) were zero. It is, in fact, not zero, and the corrections applied by Schlesinger make these Yale proper motions unfit for the determination of an inertial system. And although the postulate of zero (or, in any case, equal) right ascension averages of the proper motions for all magnitude groups is compatible with the first order theory of galactic rotation, it is not compatible with a rigorous theory of galactic kinematics that considers stars at considerable distances from the Sun. In any case, the adjustment of proper motions to fit preconceived notions as to how they should, on the average, behave makes them unfit for investigating the hypotheses upon which these motions are founded. One must admit, however, that the proper motions as published in these volumes of the Y. T. are superior, and individually more reliable than the crude, unreduced proper motions would have been, and represent perhaps an optimum of what could, under the circumstances, have been achieved.

Within the principles just explained, the proper motions were reduced to the PGC system by comparison with those common to the Yale catalogues and the PGC. For the μ_δ, this was accomplished simply by the addition of a constant within a certain right ascension interval.

In any case, the accuracy of these proper motions is rather low, corresponding to an rms. error of about 0.″02/yr.

The columns of the catalogues in Y. T., **4**, **5** and **7** contain: AGK1 number, magnitude (from AGK1), spectrum (H.D.), right ascension (oriented to 1875.0), precession and secular variation from the AGK1; declination, its AGK1 precession and secular variation, proper motion in right ascension and declination, and BD number. A minus (plus) sign in front of a BD number means that the BD zone is one degree less (more) than the degrees in the declination of the star. Stars with an asterisk have footnotes concerning them at the bottom of the page, where the epoch of the plates is also given. The footnotes refer mostly to PGC numbers, and double star data.

The following relates to the zones published in Y. T., **3**, **4** and **5**: Table 19 on p. 28 of Y. T., **5** contains reduction tables of the $+1°$ to $-2°$ zone to the PGC system. The first zone $+1°$ to $+2°$ was measured off the plates $+1°$ to $-2°$, and published in Y. T., **3** (Schlesinger et al. 1926), but since the Albany AGK1 zone was to be duplicated later anyway, only the AGK1 number, the magnitude, spectrum and the 1875 positions are listed. Epochs and remarks are at the bottom of the pages.

The zones $+50°$ to $+55°$ I and $+55°$ to $+60°$ I contain 436 stars in common which were used for the determination of systematic differences, which reach $0.^{s}027$ in α and $0.''6$ in δ, although both the Leiden and the Lick reference systems were supposed to be on the NFK system.

The *accuracies of the positions* were estimated in several ways: from the adjustment residuals, from comparison with the PGC positions and from the differences between the positions of the same stars from different plates. Generally, there was a tendency for the positions of bright and faint stars to be less accurate than for those of average magnitude. For the equatorial zone, for instance, Schlesinger estimates the accuracy of a catalogue position as in Table V-3.

Table V-3 **Accuracy of Yale Catalogue Positions in the Equatorial Zone**

mag.	<7.0	7.0–7.9	8.0–9.0	>9.0
rms. error	$0.''27$	$0.''23$	$0.''22$	$0.''23$

The figures in Table V-2 are averages. The experience gained in the measurement and reduction of the zones is reflected in that of zone $+55°$ to $+60°$ being the most accurate of the older Yale Catalogues. This may be to a large extent due to the circumstances that all measurements were carried out by the same person.

5.2.3 THE SECOND GROUP (Y. T., **9** AND **10**). For the zones $+20°$ to $+25°$ I and $+25°$ to $+30°$ I, a camera exposing a considerably larger field (10° in declination by 14° in R.A.) was used. These two zones were thus derived from measurements on the same plates and reduced simultaneously, although they were measured separately by two measurers. There was a small overlap between the measurements in order to check for systematic differences between the two measurers, but none were found that were significant.

The main reason for the change of format was the consideration that the density of reference stars required to give plate constants of adequate accuracy would be considerably lower if the plate format were enlarged. In this way it was hoped that it might become unnecessary to have special meridian observations made for the reference stars; they might even be extracted from existing catalogues, expecially the GC. This particular catalogue, however, was never used as a source of reference stars for any Yale Zone. For the two zones under consideration, the comparison stars were especially observed at the Lick transit circle.

The *camera* with which the plates for these zones were taken was located on the Yale Observatory grounds. It has a Ross objective with a focal length of 2025mm, which corresponds to a scale of 102″/mm. The effective aperture was about five inches. A guiding telescope was mounted parallel to, and was of about the same focal length as, the camera.

The format was chosen so as to make the measured field 10°20′ by 1^h at the equator. The plates used had a format of 48cm × 58cm and therefore provided an ample marginal region in which no stars were measured.

The exposures were made on plates 6mm thick which had been especially selected for planeness within 0.0001 inch. However, not all plates satisfied this rigorous criterion.

This is the first instance that an objective grating was used; the purpose was to provide means for including the very bright stars by measuring their diffraction images instead of their central images. The first order diffraction spectra are 3.4 magnitudes fainter than the central images, and almost perfectly round. With the grating, exposures of twelve minutes showed practically all stars that are listed in the corresponding zones of the AGK1, that is, the Cambridge (England) and Berlin B Zones (see Table V-1). The plates were all taken in such a way that the time of mid-exposure differed by less than two minutes from the time of the meridian passage of the plate center.

A special, large measuring machine had to be designed and manufactured for this zone to accommodate the large plates. It is still in use at the Yale Observatory station at El Leoncito. The screw has an effective length of 560mm, and the V-ways permit the carriage to be shifted up and down by the same amount.

With a measuring machine as large as this, it would have been very inconvenient if the eyepiece were moving. The light, after passing the measuring microscope (which is moved in the horizontal direction by the lead screw) is deflected to the left, collimated and directed toward a second collimator which is connected to a fixed eyepiece. The observer thus never needs to change his position during the measurements.

During the measurement of any plate, the temperature never varied by more than 5 degrees Fahrenheit. Due to the large number of stars on a plate (800 on the average) the plate had to stay in the machine for about three days before the measurements of one coordinate of all stars on it could be finished. The abovementioned narrow temperature range minimized the danger of large relative shifts between plate and measuring machine due to temperature variations; furthermore, six control points were chosen on every plate which were measured at least every hour. These points had been splashed on the film of the plate with India ink and were perfectly round. The measurements made during a sitting were rejected whenever the shift between the plate and the measuring machine exceeded one micron, which corresponds to 0.″1. This happened only in 2.5% of all sittings. These shifts were in no apparent correlation with the temperature changes during the sitting; this illustrates that a thermostatic control of the temperature in the measuring room would have been of little or no value. The odd-even scheme for the measurements (one plate direct, its neighbor reverse) was left unaltered.

On plates of this size, the experimental determination of the position of the tangential point on the plates could not have been performed with the necessary accuracy. The measurements (x, y) were therefore adjusted to the reference stars essentially by the formulas

$$\begin{aligned} \xi &= Ax + By + C + px^2 + qxy \\ \eta &= A'x + B'y + C' + p'xy + q'y^2 \end{aligned} \tag{5.1}$$

which are basically the same as equations (2.13).

The adjustment was actually performed not for standard coordinates but as follows. From the available AGK1 coordinates of the stars to be measured, approximate standard coordinates were computed, and, by multiplication with the scale factor, converted to expected measured coordinates. The left-hand sides of equations (5.1) were, in fact, the actually measured minus predicted measurement coordinates of the reference stars, called ΔX and ΔY.

This also greatly facilitated the computation of the final spherical coordinates of the field stars. Since predicted measured coordinates had been computed for the field stars from their AGK1 positions, formulas (5.1), after the determination of the plate constants through the adjustment, yielded "computed" differences between predicted and (ideal) measured coordinates of the field stars. These differences were converted to corrections to the α and δ, from which the predictions were made, by the simple formulas $\Delta\alpha = \Delta X \sec\delta$, $\Delta\delta = \Delta Y$. Only when ΔX and ΔY exceeded certain limits was it necessary to add second order terms.

On four plates, terms of the form x^3 and xy^2 in x, and y^3 and x^2y in y had to be additionally included in the adjustment model in order to produce reasonably small residuals. Without these terms, systematic residuals of up to 0.″6 would have remained on these plates.

No pre-adjustment corrections were applied either for differential refraction or for differential aberration; it is shown that the maximum error committed thereby is only 0.″04 and thus negligible. The periodic errors of the screw were below one-half micron; no corrections for these were applied. No corrections for the non-straightness of the V-ways were found necessary. Progressive errors of the measuring screw are certainly present, but the form of the error curve was so close to a parabola that the errors could be regarded as absorbed by the quadratic terms with the coefficients p and q' in the plate constants. (Equations [5.1].)

The objective used for taking the plates had considerable optical distortion; it was determined separately and applied to the comparison stars before the least squares adjustment.

The catalogue of the comparison stars was published in the Lick Observatory Bulletin No. 436 (Jeffers 1932).

The comparison stars were determined ad hoc; their (visual) magnitudes range from 8^m to $8.^m5$. In this and later zones, Schlesinger abandoned the scheme which he had followed in the former zones, namely, to employ comparison stars only in belts of 1° width at the northern and southern edges of the plates; the comparison stars for this zone are distributed uniformly over all declinations.

The corrections to the star positions, that is, the application of the reduction to standard coordinates which were found in the adjustments and the corrections due to distortion, were performed by contour line graphs. These gave accurate and reliable results faster than numerical computations would have given.

The magnitude dependent corrections could be determined very well by means of the diffraction spectra produced by the objective grating. For 1566 stars, the positions of the central images as well as those of the first order diffraction spectra were measured. The coma term coefficient (which was assumed to be the same for all plates of the series) was determined following the principles outlined above in Section 2.3.8.1. Schlesinger, who treated the problem graphically, was apparently the first one to use the grating method for the determination of magnitude dependent systematic errors affecting photographic star positions. He assumed Δx to be of the form: $\alpha(m - m_0) + \beta x(m - m_0)$, where x is the x-coordinate of the image with respect to the plate center, and an analogous expression for Δy. The observed differences in the x and y coordinates between a central image and the mean position of a corresponding pair of first order diffraction spectra produced by the same star were plotted against the x and y coordinates, respectively. This plot showed that the bright stars were displaced outward with respect to the faint ones by an amount proportional to x and to y, respectively. The amount thus obtained, namely $0.''00090$ per millimeter and magnitude in x and y, was also confirmed by a comparison with the reference stars. It was also found that the "source," that is, the point on which this coma effect (called *magnitude distortion* by Schlesinger) is zero, had the average coordinates -90mm in x and $+70$mm in y, rather far from the geometrical plate center. The corrections for the coma effect were applied to the stars after the adjustment to the reference stars. A somewhat higher accuracy for the plate constants could have been achieved if the coma correction has been applied before the adjustments; systematic errors are, however, not to be expected because the range of the magnitudes of the reference stars is only one magnitude.

The residuals of the reference star positions were again carefully investigated after the adjustment. When arranged according to the position with respect to the measuring screw, they revealed no systematic trend reaching even one micron. The treatment of the progressive screw errors as of parabolic shape and the assumption they they are accounted for by the particular model (5.1) for the plate constants was thus justified a posteriori.

When Schlesinger investigated the dependence of the residuals on the positions of the stars on the plates, he assumed that whatever systematic residuals remained after the primary adjustment between the measured and the computed positions of the reference stars would be identical for all plates. They were, therefore, collected from all plates and subjected to an adjustment in x for the additional terms

xy^2, x^2y, and y^2; and in y only an additional term xy^2. Next, the positions of the same star as obtained from two overlapping plates were compared in the sense: position from preceding minus position from following plate. These differences confirmed the systematic terms obtained from the investigation of the adjustment residuals. The small empirical terms of the third order, corrections for which were applied to the means of the coordinates from two overlapping plates by a table, cannot be readily explained. They might be due to the deviation of the form of the plates from a plane. One could argue that the plates are curved into a shape obtained by an overlap of a cylinder with a sphere; when measuring the curvature of the plates, it was found that they were predominantly concave when viewed from the film side. Schlesinger explains this by assuming that the gelatine film shrinks as it dries, and pulls the glass with it. It is, however, doubtful whether this is the correct explanation, because the same empirical terms occurred in the zones of the first group (see Section 5.2.1) with plates of only one-third of the linear dimensions of those used in the zones discussed in the present section. This demonstrates the importance of carefully investigating the residuals for any systematic trends which may not have been taken out by the model used for the relationship between measured and standard coordinates.

The curvatures of the plates were confirmed by measurements with a spherometer. From the next group on, the plates were kept flat during the exposure by pressing the back side of the plate against a flat and rigid metal plate during the exposure.

A comparison of the Yale positions in these zones with the PGC positions established a measure for the relative weights of central images and diffraction spectra. The relative weights of the central images as compared with the mean of the diffraction spectra were found. One of the results was that the central images of stars brighter than 6.0 photographic magnitude are unsuited for deriving accurate positions.

The comparison with the PGC revealed also systematic differences in the declinations (actually a term in δ^2), which were supposed to have been reduced to the PGC by application of the systematic differences between the Eichelberger and PGC systems. These systematic differences are supposedly due to failure to apply circle division corrections to the observations at the Lick transit circle, but this is not a very likely explanation.

The Tables 8 and 15 on pp. 23 and 24, respectively, of Y. T., **9** and 5, on p. 4 of Y. T., **10** will convert the positions to the system of the PGC. The proper motions were obtained by comparison with the positions in the Cambridge (England) and the Berlin B zones of the AGK1.

The "crude" right ascension proper motions, formed simply by subtracting the AGK1 positions from the Yale positions and dividing through the epoch difference, were again adjusted by subjecting them to the condition that the sum of their hourly averages in right ascension must be zero. This sum had different values for different magnitude groups. Schlesinger attributes this to the magnitude equations of the first epoch (AGK1) catalogue. This assumption agrees indeed well with the magnitude equations determined previously by Auwers. The μ_δ in Y. T., **9** were derived from the crude proper motions after application of Auwers' magnitude equation in δ to the Cambridge AGK1. The μ_δ were not so corrected for the zone in Y. T., **10** which depends on Berlin B for its proper motions. Indeed, the μ_δ should vary for stars of different magnitudes because of the variations of the effect of the solar drift.

Schlesinger expresses the opinion that the procedure used for the derivation of the μ_α insures that these proper motions refer to Struve's constant of precession. It was already pointed out that this is not so, since the sum of the hourly means of the proper motions would come out to be zero only for the true, inertially-correct, constant of precession and if there was no motion of the equinox (see Section 2.4.3.2). Also, the removal of the systematic differences between these sums for stars belonging to different magnitude groups imposes on these a certain hypothesis with regard to stellar kinematics, and thus reduces their value for investigations into stellar kinematics. There still remain systematic differences between the printed proper motions and those of the PGC. These are, for the zone between $+25°$ and $+30°$, given on p. (33) of Y. T., **9** and p. (3) of Y. T., **10**, respectively, and could be used for converting the proper motions to the system of the PGC.

The arrangement of the catalogue is essentially the same as that for the zones previously published. The positions were reduced to 1875.0 with Struve's value for the constant of precession, and, as for all of the Yale catalogues, are valid for the epochs of the plates. Not all the stars in the Cambridge AGK1 are repeated in the Yale zone, since the Cambridge AGK1 contains stars down to $9.^m2$ in the BD system, as compared to $9.^m0$ of the other AGK1 zones. Stars with an "A" after the number in Y. T., **10** refer to the AGK1 zone Berlin A. Numbers followed by "c" denote stars which are not in the AGK1. A "b" after the number indicates that central image and diffraction spectra were used to compute the positions, an "s" indicates that only the first order diffraction spectra were used for deriving the position. The numbers are the AGK1 numbers; an "a" following them refers to stars in the appendix of the Cambridge zone. An asterisk after the number indicates the presence of a note concerning this star at the foot of the page. The epoch is also at the foot of the page.

The photographic magnitudes were especially determined by J. Schilt. Those spectra which were not previously available were also determined ad hoc. The precessions and secular variations were copied from the AGK1 and thus are based on Struve's constant.

5.2.4 THE THIRD GROUP (Y. T., **11** THROUGH **22**, PT. I). The belt photographed after the one from $+20°$ to $+30°$ was that from $-10°$ to $-20°$. The overlap for every Yale zone (except that from $-70°$ to $-90°$) was one-half in right ascension and followed the edge-in-center-line pattern. The plates were measured in such a way that the catalogued position of every star depended on exactly two pairs of measured coordinates. The camera has a Ross objective of 5 inches effective opening and was mounted on the 26 inch refractor at Johannesburg in South Africa.

This objective gave better images than the one used at New Haven; its focal length was 2067.4mm, corresponding to 99.″77/mm. During the exposure, the plates, which were only 43cm square, were pressed against a very flat and rigid duraluminum plate on all four edges. Spherometric measurements have shown that this actually flattened the plates whenever they were concave at the coated side, which is almost always the case. An area of approximately 10°.5 × 10°.5 was measured on each plate. This plate format was chosen because the larger plates apparently proved too cumbersome to handle with ease. Furthermore, in view of the difficulties encountered in the zone from +20° to +30°, only excellently flat glass sheets were selected to be cut to 43cm by 43cm, and coated with film. No change was made in the layout of the measurements as compared to the first two groups, as described in Sections 5.2.2 and 5.2.3. The measurers were Miss Barney and Miss Jenkins who worked the same plates from different edges toward the center line in both coordinates. The images were measured strictly in the order of increasing coordinates.

Although Schlesinger considered it to be one of the advantages of the large plates that it is not necessary to make special observations for the reference stars, the 1117 reference stars for the zone from −20° to −10° (giving an average of 35 evenly distributed reference stars per plate) were observed at the U.S. Naval Observatory (Morgan and Lyons 1938), and are ostensibly on Eichelberger's fundamental system.

The reductions were carried out very much in the same way as for the previous zones. Cubic distortion was determined and applied to all measurements previous to the adjustment. Here again, the differences between the standard coordinates of the reference stars computed from the catalogue positions in the scale of millimeters on the plate, and their measured coordinates (corrected for distortion) were adjusted according to the model (5.1). The distortion corrections and reduction to the system of the reference stars were again applied to the field stars by means of contour diagrams, two of which had to be drawn for every plate, one for each coordinate.

For this objective, a dependence of an effective focal length on spectral type, that is, a color magnification error, was found. The corrections per 100 mm from the plate center in each coordinate were: −0.″05 for B and A, 0.″0 for F, +0.″05 for G, +0.″13 for K and +0.″18 for M stars. These corrections were applied after the least squares solutions for the plate constants had been performed.

The *coma effect* was determined by means of an objective grating (grating constant $3.^m4$), and independently by a comparison of the Yale positions with those in the GC. In this connection, it is interesting to look at Table 3 on p. (10) of Y. T., **11**. From this table it is quite apparent that the coma effect depends on the spectral type and is different in the x and the y coordinate; in the latter, it even changes sign from the B to the M stars. Analytically, the corrections to zero magnitude would be satisfactorily represented by terms of the form $(m - m_0)(\alpha + \beta c x)$ where c is a color equivalent. The corrections, however, almost never amount to over 0.″1 and influence the mean of two positions from overlapping plates only very little; so that in this group of zones, corrections for the coma effect were not applied.

Residual screw corrections (that is, exceeding those of the parabolic shape which were automatically absorbed by the plate constants) were determined as before, and again found insignificant and too small to be worth applying. Empirical terms were again determined by investigating the adjustment residuals of the reference stars over all plates in the belt, in dependence on the position of the star on the plate. In this, as well as in all other zones of this group, the adjustment residuals were averaged and collected in a double entry table against the coordinates of the stars in the plate. These residuals showed a systematic pattern. A model was then chosen to represent these systematic residuals as polynomials of the third order in x and y, the measured coordinates of the images on the plates. In every case, however, only those terms were carried in the model which were judged necessary to absorb most of the systematic trends displayed in the residuals. These models varied quite a bit from zone to zone (or rather plate belt to plate belt). Particulars are found in the introductions to the various zones concerned.

The proper motions were handled essentially as described before (Section 5.2.3). The basic principle was that the average of the hourly means of the μ_α was to be zero for all magnitude groups. Sometimes the $\Delta\alpha_\delta$ as given in the GC.I were applied to the first epoch positions, sometimes they were not. The treatment of the μ_δ was somewhat nonuniform. Corrections to the first epoch (usually AGK1) positions found by previous investigators (Auwers, GC.I) sometimes were and sometimes were not applied depending on whether corresponding trends appeared in the derived crude proper motions.

Since Struve's constant of precession was used to orient the first epoch positions to the coordinate system 1900, one may generally assume that the μ_δ in the Yale zones refer to Struve's constant of precession. As mentioned before, this is not true for the μ_α due to the particular way in which they were adjusted. For these reasons, one will have to study the introductions to the various Yale zones very carefully before the proper motions printed in them are used to investigate solar motion, galactic shearing, galactic rotation, corrections to the constant of precession and the motion of the equinox. The realization of all these handicaps is possibly what kept the German astronomers from computing proper motions from a comparison AGK2–AGK1 and including them in the AGK2, although the AGK2 had originally been initiated for this very purpose.

Tables 9 and 10 on pp. (16) and (17), respectively, give the reductions of the proper motions and positions in Y. T., **11** to the GC system. Better values are given later in Y. T., **23**. The arrangement of the catalogue is the same as before, with peculiarities explained on pp. (17) and (18) of Y. T., **11**.

Reduction tables for the proper motions and positions to the GC system, for the data in Y. T., **12** (zone $-14°$ to $-18°$) are there given on pp. (5) and (6), Pt. 1, Tables 3 and 4, respectively; for the zone $-18°$ to $-20°$ on pp. (6) and (7) of Y. T., **12**, Pt. 2.

The results of the measurements from the plates covering the declinations from $-20°$ to $-30°$ were published in Y. T., **13** and **14**. The reference stars were taken from the Third Cape Catalogue, the mean epoch of which differs by less than a year from that of the plates for this zone. The declinations of this catalogue form their own system, and the corrections to the right ascensions were applied to free them from the known periodic errors of the GC positions. Otherwise, the reductions were carried out exactly as for the plates covering the belt from $-10°$ to $-20°$, although the empirical terms in this zone are different. Proper motions were computed in the usual manner. Tables for the reduction of the positions and proper motions to the system of the GC are given on pp. (11) and (12), Tables 7 and 8 in Y. T., **13**, Pt. I; on pp. (5) and (6), Tables 3 and 4 in Y. T., **13**, Pt. 2, and on pp. (5) and (6), Tables 3 and 4 in Y. T., **14**.

The belt of plates to be measured next was centered on $-6°$. However, the measurements covered only four degrees in declination on either side of this center. The comparison stars were again taken from the Third Cape Catalogue 1925.0 as for the zone discussed above, in addition to 51 stars from other sources which were also chosen as comparison stars. The reductions followed essentially the same pattern as in the previous zones in this category; empirical terms, equal for all plates of this zone, were applied whose coefficients differed from those used in the previous zones. Proper motions were again computed by comparison with corrected AGK1 positions, but proper motions were originally not computed for stars for which there were no spectra available. The missing spectra were later determined and supplied by A. N. Vyssotsky. The positions of the stars for which no spectra were known at the time of the compilation of Y. T., **16** and **17** were recomputed and published together with their proper motions in sections A (for Y. T., **16**) and B (for Y. T., **17**) of Y. T., **23**. The positions published in these sections supersede those published originally in Y. T., **16** and **17**. All work concerning precessions was still performed with the use of Struve's constant.

The reduction to the GC system can be accomplished by Tables 7 and 8 on p. (12) of Y. T., **16**, and Table 3 of p. (6) of Y. T., **17** for the proper motions, and by Table 8 on p. (13) of Y. T., **16** and Table 4 on p. (7) of Y. T., **17** for the positions. The arrangement of the catalogue is the same as for the zones published previously.

The next belt of plates was centered on declinations $+3\frac{1}{2}°$, and a belt 11° wide was measured, covering declinations from $-2°$ to $+9°$. The positions in the belt from $-2°$ to $+1°$ were published in Y. T., **21**; those in the belt from $+1°$ to $+5°$ in Y. T., **20** which also contains the discussion of the methods, and those between $+5°$ and $+9°$ in Y. T., **22**, Pt. I. The reference star positions were taken partly from the First Greenwich Catalogue of Stars for 1950, and partly from a list of stars observed at the Cape and reduced to the Greenwich system by means of GC stars. The plates were reduced as in the other previously discussed zones, that is, cubic distortion and a correction for the color magnification error were applied. In the zone from $-2°$ to $+9°$, however, these corrections were applied before the least squares adjustment to the system of the reference stars was carried out. The calculations were to a large extent performed at the Watson Computing Laboratory of Columbia University. Empirical corrections, equal for all plates in this belt, were applied as in the other belts.

The tables for the reduction to the GC are found in Y. T., **23**; and in addition, for the zone between $+1°$ and $+5°$ on p. (13) of Y. T., **20**, Tables 7 and 8; for the zone between $+5°$ and $+9°$ on p. (8) of Y. T., **22**, Tables 3 and 4, and for the positions in the zone $-2°$ to $+1°$ on p. (6) of Y. T., **21**, Table 1. Table 1 also contains the quantities necessary for the reduction of the positions in the original Yale zones $-2°$ to $+1°$ I (Y. T., **5**) to the GC system. The proper motions published for this new zone $-2°$ to $+1°$ II were formed by combining the proper motions $-2°$ to $1°$ (new) minus Nicolayev and $-2°$ to $+1°$ II minus $-2°$ to $+1°$ (old) in the ratio 3:7, after the proper motions formed in both ways indicated above had been reduced to the GC system. The proper motions printed in this volume can, therefore, be regarded as being on the GC system; this is confirmed by Table 7 on p. (11) of Y. T., **21**, which gives a comparison of the Yale with the GC proper motions.

The proper motions of the stars between $+1°$ to $+2°$ that are included in Y. T., **20** (which covers the zone between $+1°$ and $+5°$), and which were originally computed only by comparison with the AGK1), have been recomputed utilizing the $+1°$ to $+2°$ (old), in the same manner as those published in Y. T., **21** for the zone $+1°$ to $-2°$ (new), reduced to the GC system, and are published as section F of Y. T., **23**. This reduces the rms. error of an annual proper motion in this zone to $0.''006$.

The catalogues derived from this belt of plates are arranged in the usual manner, that is, they give AGK1 numbers, AGK1 visual magnitudes, spectra, right ascension (to $0.^s001$) and the precession terms I and II computed from Newcomb's constant of precession, analogous data for the declinations (to $0.''01$), the annual proper motions in right ascension and declination to $0.''001$ and their probable errors, as well as the BD (or, when applicable, CoD) numbers, and notes as explained above. The epoch of observation is always given at the bottom of the page. The AGK1 numbers of the stars which served as comparison stars are given in appendices.

5.2.5 THE ZONES COMPUTED FROM PLATES TAKEN WITH CAMERA NO. 4 (Y. T., **18**, **19**, **22**/II, **26**/II, **27**; $+9°$ TO $+20°$ AND $+50°$ to $+60°$ II).

These belts of plates were taken at the Yale Observatory with a camera that was newly installed in 1936. It has a Ross lens of focal length 2063mm, corresponding to a scale of $99.''98$/mm, and no diaphragm between any of its components. The exposures lasted nine minutes on the average. The same procedures as in the photographic work on the previous zones were followed. The reference stars were taken from the First Greenwich Catalogue of Stars for 1950, utilizing an average of 70 reference stars per plate.

Radial (cubic) distortion, coma effect and the color magnification error had to be newly determined for this camera. Only the last of these proved significant;

the corrections for this effect are, for each 100mm distance from the center in each coordinate, $+0.''16$ for B and A stars, $+0.''08$ for F, $-0.''04$ for G and $-0.''16$ for K and M stars. These corrections were applied to the reference star positions before the least squares solutions for the plate constants were carried out. These were performed entirely in the usual manner, that is, following the model (5.1).As before, any errors in the measuring machine were assumed to have been absorbed by the plate constants and were not applied.

Empirical corrections were applied as in the previous zones. Differential refraction could be identified as the cause for the x^2 term in y, which appeared because the zenith distance for this belt (26.°8 on the average) is greater than for any of the others previously taken.

The proper motions were determined in the same way as in previous zones, by using the AGK1 zones as first epoch material after empirical magnitude corrections had been determined for them.

Besides in Y. T., **23**, the tables for the reduction of the positions and proper motions to the GC system can be found for zone +15° to +20° in Y. T., **18**, pp. (13) and (14), Tables 7 and 8; for zone +10° to +15° in Y. T., **19**, p. 8, Tables 3 and 4, and for zone +9° to +10° in Y. T., **22**/II, p. (9), Tables 5 and 6. Vyssotsky's Leander McCormick spectra, used for the determination of the color magnification correction in the volumes giving the results from the belt from +10° to +20° and which were not printed in the original catalogues, are given in section C (for Y. T., **18**) and section D (for Y. T., **19**) of Y. T., **23**.

The measuring and reduction procedures underwent rather radical changes in the zones arising from the belt covering the declinations between +50° and +60°. Although the plates were obtained in the same way as for earlier zones, they were measured on an automatic measuring machine, and the reductions performed on an electronic computer (IBM 650).

The measuring machine was the fully automatic one lead screw machine at the Watson Scientific Computing Laboratory of Columbia University described by Eckert and Jones (1954). In order to work, the machine required that the approximate positions of the stars to be measured be prepunched on cards (with extra cards where it was expected that grating images would be measured). The images were thus automatically brought into the field of view. The measurer then inspected the image and visually made note of any irregularities (oval shape, grainy background, and so on), then moved the image fairly precisely into the center of the field. The measuring principle was the following: the field containing the image was scanned by a rotating circular disc, a sector of which was light sensitive. The measuring screw was then automatically adjusted in such a way that the amount of light falling on the sector was always the same for positions 180° apart, that is, the apex of the sector was made to coincide with the center of blackening of the image. The coordinate thus measured was then punched on the card with a least count of 0.2 micron, and the coordinate, perpendicular to that being measured precisely, with an accuracy of 0.1 millimeter, was punched on the card mainly for identification purposes. The measuring process left enough control to the operator to allow him to insert extra cards if a star proved to be a double star or when an extra image needed to be measured for some other reason.

The corrections for progressive errors of the screw and those for guiding way errors were determined and applied to the measurements. They are given in Table 2 of Y. T., **26** II, p. (10). The periodic errors of the lead screw proved to be insignificant.

About sixty comparison stars were selected for every plate from the Washington Catalogue of 1952 (Watts, Scott and Adams 1952), and reduced with the proper motions given in this catalogue to the epoch of the Yale plates (the epoch difference between Yale and Washington was on the average less than two years). Corrections for the dependence of the effective focal length on the spectral type were applied before the least squares adjustment for the plate constants.

The differences Δx and Δy between the predicted (that is, the standard coordinates on the scale of the Yale plates) and the measured coordinates of the reference stars were subjected to a least squares solution following the formulas

$$\Delta x = Ax + By + C + px^2 + qxy + ry^2$$

$$\Delta y = A'x + B'y + C' + p'x^2 + q'xy + r'y^2. \tag{5.2}$$

This was a change from the previous practice, according to which no provisions were made for a y^2 term in x and an x^2 term in y. (See equations [5.1].) The coefficients r and p' of these terms, which were actually obtained in the adjustments, turned out to be rather small.

A study of systematic empirical corrections for the positions in the entire zone expressed by third order terms, which had been carried out in all previous work, showed nothing that might be regarded as significant. This could give rise to speculations concerning the reality and significance of the empirical terms applied to the measurements in all previous zones, and one wonders whether the application of the same empirical terms to the positions from plates in the same belt—but different ones to the positions from plates of another belt—was really justified; also how much the personal differences in the perceptions of the various measurers who measured on the same plate and in the same belt influenced the positions. The facts that the accidental error of a position measured with the automatic measuring machine dropped by 20%, as compared to visual measurements at the classical one screw measuring machine, and that there were no significant empirical corrections found, clearly demonstrate the superiority of the automatic measuring machine.

The proper motions in this belt were obtained by combining proper motions found from the following pairs of catalogues: AGK2 minus Yale 1916, Yale 1947 minus Yale 1916, and Yale 1916 minus AGK1, after these proper motions had been corrected for magnitude errors, reduced to the GC system, and properly weighted. The tables for the reduction of the positions and proper motions to the GC system are on pp. (14) and (15) of Y. T., **26**, Pt. II, Tables 6 and 7, and also repeated in Y. T., **27**, p. (7) as Table 1.

The catalogues are arranged in the usual way. The third term of precession is given in Table 8, Y. T., **26**, Pt. II, p. (16). Regarding the proper motions, a column of codes indicates how they were obtained.

5.2.6 THE COMBINED YALE-AGK2 CATALOGUES. The zones derived from the plate belt between +20° and +30°, published in Y. T., **9** and **10** and described in section 5.2.3, had been taken at the mean epoch of 1928.2. This is so close to the epoch of the AGK2 (1930), that the Yale investigators regarded it as worthwhile to compile the positions from these two catalogues into one combined catalogue.

The Yale positions were precessed to 1950.0 with the precession terms printed in Y. T., **9** and **10**. The small difference in the epochs was bridged by reducing the AGK2 positions to the epoch of the Yale plates by application of the Yale proper motions. Since the AGK2 is affected by systematic errors which depend on magnitude and spectral type, only the F stars of 8^m to 9^m were used for reducing the Yale positions to the system of the AGK2, which is ostensibly that of the FK3. The final $\Delta\delta$ of the Yale zones from the belt +20° to +30° is given as Table 2 on p. (6) of Y. T., **24**. The proper motions were newly determined by comparing the combination catalogue with positions from the AGK1 and other sources, and reducing them to the FK3 system. Tables for the reduction of the catalogue data to the GC system are given in the introductions to Y. T., **24** and **25**. The catalogues are arranged in the usual manner. For those positions that depend on AGK2 positions derived at Bonn, the appropriate corrections depending on the spectral type were applied.

5.2.7 THE POSITIONS DERIVED FROM PHOTOGRAPHS MADE WITH THE FIFTH CAMERA (THE NORTHERN POLAR CAP, Y. T., **26**/I). The purpose of the work on this zone was primarily to test the performance of a new eight-inch Ross camera (focal length 2063mm, corresponding to a scale of 99.″98/mm), and also to establish star positions in the neighborhood of the north celestial pole which were needed for a long-term project under work at the Yale Observatory, namely, the determination of the constant of precession.

The comparison stars were taken from the catalogue by Watts, Scott and Adams (1952) which yielded 60 comparison stars that were reduced to the average plate epoch (1957.33) by means of the proper motions in this catalogue. The region is covered by four plates centered at the pole but oriented 6^h differently in right ascension. Standard coordinates ξ and η were computed with respect to the north celestial pole as tangential point from the formulas $\xi = f \cos \alpha \cot \delta$, $\eta = f \sin \alpha \cot \delta$, with $f = 2062.6481$. The differences between the predicted (equivalent to computed standard) coordinates and measured coordinates were subjected to an adjustment according to equation (5.1). In order not to leave any systematic residuals, a term in y^2 had to be added in one instance to the Δx expression for one plate, and a term in x^2 to the expression for Δy in two other cases.

No indication for distortion could be found from the adjustment residuals, and only a slight trace of color magnification error caused by lateral chromatic aberration was noticed, but no correction for it was applied to the published positions. It is interesting that the use of a new objective system was prompted by the desire to get rid of the color magnification error, but the results from this zone are inconclusive as to whether this goal has actually been fully achieved. Also, the lens had to be stopped down considerably (because of astigmatism) before measurable images were produced in the corners of a plate of the format 17 by 17 inches.

The coma effect was found to be $+0.''00075$/mag/mm from comparison of central images and diffraction spectra produced by an objective grating. A correction for this effect was applied, and the published positions are based on the measures on all four plates.

Proper motions were computed by comparison of the Yale positions with those in the Greenwich Astrographic Catalogue (G), the AGK2 and the Pulkovo zone of the AGK2 (AGP) described above in Section 5.1.3.4. μ'' and μ' denote the proper motions obtained from $(Y\text{–}G)/\Delta t$ and $[Y - (4\text{AGK2} + \text{AGP})/5]/\Delta t$, respectively. The catalogued proper motions μ were obtained from $\mu = (3\mu'' + \mu')/4$.

A table for the reduction of the positions and proper motions to the GC system is given on p. (8) of Y. T., **26**, Pt. I, Table 3. The serial numbers of the stars in this catalogue are not AGK1 numbers since there was no AGK1 zone covering this area.

The catalogue gives positions for all stars brighter than $m_{pg} = 11.6$ and for a few that are even fainter than that. An asterisk after a number refers to a note, G indicates (as before) a GC star, A a reference star, and "s" and "b" have the same meanings as in previously published zones.

Beside the usual right ascensions and declinations, the standard coordinates of the stars with respect to the north celestial pole as tangential point, as defined before, are also given in a second part. The coefficients of the precession series for α and δ are not given for stars within 2° of the north celestial pole. Notes and epochs are, as usual, on the bottom of the pages.

5.2.8 THE ZONES SOUTH OF −30°. There are several more (essentially ten degree wide) belts of plates which were taken for the construction of the Yale Catalogues south of −30°. The results of two of these (−30° to −40° and −40° to −50°), for which the plates were exposed at average epochs of 1956 and 1942, respectively, and for which cameras 3 and 5, respectively, were used, have already appeared in print (Y. T., **28**, **29**, and **30**). The plates taken in the zone from −60° to −70° (about 1942) have been measured but not yet reduced. For the plates around the south celestial pole (−70° to −90°, taken with camera 5 in 1956), preliminary reductions have been made at Yale. The zone from −50° to −60° was never covered. Except in the zone −70° to −90°, the pattern was the usual edge-in-center-line, giving an overlap in right ascension but none in declination. The overall direction of the work on these zones has been in the hands of Dr. Dorrit Hoffleit, and was sponsored by the United States Army Topographic Command (formerly U.S. Army Map Service). Due to various circumstances, considerable pressure was exerted on the principal investigator to expedite publication. A direct result of this was that not enough time could be spent on the various trials and errors which are so necessary for finding the best procedures for producing the most accurate positions that could have been derived from the plates. There is no doubt that, without these various pressures for getting numbers out, a product of considerably superior accuracy would have been achieved. That this did not happen is certainly not the fault of the principal investigator.

The plates covering the belt from $-30°$ to $-40°$ were the first ones worked on. They were measured with the same measuring machine that had been used for most of the other zones, and which was described in Y. T., **9**. The machine had been digitized, in the meantime, but was located in a room in which the temperature was subject to rather wide variations. There was a relatively large number of different measurers. Three settings were made on every image, and x and y coordinates were measured independently. The program stars were those in the CPC (see Section 5.3). Since each plate was measured in several sittings, usually by several different measurers, the first and last ten stars of every sitting were repeated in one sitting after the entire plate had been measured. In addition, the usual alignment stars were measured at the beginning and end of every sitting. These measurements were used to check the suspected systematic differences between the observers, and also, of course, the motions of the plate in the plate holder during the period one coordinate was measured.

Over 2800 stars were available from the Second Cape catalogue for 1950, giving an average of 90 more or less uniformly distributed reference stars per plate. They were reduced to the epoch of the plates by application of the proper motions in this catalogue. Every plate was reduced according to the following model for the difference between the measured coordinates of the reference stars and those predicted from their catalogued spherical coordinates:

$$\begin{aligned}\Delta x &= C + Ax + By + Dx^2 + Exy + Fy^2 + P(m - m_0)\\ &\quad + G(m - m_0)x + H(c - c_0)x\\ \Delta y &= C' + A'x + B'y + D'x^2 + E'xy + F'y^2 + P'(m - m_0)\\ &\quad + J(m - m_0)y + K(c - c_0)y,\end{aligned} \tag{5.3}$$

where m and c are magnitude and color index, respectively, of the star and m_0 and c_0 the average magnitude and color index, respectively, of the reference stars on any one plate.

No restrictions were placed on any of the coefficients, that is, the coefficients were completely free to vary from one plate to the next, and on any particular plate neither G and J nor H and K were in any way restricted to be similar.

One fundamental criticism of the model (5.3) is the absence of a term of the form $L(c - c_0)$ in x and in y. The origin of the color magnification error, described by the terms in (5.3) whose coefficients are H and K, respectively, is thus forced to lie at the origin of the x-y coordinate system. This is an assumption which may be somewhat difficult to defend.

Also, Eichhorn (1971) has shown that there is a significant correlation between the coma effect coefficients G and J in x and y, respectively, on any one plate. This implies that large or small values of G and J have a tendency to occur together on the same plate. This proves two things: Dr. Hoffleit was justified in allowing the coma coefficients to vary from plate to plate, in contrast to the previous practice in the Yale catalogues when their effects were ignored altogether. Because, if the coma effect did not vary significantly from one plate to the next, the deviations from the mean of the coma coefficient values obtained for any coordinate on any plate would just be accidental deviations from this mean, and then there would be no reason for the deviations in both coordinates to be correlated on the same plate. Furthermore, Eichhorn noticed in the same paper that the dispersion of the individually found G and J values around their means is significantly larger than the average nominal mean error which one gets from the individual least squares solutions. This is another indication that the coma effect indeed does vary significantly between plates taken with the same optical system. J. G. Baker (1968, private communication) has stated that the coma coefficient is proportional to the temperature gradient in the components of the optics, and should therefore be correlated with the ambient temperature gradient during the exposure.

More accurate values of the coma coefficients and thus, also, more accurate values of the final positions could probably have been obtained in three ways. First, the differences between the measured coordinates of the central images and the mean of the corresponding diffraction spectra could have been employed for the determination of the coma coefficients, in the same way as they were before. The two determinations for each plate, from grating images and from the adjustment to the comparison stars, could then have been combined. Furthermore, the comparison of the positions obtained from two overlapping plates, for all stars, could have been used for determining the values of G, the coma effect coefficient in the x coordinate. Second, it might have just justifiable to keep G and J identical on every plate. Third, the values of the individual coma coefficients could have been restricted by forcing them in an appropriate way to favor a certain average value, the way it was done by Eichhorn and Gatewood (1967) for the computation of new plate constants for the Northern Hyderabad zone of the Astrographic Catalogue. (See Section 5.5.5.6.) There is a definite possibility that the uncertainty inherent in the determination of the coma coefficients from comparison with reference stars alone (because of the relatively narrow magnitude range of the reference stars) is in part responsible for the low overall accuracy of the positions in this belt of the Yale zones. Also, common systematic higher order terms were neither searched for nor applied.

The rms. error of the catalogue positions was estimated in several ways, and $0.''35$ is probably a reasonably accurate figure for its typical value.

The reference star positions were probably closer to the FK4 than to the FK3 system. A table for the corrections of the proper motions to the FK4 system is given.

The proper motions were formed by comparison with the old positions in the CPC (see Section 5.3). After correction to the FK4 system, there are hardly any deviations of the various averages for different magnitude groups of the hourly means of the μ_α, although such a trend is noticeable for the proper motions originally given in the CPC. This trend is probably due to uncompensated magnitude errors in the Cordoba meridian catalogue which served for the Cape μ_α, as first epoch material. The proper motions in declination were also reduced to the FK4 system by reducing the Yale minus CPC (μ_Y) and CPC minus Cordoba (μ_C) proper motions independently to the FK4 system, and then combining them by the formula $\mu = 0.434\mu_C + 0.566\mu_Y$. Since there are, in places, systematic trends in the differences between μ_C and μ_Y, they were compared with the curves theoretically predicted by van Rhijn and Bok (1931) in order to decide whether the μ_C or the μ_Y were to be systematically corrected. Thus, the μ_C system was reduced to the μ_Y system when the latter one agreed better with the van Rhijn-Bok curves, and vice versa. When the μ_C and μ_Y systems agreed, they were left unaltered even though in these cases they sometimes significantly differ from the van Rhijn-Bok and the Vyssotsky curves. In any case, all applied corrections were tabulated. (Y. T., **28**, Table 7) and may thus be removed from the printed values in order to restore them to those obtained originally. For investigation in stellar statistics, one should probably use the uncorrected original values of the proper motions rather than the printed ones. As usual, comparisons of positions and proper motions with the GC are also printed.

The proper motions in Y. T., **29** were treated in rather much the same way as those in Y. T., **28**, that is, the systematic trends were more or less made to agree with those predicted in the above-mentioned paper by van Rhijn and Bok, and (ostensibly) simultaneously reduced to the FK4 system. Corrections to perform these reductions were applied independently to the above-mentioned μ_C and the μ_Y. After application of the corrections, there were no noticeable systematic differences left between the μ_Y and the μ_C in the zone $-30°$ to $-35°$ (Y. T., **28**); such differences were still apparent between the μ_C and the μ_Y derived from Y. T., **29**.

The plates for the last zone which has so far appeared in print ($-40°$ to $-50°$, Y. T., **30**) had been taken in 1941–42. By the time they were to be measured during the early 1960s, the emulsions had peeled from some of the plates, thus rendering them unusable for astrometric (or any other) work. Fortunately, this had not happened in too many cases. The regions originally covered by those Yale plates which deteriorated before they could be measured were rephotographed by H. Wood in 1961 at Sydney. For this purpose, a Taylor, Taylor and Hobson lens was used which covered a field of $6° \times 6°$ (20×20 cm^2) at a scale of $116''$/mm. Objective gratings with grating constants of $3.^m0$ and $2.^m3$ for Yale and Sydney, respectively, were used at both stations. The original Yale plates were measured under the supervision of Mrs. Dorothy Eckert at the same automatic measuring machine of the IBM Watson Scientific Laboratory which had previously been used for measuring the zones from $+50°$ to $+60°$, and the Sydney plates under the supervision of A. Klemola at the Yale University Observatory.

Unfortunately, the electronics of the Watson Laboratory automatic measuring machine had deteriorated considerably since the $+50°$ to $+60°$ zones were measured, in addition to the Yale plates being in poor condition, so that the accuracy of the positions in this zone, corresponding to a rms. error of 0.″23 in both coordinates for a position based on two plates, is nowhere nearly as high as that reached in the zones previously measured with the automatic machine.

The positions from the Second Cape Catalogue for 1950 served as reference positions and were put at Dr. Hoffleit's disposal by R. H. Stoy in advance of the publication of the catalogue, and yielded an average of 105 comparison stars per plate. All CPC stars north of $-50°20'$ were put on the program. The customary odd-even scheme for the measurement of the plates was again retained. The positions of the Second Cape Catalogue, which refer to an epoch about 15 years later than those of the Yale plates, were reduced to the epoch of the latter by the application of the Cape proper motions. The reduction model employed for this zone is essentially that of (5.3), except that a term $H(m - m_0)y$ was added to the model for Δx, and one $H'(m - m_0)x$ to the Δy model. In addition to the criticism expressed for the model (5.3) above, one can observe from Table 5 on p. (9) of Y. T., **30** that these additional terms are typically insignificant and, therefore, the systematic accuracy of the finally derived positions would have been higher had these terms been left out.[3] Again, the genuine coma effect coefficients in x and y are significantly correlated, and the same comments concerning their determination that were voiced above in this section apply to Y. T., **30** also. This may perhaps be one of the factors contributing to the relatively low accuracy.

The positions as well as the proper motions were compared with those in the FK4 and the FK4 Supp. From these comparisons, $\Delta\alpha_\alpha$ and $\Delta\mu_\alpha(\alpha)$ were derived and applied to the computed positions and proper motions before they were printed. It is possible that these corrections are reliable only for stars brighter than 7^m.

Since in this zone, again, proper motions μ_C were already available from the CPC which had been computed by comparing the photographic Cape positions with those of the catalogue Cordoba D and the early Cape positions, proper motions μ_Y newly derived from comparison of the new Yale with the Cape photographic positions were compared with the μ_C in order to find systematic trends in the differences.

The μ_α display the classic behavior (the averages of the 24 hourly means should be zero for all magnitude groups and lie on a sine curve) for the bright and faint, but not the intermediary stars. The μ_α (the Y as well as C) were accordingly corrected. The systematic μ_δ were treated as described above, with the aid of the van Rhijn-Bok curves (the Y values did this even without systematic corrections), although this time the statistical parallaxes, newly derived by Vyssotsky, were employed.

Weighted means were taken of the Yale and Cape proper motions and published. The formulas according to which these means were taken vary, depending on the way in which the Cape proper motions were formed.

---

3. There is a misprint in Table 5 of Y. T., **30**. The column headed H contains the coefficient of the term $(m - m_0)x$ in Δx and not m$(m - m_0)y$ in Δx, as the formula on the bottom of page (8) in Y. T., **30** indicates.

One of the outstanding features of this group of Yale catalogues is the careful attention paid to comparing the proper motions, especially the large ones, with those available from other sources such as Luyten's Bruce Proper Motion Survey, and the often considerable effort paid off by clearing up discrepancies which might otherwise have been overlooked.

Another feature is the pointing out of *proper motions groups*, that is, groups of stars that may be rather widely scattered over the sky but still have very similar (and large) proper motions. These represent perhaps one of the most significant contributions of this group of catalogues.

The catalogues list 12876 stars for the zone $-35°$ to $-30°$, 12120 for the zone $-40°$ to $-35°$, and 17373 for the zone $-50°$ to $-40°$. The catalogues are arranged in the usual, traditional manner. The first column lists the CPC, the second the DM (CoD or CPD) number. In the third column one finds visual and photographic magnitudes, color indices and spectra (on the HD system). Next come α_{1950} and the precession terms I and II (see section 3.2.1, equation [3.1a]) to $0.^s001$, followed by the same data for the declinations (to $0.''01$). The image type (central image alone; spectra alone; both, as before) is indicated in the next column. Also, the number of plates from which the position is derived is shown, if it differs from 2 (this is especially important for the stars measured on the Sydney plates). The Yale proper motions, and the weighted means from the Yale and Cape proper motions are given, ostensibly on the FK4 system. *R* denotes a reference star. Also, there is a variety of notes. The epochs are, as customary, given on the bottom of the pages.

The zones $-70°$ to $-60°$ and $-90°$ to $-70°$, the latter having a particularly strong overlap pattern that deviates considerably from the traditional Yale pattern, have also been measured—most of the latter on a newly acquired Mann measuring machine. The measurements were sent to the United States Army Topographic Command to be reduced according to the plate overlap method, which had been perfected there by Googe, Eichhorn and Lukac (1970). Due to severe cutbacks in the budget of this agency and loss or reassignment of competent key personnel, it is rather doubtful that the reduction of these Yale zone measures will be finished at the USATOPOCOM as originally planned. Therefore, the reductions have independently already begun at the Yale Observatory where they are being conducted in the manner described in this section.

The zone from $-90°$ to $-70°$ should be particularly interesting. There are four belts of plates, centered at $-90°$, $-85°$, and $-75°$, so that (except in the zone $-75°$ to $-70°$) every star appears on at least four plates. These plates were also measured over their entire area of $11° \times 11°$. In addition to all measurable images of all CPD (not only CPC) stars, two faint AC stars per square degree were measured for the purpose of establishing faint reference material for the reduction (and, in particular, establishment of the magnitude equation) of the Melbourne zone of the AC (see Section 5.5.5.19). This will be the Yale zone with the largest star density, and the widest magnitude range. It is very much to be regretted that due to a series of unfortunate circumstances, the originally planned overlap reduction will probably not be carried out, at least not in the foreseeable future. The preliminary results will be published as Y. T., **31**.

5.3 The Cape Photographic Catalogue for 1950.0 (CPC)

5.3.1 GENERAL DESCRIPTION, EXPOSURES AND REFERENCE STARS. A program for the establishment of catalogues of the type of the Yale zones and the AGK2 south of $-30°$ declination was finished in 1968 at the Royal Observatory at the Cape of Good Hope (Jackson and Stoy 1954, 1954a, 1954b, 1958, and Stoy 1966, 1968). It was initiated by the late Astronomer Royal (then His Majesty's Astronomer at the Cape of Good Hope), Sir Harold Spencer Jones, and carried through by his successors J. Jackson and R. H. Stoy. Plates were taken between 1930 and 1953. (See also Table V-4.) No plates were taken in the zone between $-40°$ and $-52°$ since this was already covered by the Cape Astrographic Catalogue and positions and proper motions of stars in this belt were published earlier by Cape Observatory astronomers (Gill and Hough 1923, Spencer Jones and Jackson 1936). The purpose of the catalogue was a uniform coverage of the zones concerned by about 10 stars per square degree, with magnitudes chiefly between 7^m and 10^m.

The *camera* was mounted on the normal astrograph until early 1953, and after that on the Victoria telescope. The optics of this camera were quite different from the symmetrical objectives used for the Yale zones and the AGK2-3. They were a five-inch triplet by Taylor, Taylor and Hobson of about 2070 mm focal length, corresponding to a scale of 103.″2 per mm. The camera was used with full opening only in the zone from $-30°$ to $-35°$, but since the quality of the images deteriorated toward the corners, it was stopped down for the exposures of all other zones to an effective aperture of about $4\frac{1}{2}$ inches. The objective was carefully centered and the plates were carefully squared on. The plates for the zone from $-30°$ to $-35°$ were of normal thickness, but for all other zones the exposures were made on plates of selected glass $2\frac{1}{2}$ millimeters thick. The plate pattern was essentially the same as that employed for the Yale Catalogues. The plates were taken in the "edge-in-center-line" overlap pattern in such a way that the right ascension of every star depends on exactly two images per zone. Between adjacent zones, there was always a slight overlap in declination. For the first two zones to be explored, the format was thus $4.°8 \times 5°$, but in the other zones, the format was cut to essentially $4° \times 4°$. In the individual cases, the width of the measured area was chosen in α so as to provide a complete overlap, and in δ, $4.°3$ degrees. In the zone from $-56°$ to $-80°$, the declinations of the plate centers were $4°$ apart, but in the polar zone ($-80°$ to $-90°$), the plates were centered on the declinations $-82°$, $-85°$, $-88°$ and $-90°$ in order to provide an overlap of about $1°$ in declination also. The plate area measured in the polar zone was again about $4° \times 4°$, so that there the position of a star frequently depends on more than two, even up to 10 plates.

The *exposures* were of varying duration between 8^m and 20^m. Except in the zone from $-30°$ to $-35°$, where each plate was given three exposures, two exposures were put on each plate, one for measuring and the other one to identify spurious images. On each plate, only one set of images was measured.

The stars to be measured were selected from the AGK1 type meridian catalogues observed at Cordoba and La Plata, other available old meridian catalogues, the CoD, CPD and other sources, such that the average density of about 10 stars per square degree was reached.

The *measuring machine* (made by Hilger) is of the Schlesinger type with a lead screw 180mm long. When the measurements of the zone $-30°$ to $-35°$ were about half finished, a projection device was installed. The readings were made to a tenth of a micron; every plate was measured in direct and reverse position. Corrections for the errors of the measuring screw and the V-way were applied to the measures. Also, for the purpose of applying diameter dependent corrections to the positions, the diameters of the star images were determined during the measurements except in the zone $-30°$ to $-35°$. The zone $-56°$ to $-60°$ was measured at the Royal Greenwich Observatory on a machine identical to that at the Cape. The measurements at the Royal Greenwich Observatory required neither screw nor V-way corrections.

The *reference stars* were specially selected, and before 1939 were observed with the 8-inch Airy Transit Circle; afterwards, that is for the zones from $-56°$ south, with the 6-inch Gill Reversible Transit Circle rather contemporaneously with the exposures of the plates. Their density was on the average 1.25 stars per square degree; their average visual magnitudes, mostly between 8^m and 9^m. For the zones between $-30°$ and $-56°$, the reference star positions were determined differentially with respect to the First Cape Catalogue for 1925.0 whose positions were brought to the epoch of the observations by the application of proper motions. For the zones south of $-56°$, the FK3 itself, when necessary augmented by stars from the GC, reduced to the FK3 system by means of Kopff's (1939a) tables, served as the fundamental reference catalogue for the observations. In the polar zone, the reference star observations were reduced fundamentally and form part of the Second Cape Catalogue for 1950.0.

There is a slight preference for spectral types G and K. The observations at the Airy circle are affected by a small magnitude equation, which was later detected and corrected for.

The *magnitudes* were determined by a special program. The spectral types were partly taken from the Henry Draper catalogue, and when not available there, were especially determined at the Harvard College Observatory by Dr. Hoffleit and later by Mrs. Mayall.

5.3.2 THE DERIVATION OF THE POSITIONS IN THE CATALOGUE. The reductions to the system of the reference stars and further to that of the FK3 were executed in steps. First, six linear plate constants, following equations (2.14a), were determined individually for each plate from comparisons with the reference stars by the so-called Christie-Dyson method, that is, not by a rigorous least squares adjustment, but by combining all (reference) stars in any one plate quadrant into a normal position. This method was originally proposed by Christie and Dyson (1894) and is also well described in a paper by Walther (1946). The plate constants are determined from the four normal positions. This method for obtaining the plate constants was frequently used in the various zones of the astrographic catalogue. Some astronomers, mostly in the British Commonwealth, feel that this method is not essentially inferior to a least squares adjustment. This author does not agree with them.

The *reference star residuals from this adjustment* were then collected and it was assumed that, for all plates of each zone, they could be well represented in x and y, respectively, by a pair of complete polynomials of the third order in x and y. A collection of the coefficients of the terms in these polynomials, which came out somewhat differently for different zones, is given on p. ii of the Annals of the Cape Observatory, Vol. XXI. The coefficients in this polynomial can, of course, also be determined from an analysis of the systematic residuals of star positions as obtained on the "preceding" minus that from the "following" plate. However, this principle was used only in the zone $-30°$ to $-35°$, where the coefficients of the corrected polynomial were determined by a combination of an analysis of the reference star residuals and the systematic differences between the star positions obtained from overlapping plates. The correction polynomials were not determined in the polar zone where the material was scarce. There the corrections applied were those whose coefficients are the averages of the corresponding ones in the zones from $-64°$ to $-80°$. The residuals which still remained after the application of the third order polynomial corrections, depended not only on the position of the image on the plate, but also on magnitude and color. Their systematic effects on the positions were removed by *subsidiary corrections*. These were applied to the positions in all zones except that from $-30°$ to $-35°$.

The subsidiary corrections are modeled in the form $\Delta x = f(d, \lambda)x$ and $\Delta y = f(d, \lambda)y$, common for all plates of the same zone. They are thus assumed to be radial and to vary with the first power of the distance from the geometric plate center, which is practically identical with the tangential point. In the function $f(d, \lambda)$, d means the image diameter and λ is a kind of effective wave length which depends on the spectral type. A table for the conversion of spectral types to effective wave lengths λ is given in Cape Annals **18** (Jackson and Stoy 1954a), page v. The color index might also have been usable as an argument instead of λ, and perhaps would have been more significant since interstellar absorption frequently disturbs the strong correlation between a star's color and its spectral type.

Two principles are available for the determination of $f(d, \lambda)$: the investigation of reference star residuals (after application of the corrections which depend on position only and which are expressed in the form of third order polynomials), and the investigation (again after application of the corrections just mentioned) of the systematic differences between the positions of the same stars obtained from overlapping plates. Only in the zone from $-35°$ to $-40°$ were the reference star residuals utilized for the determination of $f(d, \lambda)$. Even so, only that part of $f(d, \lambda)$ which depends on λ only, that is, which describes purely a color magnification error) was determined from the reference stars. This is so because the narrow magnitude range of the reference stars makes it impossible to investigate, by means of their residuals, the magnitude dependent character of $f(d, \lambda)$. In the zones from

$-35°$ to $-40°$ and $-52°$ to $-56°$, the model

$$f(d, \lambda) = a(\lambda - \lambda') + b(d^2 - d'^2)(\lambda - \lambda'')$$

was used; in the zones from $-56°$ to the south celestial pole, $f(d, \lambda)$ was applied numerically from a double entry table. a and b were empirically determined coefficients, and d', λ' and λ'' were certain mean values.

The application of these subsidiary corrections led to the "measured positions" which were converted to right ascensions and declinations. From these, the proper motions were determined, and to these the final corrections were applied to bring them to the FK3 system before they were printed in the catalogue.

5.3.3 THE REDUCTIONS OF POSITIONS AND PROPER MOTIONS TO THE FK3 SYSTEM. The "measured positions" were first put onto the FK3 system by application of the corrections of the system of the reference stars to the GC, and the systematic corrections from the GC to the FK3 system. The data for correction to the GC system were always extracted from the tables in the GC. I; those for correction from GC to FK3; partly from the tables in the GC and later from those of Kopff (1939a). Because of the uncertainties of the corrections of the reference stars to the GC system, and possible distortions introduced by the photographic process, the above described reduction to the FK3 system was regarded as only a preliminary one.

The proper motions were determined by a comparison of the newly determined position with those in available old sources, after they had been corrected to the FK3 system (by use of the tables in GC.I and Kopff [1939a]). These sources were the Cordoba and La Plata AGK1 type zones, the Argentine General Catalogue, the Melbourne meridian catalogues Nos. 3, 4, and 5, Gilliss' catalogue, the Melbourne zone of the Astrographic Catalogue and the polar catalogue in the Annals of the Cape Observatory **11**/II (Hough 1905). When there was more than one old position for a star, the proper motions were determined individually, and their weighted means were taken and published.

The final reductions to the FK3 system were then performed as follows: The positions of the stars which the catalogue had in common with the GC were compared with the GC positions after they had been reduced to the FK3 system, and updated to the epochs of the CPC positions by means of the just derived proper motions which were, ostensibly, already on the FK3 system. There were about 1000 of these in each zone. From this comparison, analytical expressions (in the zones from $-30°$ to $-56°$) or double entry tables were constructed to perform the final reductions to the FK3 system. The analytical expressions, as well as the double entry tables, depended on the magnitudes as well as the measured y coordinates.

In δ, there was a coma term in y in two of the zones ($-30°$ to $-35°$, and $-35°$ to $40°$), as well as a quadratic magnitude term. In α, the corrections were essentially linear in y and in m. The latter corrections were applied, and the positions obtained in this way are the ones published.

How well the positions thus established represent the FK3 system was checked by comparisons with the GC on the FK3 system, reduced to the epoch of the CPC positions by means of the proper motions in the GC, which had also been reduced to the FK3 system, and also through comparisons with other catalogues which had been treated likewise. Generally, it was found that the CPC positions represent the FK3 system quite well, although systematic errors of up to 0.″3 will occur in places, especially for the more southern zones. It is, of course, quite possible that systematic errors in the proper motions are to some extent responsible for these still remaining systematic differences.

5.3.4 DESCRIPTION OF THE CATALOGUES.

Table V-4 presents a summary of the zones of the CPC with some of the relevant data. There is a general introduction to the entire enterprise in Volume **17**. Every volume starts with a detailed introduction in which the procedures that led to the data in that volume are described. These introductions ought to be read by everybody who intends to critically use the positions and proper motions in the CPC zones. A part of these introductions is the detailed description of the methods used for the determination of the photographic and photovisual magnitudes of the program stars, which was carried out mainly by R. O. Redman from plates especially taken for this purpose. Also, every introduction carries a description of the catalogue of the zone (or zones) in the volume concerned. One table always gives the dates when the plates were taken.

Except in the polar zone, the data are arranged in the order of right ascension within the belts of plates on which they were measured. The columns of the catalogues typically contain: Serial number, CoD or CPD number, right ascension (to 0.ˢ01) for the orientation 1950, ostensibly on the FK3 system, the first two coefficients I and II of the precession series for use in equation (3.1a), computed with Newcomb's constants, and analogous data for the declinations (to 0.″1), the cen-

Table V-4 The Zones of the Cape Photographic Catalogue for 1950.0

Zone	*Range of Epochs*	*Number of Stars*	*rms. error* $\alpha \cos \delta$	δ	*Avg. No. of Ref. Stars per Plate*	*Pub. in Cape Ann.*
−30° to −35°	1931–1933	12846	0.″22	0.″22	26	17
−35 −40	1935–1937	12115	.15	.21	27	18
−52 −56	1938	9215	.15	.18	21	19
−56 −64	1945–1947	14710	.15	.15	21	20
−64 −80	1947–1948	16599	.18	.21	23	21
−80 −90	1953–1955	2981	.15	.21	18	22

tennial proper motions: in α to $0.^s01$ on the parallel and to $0.''1$ on the great circle, in δ to $0.''1$, and the probable error (in units of $0.''1$) of the proper motion components. (The latter are missing in the zone $-30°$ to $-35°$.) Proper motion data are, of course, given only for those stars for which early epoch precision positions were available. An M in the proper motion error column indicates that the printed position was derived not from photographic measurements but from meridian observations, and was reduced to the epoch printed at the bottom of the page. GC in this column indicates that the positional data are from the GC, reduced to the system of this catalogue. The next three columns give photographic and photovisual magnitude, and color index, respectively. Underlined data were not determined in the regular series made for this purpose but are from other sources. The last column shows the spectral type, usually from the HD or the Henry Draper extensions. An underlined spectrum means that it was determined at the Harvard College Observatory for the purposes of this catalogue, first by Dr. Dorrit Hoffleit, later by Mrs. Margaret Mayall.

Asterisks after the serial numbers refer to footnotes which are found at the bottom of the page, where the epochs are also listed. The arrangement is different in the polar zone, $-80°$ to $-90°$. There, in addition to the data given for the other catalogues, one finds the terms III_α and III_δ; the epochs are given (to $0.^y1$) for every star, as well as the number of images from which the ascribed position was derived, the HD numbers, when available, the components of the vector $\hat{\mathfrak{x}}(\alpha, 90° - \delta)$ to eight figures (there called direction cosines and denoted by X, Y and Z). Note that the error data for this zone in Table V-4 are valid only for positions which depend on exactly two images. The catalogues also contain the meridian positions of the reference stars and, in appendices, contain corrections to the first epoch catalogues used for the derivation of the proper motions, or a list of stars in the first epoch catalogues that were simply discordant and for which a correction was not obvious. Appendix III of the zone $-80°$ to $-90°$ also contains a list of errata in the previous zones.

5.4 Other Photographic AG Type Catalogues

5.4.1 THE CATALOGUES BASED ON COMPOSITE PLATES (ZI-KA-WEI AND PARIS). One of the most important conditions for the derivation of reliable star positions by photography is the availability of a sufficient number of reference stars on the plate. If there are not enough reference stars, the positions derived from the measurements will be affected with intolerably large systematic errors.

There are, in principle, two ways to achieve a large number of reference stars on a plate. The most obvious of these is to make the plate large enough so that there will be the required number of reference stars on it. This is the way first followed by Schlesinger in the construction of the Yale Catalogues (Section 5.2), and later by the makers of the AGK2 (Section 5.1.3), the AGK3 (Section 5.1.4) and the Cape Photographic Catalogue (Section 5.3). While the first large systematic photographic astrometric program, namely the Astrographic Catalogue (Section 5.5), used plates that covered an area of somewhat less than 5 square degrees, the areas covered by the individual plates from which the positions in the catalogues listed above were derived, covered areas from 25 to 140 square degrees which contained a correspondingly larger number of reference stars.

However, there is a possibility to have more reference stars available, even on plates with a small field, by the construction of plate complexes which are then treated for the purpose of orientation against the sky as a single plate. There are several ways in which such complexes can be constructed by finding the constants of orientation of two (or more) overlapping plates against each other. This would, in fact, be a simple version of the plate overlap method, and has in the past been applied with good success to several problems.

There is a particularly ingenious version of this method which has so far been applied to construct two photographic catalogues, the "Tour de l'Équateur sur 12 plaques photographiques portant 300 clichés photographiquement reliés" (S. Chevalier 1921, 1927, 1928) in Vols. **12**, **14** and **15** of the Zô-Sè annals, and the "Catalogue de 11755 Étoiles" (Baillaud 1950) which was intended to serve as catalogue of reference stars for the determination of proper motions of the stars in the Paris zone of the AC (Section 5.5.5.9).

The basic idea of this method is as follows. On the same plate (that is, piece of glass) one takes several (16 to 25) fields which partially overlap each other, following the "edge-in-center-line" pattern. The exposures are made in such quick succession (about 5^m apart) that the telescope always remains at the same declination and scans the same hour angle range during each of the exposures on the same plate. Also, the plate is then left completely undisturbed in the plate holder so that the orientation of the plate with respect to the telescope presumably does not change during the series of exposures that go on the same plate. One will, therefore, be able to assume that the orientation angle and the scale factor, namely the quantities ρ and σ in equation (2.11a), are the same for all plates or are at least connected by a known relationship. Under the circumstances, barring significant changes of the relevant parameters during the exposures, refraction will affect all exposure sets on one plate in the same way, and the way in which the effects of aberration distort the positions from different exposure sets differently can be accurately computed. Baillaud, in the work quoted above, discussed in detail the relationship between the ρ and σ belonging to the various exposure sets on the same plate, under the assumption that the plate itself is left undisturbed in the plateholder during all exposures. The zero points (that is, the parameters corresponding to C and D in equations (2.11a) of the positions derived from the various exposures relative to each other, that is, the amounts by which the plate was shifted in right ascension and declination, respectively, between successive exposures, are determined from the investigation of the relative positions of the same stars in different exposures. This requires, of course, no reference stars, and is a utilization of the plate overlap. The only parameters to be determined by comparison with reference stars are then scale and orientation angle, as well as the zero points of the entire strip-like plate complex.

At Zô-Sè, the equator region ($-50'$ to $+50'$) was covered with 300 overlapping exposures on twelve plates. The catalogue gives in this region the positions of 14268 stars for an epoch in the early 1920s with an rms. error of about $0.''15$ in $\alpha\cos\delta$ and $0.''10$ in δ. The density is about 20 stars per square degree, thus rather high. This catalogue is, therefore, a valuable source of relatively early epoch star positions of high accuracy around the equator. An investigation of systematic corrections to any well established system seems to be lacking.

The Paris Catalogue contains the positions of 11755 stars with magnitudes between $9.^m5$ and $10.^m5$ in the zone between $+17°$ and $+25°$, and epochs between about 1930 and 1936 ostensibly on the system of the GC. The rms. error of a catalogue position, which typically depends on measurements on two plates, is about $0.''2$ or slightly less in both coordinates. The catalogue itself contains the positions in zones that are 2° wide. Within these zones, the stars are numbered consecutively in the order of increasing right ascension. Underlined numbers indicate reference stars. Also given are the numbers on one plate of the Paris Astrographic Catalogue zone, the AC magnitude, the 1925.0 right ascension (to $0.^s001$) ostensibly on the system of the GC, the first and second term of the precession series, the centennial proper motions (only for a selection of stars) to $0.^s1$ from comparison with the Paris AC, the epoch $t - 1925.0$ in years, the declination 1925.0 to $0.''01$, the precession terms in declination and, finally, the centennial proper motion in declination to $0.''1$.

5.4.2 THE NEW REDUCTION OF THE CPC ZONES $-52°$ TO $-56°$ AND $-56°$ TO $-60°$. At the United States Army Topographic Command (USATOPOCOM, formerly United States Army Map Service), a project was initiated to re-reduce by means of a sophisticated version of the plate overlap method (Googe, Eichhorn and Lukac 1970) all the CPC and Yale zones. Due to untoward circumstances, not the least of which were budget cuts, the only part of the project that was realized was the development of the theory referred to above and the re-reduction of the CPC zones $-52°$ to $-56°$ and $-56°$ to $-60°$. In contrast to the original reduction, this new reduction (Lukac, Murphy, Googe and Eichhorn 1971) was not carried out in steps, that is, by adding several corrections together which were derived by different procedures in order to get the positions, but by first establishing a model for the reduction and then simultaneously determining the constants in this model. For the establishment of the model, the terms whose existence had been previously established by the Cape astronomers yielded valuable guidance. As a special feature of the new reduction, a criterion was developed and used to determine how much the coefficient of a certain term changes from plate to plate. Some of these coefficients proved to be virtually constant in the whole zone, others were completely uncorrelated on neighboring plates and some varied in a restricted fashion from plate to plate.

In terms of the accuracy measured by the internal agreement between the star positions on overlapping plates, the new reduction was successful. The new rms. errors of a catalogue position in the zone $-52°$ to $-56°$ are $0.''15$ in $\alpha\cos\delta$ and in δ, as compared to $0.''15$ and $0.''18$ in the original reduction, while in the zone $-56°$ to $-60°$ the rms. errors of a catalogue position, after the new reduction, are $0.''10$ in both $\alpha\cos\delta$ and δ, while the corresponding figure in the original reduction was $0.''15$. It is worth mentioning, however, that the positions obtained through the new reduction sometime differ by several tenths of a second of arc from those originally obtained. The newly reduced positions are also on the system of the FK3. No proper motions were derived.

5.4.3 THE SYDNEY ZONE $-48°$ TO $-54°$. Plates covering this belt had been taken in 1965 at the Sydney Observatory by Harley Wood following, essentially, the edge-in-center-line overlap pattern in right ascension, as did the plates taken for the Yale Zones and the CPC zones. The Taylor, Taylor and Hobson camera used gave a scale of about $116''$/mm. The plates were measured at the Observatory of the University of South Florida (in direct and reverse positions) under the supervision of H. Eichhorn, and reduced at the USATOPOCOM under the supervision of H. Eichhorn and W. D. Googe. The program covered most CPD stars in the area and a selection (usually two per square degree) of faint AC stars for the purpose of providing material for establishing the magnitude equation of the AC zones involved.

The reductions were made following the plate overlap method as described by Googe, Eichhorn and Lukac (1970) in the same way as the re-reductions of the Cape zones. The entire field of the plates was measured, so that some positions depend on more than two images. The catalogue is finished and ready to be printed; the accuracy of a position depending on two images corresponds to a rms. error of about $0.''14$.

The positions are ostensibly on the system of the FK4. One of the particularly difficult problems in the reductions of the measurements in this zone was the fact that there was no homogeneous source of reference stars covering all declinations in the zone. This resulted in the fact that the reference material had to be collected from more than one catalogue. It looks, however, as if undue systematic errors have been avoided. The catalogue also contains only positions and no proper motions.

5.4.4 WORK PLANNED AT WASHINGTON AND AT SYDNEY. In connection with the SRS program, or rather as its utilization, the U.S. Naval Observatory in Washington plans the establishment of a photographic AG type catalogue for the entire southern hemisphere of the sky. Investigations are currently under way that are to decide on optimum procedures. At the Sydney Observatory, a large region from $-43°$ to $-90°$ is being photographed currently with the Taylor, Taylor and Hobson lens. There, neither measurement nor reduction have yet begun.

5.5 The Astrographic Catalogue

5.5.1 DISPOSITION AND HISTORY. Soon after the introduction of dry emulsion plates, a number of pioneering investigations demonstrated the great potential of photographic methods for obtaining accurate star positions.

Around 1880, two astronomers and instrument makers at the Paris Observatory, the brothers Paul and Prosper Henry, had built a telescope for the photography of stars, which was later to be called a *normal astrograph.*

It is basically a combination of two telescopes; one, a photographically corrected lens system with an aperture of thirty-three centimeters and a focal length of 3438 mm, producing a scale of $1' = 1$ mm, which is used on a field about $130' \times 130'$, and the other, a telescope with a visually corrected lens system of about 25 centimeters aperture and a focal length of about 3650 mm. The two lens systems are mounted practically in one tube, so that they are rigidly connected as possible. The photographs are taken with the photographic telescope functioning as a camera, while the visual one is used for guiding. The Henry brothers' experience with this instrument was very encouraging. Therefore, in April of 1887, the French Academy of Science, at the urging of Admiral Mouchez, who was then the Director of the Paris Observatory, invited astrometrists from many countries to a conference at Paris to discuss the formation of a catalogue of star positions by means of photographs taken at instruments patterned after the prototype at Paris.

This was the beginning of the biggest astronomical enterprise ever undertaken by international cooperation: *The Astrographic Catalogue* (AC) (German: Photographische Himmelskarte [PHK]; French: Carte du Ciel [CdC]).

One must realize how revolutionary the idea of photographic star positions was at the time. Up to then, most star positions were obtained by observations on transit circles. This imposed a limit of about the 9th magnitude on stars accessible to observation; and (at that time) the typical rms. error of a meridian position was in the neighborhood of about $0.''5$ in either coordinate.

The photographic method then under consideration made it practical (without too much effort) to yield positions with a rms. error of about $0.''3$ in either coordinate for stars as faint as 13th or even 14th magnitude. Although there was considerable resistance against the project, in part by competent and very influential astrometrists,[4] the plans were approved and work started almost immediately.

The catalogue was to give accurate rectangular coordinates, as measured on the plates, of all stars brighter than eleventh magnitude. Some zones give the coordinates of stars as faint as 13^m. On the average, the coordinates of about 200 images were measured on each plate, corresponding to a density of approximately 40 stars per square degree. The measuring accuracy should not be less than that corresponding to a rms. error (standard deviation) of $0.''3$ in either coordinate.

4. The eminent German astrometrist Artur Auwers, for instance, is said to have opposed the Astrographic Catalogue project because he claimed that the application of photographic methods in astronomical research was "against nature."

This aim has, with one or two exceptions, been attained and even surpassed. Unfortunately, very little was known at the time about the difference between measuring accuracy, that is, the accuracy with which the position of the image on the plate can be measured, and the actual position accuracy, that is, the accuracy with which the position of the image on the plate represents the position of the star on the sky. We shall call the latter the *intrinsic accuracy* because it cannot be increased no matter how accurate the measurements—or the settings on the images of the stars on the plate—are made. The total accuracy of the rectangular coordinates is thus a combination of the intrinsic accuracy and the measuring accuracy.

The catalogue was also to contain plate constants for the conversion of measured into standard coordinates (in units of minutes of arc and oriented to 1900) from which the spherical positions (that is, right ascensions and declinations) of the stars could be derived with an rms. error of not more than $0.''5$. This figure was apparently suggested because it corresponded (at that time) to the accuracy of good relative meridian observations. There may have been some doubt as to whether photographic positions could actually be more accurate, or it may simply have been a case of scientific inertia, of which the history of the Astrographic Catalogue is so full. Economic considerations may also have played a role, as it is obvious that too rigorous an accuracy requirement would have led to a highly increased cost. It now seems, in retrospect, that this figure for the maximum acceptable rms. error is responsible for much of the inhomogeneity characteristic of the finished product. While practically all zones are capable of yielding positions with a rms. error smaller than $0.''5$, the mean error of a star position obtainable from the AC varies between about $0.''17$ and $0.''40$.

The photographs were to be arranged in such a way that every region of the sky was covered by at least two plates. The format of the plates was 16×16 cm^2 and there were to be 2°10′ wide "belts" (or zones) of plates centered on every parallel of declination (equator 1900) corresponding to a full degree. As far as the convergence of the meridians permit this, the right ascensions of the plate centers should lie between those of the neighboring zones, so that the entire sky is covered by a roof-tile like pattern of plates in which the center of each plate coincides with points close to corners of four other plates. This is the so-called corner-in-center overlap pattern. Fortunately, this particular overlap pattern (where every region is recorded on at least two and, at the most, five plates) lends itself very well for the determination of certain types of systematic errors and is, for this reason, to be preferred to the "edge-in-center-line" overlap pattern where there would be no appreciable overlap between plates whose centers are at different declinations. Unfortunately, in some zones, only the measures of images on not more than two adjacent plates were published.

Due to this arrangement, stars which were near the center of one plate could be measured on as many as five plates. However, faint stars are frequently measured on one plate only. This usually happened when the region containing them was photographed on only two plates and when their images on one plate were too faint to be measured.

Table V-5

The Zones of the Astrographic Catalogue

Observatory	*Plate Centers* *From*	*To*	*Plate Epochs* *Earliest*	*Latest*	*Coord's. of Plate Center* x	y	*Units of Meas. Coord's.*	*Mode of Meas.*	*Remarks*
Greenwich	+90°	+65°	1892	1905	14	14	5′	S	
Vatican	+64	+55	1898	1922	13	13	5′		1
Catania	+54	+47	1898	1926	0	0	1′	M	
Helsingfors	+46	+40	1892	1903	0	0	1′	M	
Potsdam	+39	+32	1893	1900	0	0	5′	M	2
Hyderabad N	+39	+36	1928	1936	13	13	5′	S	
Uccle	+35	+34	1940	1950	0	0	1′	M	3
Oxford II	+33	+32	1932	1936	13	13	5′	S	
Oxford I	+31	+25	1892	1904	13	13	5′	S	
Paris	+24	+18	1891	1906	0	0	1′	M	
Bordeaux	+17	+11	1893	1913	0	0	1′	M	
Toulouse	+10	+5	1893	1935	0	0	1′	M	
Algiers	+4	−2	1891	1903	0	0	1′	M	
San Fernando	−3	−9	1894	1903	0	0	1′	M	4
Tacubaya	−10	−16	1900	1912	0	0	5′	S	
Hyderabad S	−17	−23	1914	1928	13	13	5′	S	4
Cordoba	−24	−31	1909	1913	14	14	5′		6
Perth	−32	−37	1910	1919	13	13	5′	S	
Perth-Edinburgh	−38	−40	1910	1919	14	14	5′	S	4
Cape	−41	−51	1892	1910	0	0	1′	M	
Sydney	−52	−64	1891	1918:	14	43	5′	S	5, 7
Melbourne	−65	−90	1891		0	0	1′	M	

In the column "*Mode of Measurement,*" S means an eyepiece scale, and M a micrometer screw. x always increases toward the east and y toward the north, unless otherwise indicated in the remarks.

Remarks:

1. Special measuring machine (eyepiece grid, see Section 5.5.5.2).
2. Zone incompletely published. All plates were taken, but may have been destroyed during the events of World War II.
3. The measurements were carried out at Paris.
4. y increases toward the south.
5. x increases toward west.
6. 12% of the measurements, M, 88% S.
7. Plates measured before 1912 were measured at Melbourne on the M-type machine of that observatory

The photographs were made for two purposes: (a) To establish a photographic record of the sky complete to about 15th magnitude, and (b) to establish a catalogue of star positions. The purposes set forth in (a) and (b) were to be accomplished by different sets of plates. We are concerned here with the catalogue part, that is, (b) only, and therefore will not further discuss (a).

The work of determining roughly one-and-a-half million star positions was to be divided among about twenty observatories, none of them, curiously, in the United States; see Table V-5. Each of them was assigned a zone, that is, a belt between two parallels of declinations (equator 1900), in which they were to take the necessary photographs and to measure and reduce them. As must be expected in a project of this size, not all observatories which originally had intended to participate went through with their plans, so that some of the zones had to be reassigned. This was especially serious in the case of the Astrophysikalisches Observatorium at Potsdam, which stopped work on the Astrographic Catalogue after the First World War, after a good deal of their work had been done and published. In order not to leave the zone, which was originally Potsdam's responsibility, a fragment, it was reassigned to the observatories at Hyderabad and at Oxford. When H. H. Turner, director of the Oxford Observatory died, Oxford stopped the work after completing half of its share of the reobservation of the Potsdam zone. The rest of the Potsdam zone was finally photographed at Uccle (near Brussels) and measured and reduced at Paris. The volumes containing the results of the Uccle-Paris effort appeared in 1962.

The exposures were to be made on normal astrographs as described above. Every plate was to be given a set of three exposures, mostly lasting 6^m, 3^m, and 20^s, respectively.

In order to facilitate measuring (at least according to the state of the art at that time), and to have a check on the film shifts that one feared might occur during the processing and thus introduce serious errors, a réseau was to be exposed on the plates either before or after the sky was photographed.[5] This has an advantage and a disadvantage. The advantage is that one star or a few stars can be remeasured on the plate at any time without having to remeasure the reference stars, because all measurements are made with reference to that réseau square which contains the star. The disadvantages are that the measurements are more time consuming, since the coordinates of not only the stars but also the réseau lines must be meas-

---

5. A réseau is a silvered glass plate. Into the silver film are scratched fine lines forming a quadratic net with the lines 5 mm apart. To copy the réseau on the photographic plate, the silver film and the photographic emulsion are brought almost into contact and then shortly exposed to light which is arrested by the silver film, but passes through the lines. The measurements of the positions of the stars' images were then made with respect to the photographic images of the réseau lines. In the meantime, astrometrists have found that the use of a réseau is of very doubtful value. The accidental film shifts were shown (independently by several investigators) to have a dispersion of about 2 microns. In general, while the intrinsic error of a photographic position is not decreased by the use of a réseau, the measuring labor is at least doubled.

ured, and secondly, that the measurements lose accuracy because they are actually the differences between the settings on the réseau line and on the stars.

Measuring was, therefore, at that time a much more laborious and time consuming process than it is today. Since setting on the réseau line was an integral part of the measurements, long screw measuring machines were not used, perhaps because the men in charge did not trust that long measuring screws could be manufactured with the necessary accuracy, or because long screw machines were judged to be inconvenient. We know now that it would have been faster and more accurate if all the plates for the Astrographic Catalogue had been measured on long screw measuring machines.

Today, the early epoch of most of the AC and the high achievable accuracy makes the information, embodied in the volumes of this gigantic enterprise, one of the most powerful tools available for the determination of accurate proper motions.

5.5.2 PHOTOGRAPHY, MEASUREMENT AND REDUCTION. Only few guidelines were given which could not be deviated from: type of telescope, plate format, overlap pattern, minimum measuring accuracy, use of réseau and length of exposures. Except for these guidelines, which were more or less followed,[6] the various observatories had complete freedom regarding the processing of the plate the measurement of the images, the reduction of the measurements, and the format of publication.

As might be expected under these circumstances, the various phases of the work were executed differently at the several observatories participating in the program. This lack of uniformity resulted in a considerable variety of formats, accuracies, and so forth, causing quite a nuisance to any user of the Astrographic Catalogue.

Normal astrographs are so designed that during the *photography*, the eyepiece of the visual guiding telescope is carried by a micrometer plate and can be moved in all directions up to usually about one degree off the optical axis. The large size of the area which can be scanned by the eyepiece of the guiding telescope is intended to increase the chance that a sufficiently bright guiding star can be selected.

In all normal astrographs used to expose the plates for the AC, there were more or less elaborate provisions to carefully define and adjust the plane of the plate perpendicular to the optical axis, to have the edges of the plate parallel to the parallels of declination, to focus the plate, and to properly center the objective. Adjustments to achieve ideal conditions were always made so that, theoretically, corrections for tilt or for decentering distortion would in no case be necessary. This aim has probably not always been achieved. Most authors of the introductions to the various AC zones complain that the focal surface is not plane, and most plates were taken with the sharpest focus 30′ to 50′ from the center of the plate.

The coordinates of the stars on the *plates* were measured with two fundamentally different types of *measuring machines*. In one, the distances between the

6. The only observatory at which the plates were not exposed on a normal astrograph is Hyderabad.

lines of the réseau and the images of the stars were measured with a short eyepiece micrometer screw. This approach was followed by all the participating French observatories, and some others. These are denoted by M in Table V-5. In the other, these distances were measured by means of a cross of scales that appeared in the eyepiece of the measuring microscopes; this method was preferred by most observatories within the British Empire. The procedure with measuring machines of this type was as follows: The center of the cross of scales (each half beam of which is divided into 100 parts, reading from 0 to 100 or -100, and matches a réseau interval in length as accurately as possible) was centered on the star image to be measured, and the four places on the cross of scales where the réseau lines intersect were recorded in units of one-thousandth of a réseau interval which corresponds to to 5μ. Zones measured by this method are indicated by S in Table V-5. (A still different method was employed for the measurement of the coordinates in the Vatican zones.)

It has been argued that this method cannot be regarded as a measurement, but must rather be considered only an estimate of the coordinates of the star images. This view is hardly justified. If the setting is perfect, no error larger than $2\frac{1}{2}\mu$, corresponding to 0.″15, can be made. Even this error can occur only in extreme cases (provided, of course, that the setting is indeed perfect), and thus the standard deviation of the recorded coordinate will be about 0.″10. Since most plates were measured in direct and reverse positions, this may even further be reduced by a factor of $\sqrt{2}/2$ to 0.″07. Considering that the intrinsic error (depending on the quality of the plate, and so on) is about 0.″15 to 0.″20 for good plates, the limitations in recording accuracy will thus produce total standard deviations of the recorded positions of from 0.″18 to 0.″22. If—and experience has indeed shown this—zones measured with the eyepiece scale(S) method yield somewhat less accurate positions than zones measured with short screw micrometers (M), this cannot be due to the failure to record the S measurements to 1μ (or even 0.1μ as some of the M zones have done) but to the fact that the settings are made faster and less carefully with the eyepiece scale cross than with a micrometer screw. The opinion that the eyepiece scale micrometers will produce positions with a rms. error of 1″ is untenable: rather, 0.″3 is a more realistic figure (see, for instance, Eichhorn and Gatewood 1967, Standish 1962) as compared to 0.″2 for the excellent screw micrometer zones.

The measuring machines preferred by the French observatories, and some others which participated in the project, were equipped with small screw-type eyepiece micrometers which permitted one to read off the distance between a star image and a réseau line to 0.1μ. Several observatories published their measurements to 0.0001 mm. The last figure (that is, 0.1μ) is of course entirely insignificant, because the intrinsic errors are usually of the order of two to three microns. If the measurements had been published with microns as last figures, the published measurements would have given just as accurate positions and the printing costs would also have been noticeably reduced. At the different observatories, the workers went about measuring in different ways. Some measured two images (the 6^m and the 3^m exposure), some only one, some measured the plates in two positions turned relatively to each other by 180° (that is, in "direct" and "reverse" position), some in only one position (per coordinate), some measured all four réseau lines of the square surrounding a star, others only two, and so on.

One of the reasons why even today the measured rectangular coordinates of the stars, which are printed in the various volumes of the AC, will not yield spherical positions with the accuracy that they would be ultimately capable of, is the fact that almost *all of the originally derived plate constants were only provisional*, and besides, they did not yield positions on a well-defined system. We shall see below that these inaccurate constants are only gradually being replaced by more definitive ones, but even so, there is still a great deal of work to be done before the full potential accuracy of the AC will have been realized.

Remembering formula (1.8c), we realize that inaccurate plate constants will give rise to inaccurate positions by producing a large plate constant variance. Naturally, this quantity depends mainly on the position of the image on the plate and on the model chosen for the conversion of measured to standard coordinates. Eichhorn and Williams (1963) have illustrated the behavior of the plate constant variance for a variety of conversion models of $\{x, y\}$ to $\{\xi, \eta\}$, although in their paper they did not consider the possible dependence of the standard coordinates on magnitude and color as well as on the position of the image.

Since one can, at least approximately, always express the standard coordinates in terms of the measured coordinates through the formula (2.14a), the contribution of the variance of the measured coordinates x and y, σ^2_{xx} and σ^2_{yy}, respectively, to the variances $\sigma^2_{\xi\xi}$ and $\sigma^2_{\eta\eta}$, will be $A^2\sigma^2_{xx}$ and $A^2\sigma^2_{yy}$, respectively. Nothing, short of a remeasurement of the plates, will diminish these components. They thus represent an upper limit of the achievable accuracy. On the other hand, the choice of a more adequate model for the computation of the standard from the measured coordinates can very well reduce the contribution of the plate constant variances to the total variances $\sigma^2_{\xi\xi}$ and $\sigma^2_{\eta\eta}$.

None of the observatories at which any of the astrographic catalogues were measured and reduced at that time had modern calculating aids, or even desk calculators. Under these circumstances, the adjustment model for the computation of ξ, η from (x, y, m, c) had to be kept as simple as possible. m and c were originally not taken into account anywhere. In almost all cases, either equations (2.11) or (2.14a), (that is, linear four- or six-constant approach) were used. When equations (2.11) were chosen as the reduction model, appropriate corrections for aberration and refraction were computed and applied, so that even then in the final formulas for computing ξ and η from x and y, $A \neq E$ and $B + D \neq 0$. Only in a few instances were the plate constants obtained by least squares adjustments. Usually, four fictitious reference stars were created by averaging the data in each quadrant of the plate, and the plate constants were computed from them by an adjustment process known as the Christie-Dyson method mentioned above in Section 5.3.2. The number of the reference stars per plate and the accidental and systematic accuracy of their positions varies greatly from zone to zone, and even within a zone. For some plates, the reference stars were taken from available catalogues sometimes updated to the epoch of the plates by application of proper motions; for others, the reference stars were especially observed on transit circles. In no case are the covariance matrices of the plate constants given nor are they even easily obtainable from the materials published in the various volumes of the AC. Thus, an accurate estimate of the influence of the inaccuracies of the plate constants on the computed standard coordinates cannot be given. The plate constant variances could, however, be rather accurately estimated from the curves given in the above-mentioned paper by Eichhorn and Williams (1963).

The principal sources for *reference stars* were the volumes of the AGK1 (Catalog der Astronomischen Gesellschaft) and some other catalogues, some of which were observed ad hoc for the purpose of providing reference stars for the Astrographic Catalogue. Almost all of these reference star positions were of course affected by systematic errors. The reference star catalogues were observed on various fundamental systems and were, of course, also affected by systematic deviations from the system on which they were observed (as were almost all meridian catalogues, especially at that time). In some cases, the reference stars were taken from several different catalogues which showed systematic differences against each other. Additional systematic errors were frequently introduced by neglecting the effect of proper motions during the interval between the epochs of the plates and the reference star catalogs, which sometimes differed by several decades. In such cases, the systematic errors were in part due to the systematic component in the proper motions.

All systematic errors of the reference stars were, of course, propagated into the positions computed with plate constants obtained by using these reference stars.

5.5.3 DEFINITIVE REDUCTION PARAMETERS FOR THE ZONES OF THE ASTROGRAPHIC CATALOGUE.

5.5.3.1 THE DETERMINATION OF TERMS THAT DEPEND ON MAGNITUDE ALONE. As was pointed out in the previous section, a linear model was always used for the establishment of the original formulas for the conversion of the measured AC coordinates to standard coordinates. Terms of order higher than one were originally included in only one case in the reduction models; in addition, no attention was paid to the fact that the positions of the stars might depend not only on the measured positions of the images alone, but also on the colors of the stars and especially their magnitudes. The possibility that these last two parameters might influence the positions was completely unforeseen, and therefore no intentional precautions were taken initially, when the plates were exposed, to provide means for an eventual determination of their effect on the positions. No grating (see Section 2.3.8.1), for instance, was ever used on an AC photograph. As a result, it is not, at this time, possible to determine from the published measured coordinates, systematic errors in the AC which depend on the magnitude alone because there just are no positions of sufficiently faint stars available which might serve as a standard against which the AC positions could be calibrated.

This is unfortunate because magnitude dependent position errors are especially dangerous. Terms in the model for converting measured to standard coordinates which depend on magnitude alone cannot be determined from overlaps. Such magnitude errors will propagate themselves into the proper motions that will eventually be derived from them, and thus will lead to erroneous values for secular parallaxes whose determination is, after all, one of the reasons for deriving proper motions. For this reason, it is important that enterprises for the establishment of catalogue positions, say, of the type and coverage of the AG catalogues, include in their programs a sufficiently large number of faint AC stars so that these may sometime be used as reference stars for the determination of definitive reduction parameters of the AC measurements, in particular of coefficients of those terms in the reduction formulas that depend on the magnitude only. Even so, they will have to be observed at two epochs so that their own proper motions may become known. So far, stars of this type have been included only in the Yale zone $-70°$ to $-90°$ (Section 5.2.8) and the Sydney zone $-48°$ to $-54°$ (Section 5.2.4).

The only chance to determine from existing material those errors in the AC positions that depend on magnitude alone might be provided by the fact that in almost all zones, and on every plate, at least two unequally long exposures were taken. This means that the image of every star is accompanied by a fainter image of the same star in such a way that the ratio of total incident light is constant for each pair of images. If the magnitude errors were intrinsically produced by the camera system and not through the peculiarities of guiding, they would be very similar on all plates taken with the same optical system. In this case, a significant dependence of the differences between the coordinates of the strong and faint images of the same stars with, say, the total blackening of the strong image would indicate that the magnitude equation is not linear. (One can easily see that a linear magnitude equation would just add a constant to the intrinsic coordinate difference, provided that the magnitude difference between strong and faint image is constant; but then, a purely linear magnitude equation could, of course, be established by comparison with reference stars whose magnitude range is less than the magnitude range of the field stars.) Unfortunately, the investigation of a single plate according to this method has no significance because the magnitude equation depends rather strongly on the peculiarities of guiding on this plate. If, however, a combination of samples of pairs of images from all plates were investigated, the peculiarities of the individual plates could be regarded as random, and the question of whether or not nonlinear magnitude equation exists for the whole zone could be answered. Once this is established, the behavior of the magnitude errors on different plates could be established in principle by a careful investigation of plate overlaps. The availability of electronic computers and of modern measuring machines would make an investigation as described above a much less formidable enterprise than it might seem at first sight. Of course, the cooperation of the observatories in whose archives the plates are stored (if they have not yet deteriorated) would be a necessary condition for success.

Investigations of the systematic differences in the 1° wide belts that are covered by two adjoining zones are also useful material for the investigation of the magnitude equation of the positions derived from the measurements in any one particular zone. It was in this way that Heckmann, Dieckvoss and Kox (1949) established the magnitude equation of the Bordeaux zone (see Section 5.5.5.10).

5.5.3.2 EXISTING DETERMINATIONS OF DEFINITIVE PLATE PARAMETERS. The chief reason why definitive plate constants for the AC zones were not derived soon after the plates were measured was that contemporaneous and adequately accurate positions of a sufficient number of reference stars were not available. Frequently, the adjustment was by nature only approximate, so that systematic errors caused by inaccurate plate constants were introduced into the astrographic positions in addition to the systematic errors of the reference star system (if this can indeed be defined).

Therefore, the accuracy of Astrographic Catalogue positions computed from the measurements with the original plate constants and referred to a modern fundamental system, say the FK4, is on the order of one or two seconds of arc, while it is now known that the accuracy with which Astrographic Catalogue positions represent the system of the reference stars can practically be as good as the relative accuracy. For this purpose, it is necessary that the plate constants be determined by rigorous least squares ajustments which involve enough reference stars whose positions are known with adequate accuracy.

The determination of new linear plate constants, however, is not always the only step necessary to realize the full potential accuracy of Astrographic Catalogue positions. Frequently, an affine transformation will not adequately represent the relationship between the standard and the measured coordinates. In these cases, more complicated transformations, which in some cases involve higher order terms as well as the magnitudes and colors of the stars, are necessary.

The Astrographic Catalogue thus potentially contains the positions of almost one-and-a-half million stars, complete down to at least 11th magnitude, with an accuracy corresponding to a rms. error of $0.''25$ or better. Considerable effort is necessary, however, to realize this potential. The Astrographic Catalogue is at the same time the only source for these positions, and particularly valuable to astronomy because of its early epoch. (This very same early epoch, however, might be considered a disadvantage by those who need positions at a recent epoch.) The reason why only linear reduction models were chosen, and why the rectangular coordinates were not (except in two zones) converted to spherical coordinates by the use of the provisional plate constants is the enormous calculating effort that would have been necessary to determine the parameters in more complicated models, and independently of this, to convert the measured rectangular coordinates to spherical coordinates. In retrospect, one can now say that the publication of the AC in the present form, that is, as lists of rectangular plate coordinates of the images, lends itself much better to the derivation of definitive reduction parameters and of positions, now that computers are available, than if the spherical coordinates of the stars had been published. At the time when the format of the publication of the AC was decided, all this could hardly have been foreseen. Today, however, with electronic computers available, there is no longer the impediment of the forbidding amount of numerical calculations to the optimum reduction of the measurements published in the AC. A new reduction is, however, indicated only if reference star positions are available at the epochs at which the plates were taken. In this respect, the situation is becoming more and more favorable. On the northern hemisphere, positions and proper motions contained in the AGK2-3 provide reference material for all the AC zones in this region. These were used by Günther and Kox (1970, 1970a, 1971) to derive improved plate constants for the AC zones Greenwich, Vatican, Catania, Helsingfors, Potsdam, Hyderabad N, Uccle, and Oxford. Their method is described in their 1970 paper, while the constants themselves were published in their 1970a and 1971 papers. The reduction formulas are a variation of the equations (2.14a) in the sense that for each plate the right ascension α_0 and declination δ_0 (1950, FK4 system) of the tangential point is given, with respect to which the standard coordinates in units of minutes of arc are computed by

$$\begin{pmatrix}\xi\\ \eta\end{pmatrix} = \begin{pmatrix}A & B\\ C & D\end{pmatrix}\begin{pmatrix}x + \mathrm{d}x\\ y + \mathrm{d}y\end{pmatrix}, \tag{5.4}$$

where x and y are the measured coordinates approximately in units of minutes of arc and reckoned from the geometric plate center, and A, B, C and D the newly determined plate constants. To find the $\mathrm{d}x$ and $\mathrm{d}y$, these investigators assumed, as did the astronomers responsible for most of the Yale zones and those who constructed the CPC, that higher order and magnitude dependent terms $\mathrm{d}x$ and $\mathrm{d}y$ are of the same form for all plates in each zone. Günther and Kox assumed these corrections to be of the form

$$\begin{aligned}\mathrm{d}x &= a_1x^2 + a_2xy + a_3y^2 + a_4x^3 + a_5x^2y + a_6xy^2 + a_7y^3 + a_8d\\ &\quad + c_1dx + c_2d^2x + c_3dxr\\ \mathrm{d}y &= b_1x^2 + b_2xy + b_3y^2 + b_4x^3 + b_5x^2y + b_6xy^2 + b_7y^3 + b_8d\\ &\quad + c_1dy + c_2d^2y + c_3dyr,\end{aligned}$$

where d is the measured diameter of the image, and $r^2 = x^2 + y^2$. Coma type aberrations (those whose coefficients are c_1, c_2 and c_3) were thus assumed to be radially symmetric. All a_i, b_i and c_i were further assumed identical for all plates of the same zone. Note that, obviously due to the insufficient magnitude range of the reference material, no provisions were made for a term in d^2 in either $\mathrm{d}x$ or $\mathrm{d}y$. As emphasized before, it is exactly these terms whose presence could endanger the correctness of secular parallaxes derived from the material in the AC. Also, Günther and Kox made no use of overlap conditions in their derivation of plate constants and higher order corrections. Their constants, while certainly the best available at this time, are not the best ones which could be derived by a more ambitious effort.

New plate constants, in the form

$$\xi = \alpha_1 x + \alpha_2 y + \alpha_3 + \alpha_4 d + \alpha_5 dx$$
$$\eta = \beta_1 x + \beta_2 y + \beta_3 + \beta_4 d + \beta_6 dy \qquad (5.5)$$

were computed and published by Eichhorn and Gatewood (1967, 1967a) for the northern Hyderabad zone who used the practically contemporaneous AGK2 positions as reference positions. The coma effect constants α_5 and β_6 were allowed to vary only by a restricted amount around a common value, and some use of the overlap method was made in the derivation of the α_i and β_i. There is evidence, however, (Eichhorn 1971) that the model (5.5) is not quite adequate for the description of the dependence of ξ and η on x, y and d, especially when the α_i and β_i are determined by the plate overlap method.

The pioneer investigations concerning the determination of new, improved plate constants were carried out by Heckmann, Dieckvoss and Kox (1949, 1954). In their first paper, they computed improved plate constants for a selection of the plates in the AC zones between +14° and +25°. This includes the declination zone covered by the AGK1 zone Berlin B which is the most accurate zone of the AGK1, as demonstrated by the figures in Table V-1. Reference star positions for the AC zones in this declination range were obtained by first deriving proper motions of the reference stars from their positions in the AGK2 and the Berlin A and B zones of the AGK1, respectively,—the latter having been reduced to the FK3 system. With the aid of these proper motions, the reference star positions were computed for the epochs of the AC plates, in the FK3 system, and the new plate constants were obtained by a least squares adjustment following the formula (2.14a). In their first paper, (1949) the authors solved for six constants A, B, C, D, E, F; but C and F were converted to corrections to the tangential point so that they give A, B, D, E, and the coordinates of the tangential points, α_0 and δ_0, precessed to the orientation 1950. In their second paper (1954), the nonorthogonal terms of refraction were computed theoretically, and the adjustment followed the model of equations (2.11). There they give new plate constants for all plates in the AC zones with centers +21° to +24°. The formulas for the computation of standard coordinates, oriented to the coordinate system 1950 from measured coordinates (counted from the plate center) are given also in the form of equation (5.4) although without the dx and dy. To replace C and F by α_0 and δ_0, respectively, makes it somewhat simpler to use the formulas, since the additive constants are no longer there. This procedure makes it mandatory, however, that the coordinates x, y be reckoned from a point that is reasonably close to the actual geometrical tangential point, otherwise serious systematic errors would creep in. For the AC, it is always sufficient to reckon, for this purpose, x and y from the geometric center of the plate, which may always be assumed to be sufficiently close to the tangential point, without thereby introducing avoidable systematic errors. See also Herget's (1973) work.

5.5.3.3 Other problems concerning the determination of definitive plate parameters. At Strasbourg, new, improved plate constants are being determined by Lacroute and his collaborators for the zones Paris, Bordeaux, Toulouse, and Algiers, all with the aid of reference star positions computed from the proper motions in the AGK3. The French investigators are using a plate overlap method, but nothing is known at this time about the details of the model which they employ.

Lacroute (1971) has pointed out that the unavoidable systematic errors in the AGK3 proper motions will lead to considerable systematic errors in the AC positions of the order of half-a-second of arc. These could be made considerably smaller if other catalogues, besides AGK3 and AGK2, were used as sources for the computation of reference star positions at the epochs of the AC plates. The efforts of Günther and Kox on the one hand, and of Lacroute and his collaborators on the other, lead to constants which are better than those available up to now, and will supersede all previous work, although it is doubtful whether they may be termed definitive. They will allow one to compute, from the published rectangular coordinates, the positions of the stars on a system not too different from that of the FK4.

On the southern hemisphere, proper motions whose accuracy is comparable to that of the proper motions in the AGK3 will not be available for some time, and one may therefore assume that the computation of improved (let alone definitive) plate constants for the zones south of $-2°$ will only be accomplished sometime in the future.

It was already mentioned above that even the newly determined plate parameters for the northern hemisphere AC zones are not the best ones which could, with the appropriate effort, have been obtained at this time. Günther and Kox made no use of overlap conditions, and assumed the higher order terms in the coordinates and the magnitude terms to be identical for all plates obtained at any given observatory. Undoubtedly, the use of the plate overlap method would have yielded plate parameters and thus, in the end, positions of superior accuracy. Furthermore, it is a quite well established fact that at least some of the correction terms (that is, the coefficients in the expressions for dx and dy) vary significantly from one plate to the next among the plates taken on the same telescope. An algorithm for the proper consideration of this circumstance was published by Googe, Eichhorn and Lukac (1970), and applied in the new reduction of the CPC measurements (Section 5.4.1) and the establishment of the Sydney zone $-48°$ to $-54°$ (Section 5.4.2). All these procedures do, however, require considerable effort, and different investigators may be of different opinion as to whether the rather significantly increased effort is really worth the possibility of only a slight increase in accuracy.

One point, however, which for some time to come is bound to defy the efforts of the astrometrists, is the *color dependent systematic errors* of the positions derived from the measurements in the AC. According to the experience gathered almost universally at the observatories that produced photographic AG-type star catalogues, for instance Yale, Cape, Hamburg, Bonn and Sydney, the measurements must be corrected for color effects, notably color magnification error whose magnitude often even depends on the intensity of the stellar images on the plate. Eichhorn (1962, unpublished) noticed that such effects do exist in the Northern Hyderabad zone. On the other hand, corrections for them cannot be applied unless one knows the color equivalents of all AC stars. Needless to say, these are nowhere near yet available in their entirety. Thus, the determination of color dependent effects in the AC positions—and they certainly do exist in some zones—is, at this time, somewhat of a futile effort, at least on a comprehensive scale.

5.5.4 THE HANDLING OF THE COORDINATES PUBLISHED IN THE AC.

5.5.4.1 PRECESSION OF PLATE CONSTANTS. The available published plate parameters (constants) yield spherical coordinates oriented to either 1900 or 1950. Frequently, the positions will be required at different orientations. If this concerns only one or two stars on any one plate, one could precess the star positions obtained from the available constants, by the standard methods. If the positions of a considerable number of stars on a plate are required for an orientation different from that which the available plate constants would yield, much computing will be saved by finding, from the available plate constants, those which will immediately yield the positions for the required orientations without precessing each star individually.

Assume that, with respect to the tangential point $\alpha_1\delta_1$ oriented to the coordinate system at the epoch t_1, the standard coordinates (ξ_1, η_1) are obtained from the measured coordinates (x, y) by the model (2.14a), namely,

$$\begin{pmatrix} \xi_1 \\ \eta_1 \end{pmatrix} = \begin{pmatrix} A_1 & B_1 & C_1 \\ D_1 & E_1 & F_1 \end{pmatrix} \begin{pmatrix} x \\ y \\ 1 \end{pmatrix}.$$

The standard coordinates (ξ_2, η_2) oriented to the coordinate system at the epoch t_2 will refer to the same tangential point whose spherical coordinates have been precessed to the orientation at t_2, and are now α_2, δ_2.
We have, generally

$$\begin{pmatrix} \xi_2 \\ \eta_2 \end{pmatrix} = \begin{pmatrix} \cos\Delta P & \sin\Delta P \\ -\sin\Delta P & \cos\Delta P \end{pmatrix} \begin{pmatrix} \xi_1 \\ \eta_1 \end{pmatrix} \tag{5.6}$$

and, with the model above,

$$\begin{pmatrix} \xi_2 \\ \eta_2 \end{pmatrix} = \begin{pmatrix} A_2 & B_2 & C_2 \\ D_2 & E_2 & F_2 \end{pmatrix} \begin{pmatrix} x \\ y \\ 1 \end{pmatrix}$$

with

$$\begin{pmatrix} A_2 & B_2 & C_2 \\ D_2 & E_2 & F_2 \end{pmatrix} = \begin{pmatrix} \cos\Delta P & \sin\Delta P \\ -\sin\Delta P & \cos\Delta P \end{pmatrix} \begin{pmatrix} A_1 & B_1 & C_1 \\ D_1 & E_1 & F_1 \end{pmatrix}. \tag{5.7}$$

ΔP is the change of the position angle at the tangential point due to precession, and is given by

$$\tan\Delta P = \frac{\sin J \sin(\alpha_2 - z)}{\cos J \cos\delta_2 + \sin J \sin\delta_2 \cos(\alpha_2 - z)} \tag{5.8}$$

in terms of α_2, δ_2, or by

$$\tan\Delta P = \frac{\sin J \sin(\alpha_1 + \zeta_0)}{\cos J \cos\delta_1 - \sin J \sin\delta_1 \cos(\alpha_1 + \zeta_0)} \tag{5.8a}$$

in terms of α_1, δ_1, where ζ_0, z, and J are the quantities previously defined by equations (3.4).

Equations (5.6) are rigorous, and therefore it is easy to see how equations analogous to equation (5.7) can be derived even if the transformation of measured to standard coordinates is not the simple affine model to which equation (5.7) relates.

5.5.4.2 IDENTIFICATION OF IMAGE COORDINATES. Often the following question arises: Which will be the measured coordinates of the image of a star with the spherical coordinates (α, δ), on a plate whose tangential point is at (α_0, δ_0)? For this purpose, we assume that the relationship between the star's standard coordinates and the measured coordinates of its image on the plate is expressed in the form of an affine transformation by equation (2.14a). We must also assume that at least approximate values for the plate constants $A, \ldots, F$ are known. With these assumptions, one obtains by solving equations (2.14a) for x and y:

$$\begin{pmatrix} x \\ y \end{pmatrix} = \frac{1}{AE - DB} \begin{pmatrix} E & -B & BF - EC \\ -D & A & CD - AF \end{pmatrix} \begin{pmatrix} \xi \\ \eta \\ 1 \end{pmatrix} \tag{5.9}$$

ξ and η are, of course, the standard coordinates of the star with respect to the tangential point of the plate. They must be pre-computed for this purpose by means of the formulas (2.9).

5.5.4.3 COMPARISON OF THE COORDINATES OF THE SAME STAR ON DIFFERENT PLATES. Assume that the same star is imaged on two plates, P_1 and P_2 whose tangential points are (α_1, δ_1) and (α_2, δ_2), respectively, and whose affine plate constants for the computation of standard coordinates from measured coordinates are $A_1, \ldots, F_1$, and $A_2, \ldots, F_2$, respectively. On plate P_1, the measured coordinates of the image of the star will be (x_1, y_1); on plate P_2, they will be (x_2, y_2).

How can one ascertain that the two pairs of coordinates (x_1, y_1) and (x_2, y_2) belong to the same star, that is, what is the relationship between (x_1, y_1) and (x_2, y_2)?

To answer this question, we remember that the standard coordinates (ξ_1, η_1) and (ξ_2, η_2), in units of radians, of the same star with respect to the different tangential points (α_1, δ_1) and (α_2, δ_2) are related by

$$\begin{pmatrix} \Xi_2 \\ \mathrm{H}_2 \\ \mathrm{Z}_2 \end{pmatrix} = \mathbf{S} \begin{pmatrix} \xi_1 \\ \eta_1 \\ 1 \end{pmatrix}, \quad \text{and} \quad \xi_2 = \frac{\Xi_2}{\mathrm{Z}_2}; \quad \eta_2 = \frac{\mathrm{H}_2}{\mathrm{Z}_2}, \tag{5.10}$$

$$\text{where } \mathbf{S} = \mathbf{R}_1(-\delta_2)\mathbf{R}_2(\alpha_2 - \alpha_1)\mathbf{R}_1(\delta_1) = \begin{pmatrix} a_1 & a_2 & a_3 \\ b_1 & b_2 & b_3 \\ c_1 & c_1 & c_1 \end{pmatrix}, \tag{5.11}$$

or in detail:

$$\begin{pmatrix} a_1 & a_2 & a_3 \\ b_1 & b_2 & b_3 \\ c_1 & c_2 & c_3 \end{pmatrix} =$$

$$\begin{pmatrix} \cos\Delta\alpha & \sin\delta_1\sin\Delta\alpha & -\cos\delta_1\sin\Delta\alpha \\ -\sin\delta_2\sin\Delta\alpha & \cos\delta_1\cos\delta_2 + \sin\delta_1\sin\delta_2\cos\Delta\alpha & \sin\delta_1\cos\delta_2 - \cos\delta_1\sin\delta_2\cos\Delta\alpha \\ \cos\delta_2\sin\Delta\alpha & \cos\delta_1\sin\delta_2 - \sin\delta_1\cos\delta_2\cos\Delta\alpha & \sin\delta_1\sin\delta_2 + \cos\delta_1\cos\delta_2\cos\Delta\alpha \end{pmatrix},$$

where we have introduced

$$\Delta\alpha = \alpha_2 - \alpha_1. \tag{5.12}$$

If the standard coordinates are expressed in units other than radians, then

$$\mathbf{S}' = \begin{pmatrix} a_1 & a_2 & fa_3 \\ b_1 & b_2 & fb_3 \\ \dfrac{c_1}{f} & \dfrac{c_2}{f} & c_3 \end{pmatrix} \tag{5.11a}$$

where f is the number of the units per radian, in which ξ and η are expressed. For ξ and η in seconds of arc, we have, for example, $f = 206264.806$, and for ξ and η in minutes of arc, $f = 3437.74677$.

The affine relationship between measured and standard coordinates, equation (2.14a), which is, at least for the purpose at hand, always adequate, may be written in the form

$$\begin{pmatrix} \xi \\ \eta \\ 1 \end{pmatrix} = \mathbf{A} \begin{pmatrix} x \\ y \\ 1 \end{pmatrix}$$

with

$$\mathbf{A} = \begin{pmatrix} A & B & C \\ D & E & F \\ 0 & 0 & 1 \end{pmatrix}.$$

Therefore

$$\mathbf{A}^{-1} = \frac{1}{AE - DB} \begin{pmatrix} E & -B & BF - EC \\ -D & A & CD - AF \\ 0 & 0 & AE - DB \end{pmatrix}.$$

If we now put $(\xi_1, \eta_1, 1) = (x_1, y_1, 1)\mathbf{A}_1^T$ and $(\xi_2, \eta_2, 1) = (x_2, y_2, 1)\mathbf{A}_2^T$, we get, as the solution of our problem for the relationship between x_1, y_1 on the one plate, and x_2, y_2 on the other plate,

$$\begin{pmatrix} X_2 \\ Y_2 \\ Z_2 \end{pmatrix} = \mathbf{A}_2^{-1}\mathbf{S}'\mathbf{A}_1 \begin{pmatrix} x_1 \\ y_1 \\ 1 \end{pmatrix}, \quad \text{and} \quad x_2 = \frac{X_2}{Z_2}, \quad y_2 = \frac{Y_2}{Z_2}. \tag{5.13}$$

5.5.4.4 PLATE LIMITS. Frequently one must decide whether a certain star's image is to be expected on a particular plate. Let the spherical coordinates of the corners of this plate be (α_1, δ_1), (α_2, δ_2), (α_3, δ_3) and (α_4, δ_4), respectively, and counted clockwise as seen from the center of the sphere, that is, normally with the film toward the viewer (except for plates taken through the glass). Since we assume the measured coordinates of these corners, and the plate constants of the plate to be known, the $\alpha_1, \ldots, \delta_4$ can be computed from the standard coordinates through equations (2.10). The unit vector to the i-th corner is $\hat{\mathbf{x}}_i(\alpha_i, 90° - \delta_i)$, whose components are computed according to equation (1.1). The unit vector in the direction to the star at (α, δ) is $\hat{\mathbf{x}}(\alpha, 90° - \delta)$. This star's image will be within the area imaged on the plate, if all four of the following inequalities are satisfied simultaneously:

$$\begin{aligned} &\hat{\mathbf{x}} \cdot \hat{\mathbf{x}}_1 \times \hat{\mathbf{x}}_2 > 0, \\ \text{and } &\hat{\mathbf{x}} \cdot \hat{\mathbf{x}}_2 \times \hat{\mathbf{x}}_3 > 0, \\ \text{and } &\hat{\mathbf{x}} \cdot \hat{\mathbf{x}}_3 \times \hat{\mathbf{x}}_4 > 0, \\ \text{and } &\hat{\mathbf{x}} \cdot \hat{\mathbf{x}}_4 \times \hat{\mathbf{x}}_1 > 0. \end{aligned} \tag{5.14}$$

5.5.5 THE ZONES OF THE ASTROGRAPHIC CATALOGUE. In this section will be given the information which is peculiar to the individual zones, and references relevant to additional investigations or the improvements of the positions derived from the measurements in the zone.

5.5.5.1 GREENWICH (+90° TO +65°). An introduction to the published rectangular coordinates and description of the photographic measuring and reduction methods is given by W. H. M. Christie in the introduction to Vol. I, supplemented by additions in the introduction to Vols. II and III.

The plates were taken between 1892 and 1905, but most of them before 1898, so that the average of the plate epochs is somewhere around 1895.

The measuring machine deserves special attention. It was a machine with two separate measuring microscopes, and eyepiece scales which permitted the simultaneous measurement of two plates. These plates were lined up in the machine in such a way that both measuring microscopes (which are moved while rigidly connected to each other) always pointed at images of the same star on the two plates. This was done to make sure that every star was measured on exactly two plates. This means that the plates were not measured to the edges. However, many faint stars which showed on one but not on the other plate were measured on only one plate. Only reference stars were measured on more than two plates when this seemed desirable.

Both the 6^m and the 3^m images were measured, in both direct and reverse positions (that is, the plate was rotated 180° between the series of measurement). The 6^m and 3^m images were measured by different measurers, but (with the exception of the zones +64° to +67°) the same measurer measured the same images in the direct and the reverse positions. The above described arrangement, namely to have always two plates in the measuring machine, made it impossible for a plate to remain in the machine until all of it was measured. Each plate was measured in two sections: southern half and northern half. This can be done without having to use two sets of plate constants for every plate because the rectangular coordinates of the star's images on the plate are not reckoned with respect to the measuring machine (as it would be if the measurements were made with a long-screw measuring machine), but with respect to the lines of the réseau which are intrinsic to the plate. This is one of the (few) advantages of using a réseau.

The plate constants originally communicated were determined on the basis of rather unconventional assumptions, but are now obsolete anyway due to the above-mentioned work of Günther and Kox (1970, 1970a). The scale value ρ (that is, number of millimeters on the plate corresponding to seconds of arc) was assumed to be constant, so that only the orientation and the zero points in x and y were considered unknown in the equations. It is claimed that this gives a more accurate scale value than the one individually determined from each plate. The influences of refraction and aberration were determined theoretically. The plate constants were derived by comparison with reference stars taken from the Greenwich Second Nine Year Catalogue (1900). They are given for the plates in the zones +72° to

$+90°$ on pp. xxiv to xxvii of Vol. II, and for the zones $+65°$ to $+71°$ on pp. Bix to Bxvi of Vol. III. (The latter supersede the values communicated on pp. xlvii–liv of Vol. I.). Standard coordinates (ξ, η) in units of minutes of arc, oriented to 1900 and with respect to the plate centers (α_0, δ_0) are obtained from the measured rectangular coordinates (x, y) by the equations

$$\xi = 5(x + ax + by + c - 14), \qquad \eta = 5(y + dx + ey + f - 14).$$

Needless to say, standard coordinates derived by the formulas of Günther and Kox are more accurate. From the stars within 3° of the pole, standard coordinates with respect to the pole are given in Vol. III on pp. A9 through A29. Their functional dependence on α and δ is explained in the same volume on p. A4.

The catalogues (we refer here to Vols. I and II only) give the measurements (x, y) in declination zones 1° wide, in R.A. groups which correspond in width to about one degree. The measurements of the same star on different plates are printed side by side.

At the "head" of every group of measurements are given α_0 and δ_0, numbers and exposure dates of the plates on which the measurements were made.

The catalogue assigns serial numbers (starting with 1 in every zone) and gives diameters and measured coordinates from both plates, the latter to 0.0001 of a réseau interval (corresponding to 0.″03). At the end of the R.A. groups are printed the "extra" measurements of reference stars.

The α and δ for 1900 for all AGK1 stars and stars brighter than 9^m (on the BD system) as computed from the Greenwich AC measurements are given in Vol. III. Vols. IV, V and VI contain proper motions for the stars in Vols. I, II, and III, respectively, with an accuracy of about 0.″017 (per year) in Vol. IV, and 0.″012 in Vols. V and VI. According to the Greenwich investigators, the measuring accuracy of a published Greenwich rectangular coordinate corresponds to a rms. error of about 0.″19, and the intrinsic error of a coordinate (including every error source except the plate constants) is quoted as 0.″32.

5.5.5.2 Vatican ($+64°$ to $+55°$). An introduction to the "Catalogo Astrografico 1900.0, regione Vaticana" is given in Vol. I, partly by G. Lais and I. G. Hagen (history, photography and measurement) and partly by H. H. Turner (magnitudes and reduction, which were done at Oxford).

The plates were exposed with a standard type normal astrograph in the usual manner between 1898 and 1922, most of then between 1900 and 1910.

The Vatican observatory suffered from lack of personnel, and the simplest and fastest measuring methods had to be employed if the task was to be completed in a tolerable time. The measuring machine showed in the eyepiece a "diaframma," that is, a network of coordinate lines three seconds of arc apart, which were lined up with a réseau square. Thus, the positions of all stars in a square could be read off (with the last figure corresponding to 0.″3). Only the images produced by the middle exposure (usually 3^m) were measured, always by the same measurer on the same machine in direct and in reverse positions. Whenever the discrepancy between direct and reverse measurements was larger than 1.″5 (that is, five units), the measurements were repeated. The errors of the réseau (of which twelve different ones were used in the course of the work) and those of the "diaframma" were regarded as accidental and not corrected for in the measurements.

The original plate constants were derived at Oxford under Turner's supervision by using as reference star positions those in the AGK1 Helsingfors, precessed to 1900. No least squares adjustment, but "averaging" over the four quadrants of a plate (that is, the Christie-Dyson method), was used to compute the six independent plate constants, which connect standard coordinates (ξ, η) in units of minutes of arc and oriented to 1900, and measured coordinates (x, y) as follows:

$$\xi = 5(x - 13 - Ax - By - C), \qquad \eta = 5(y - 13 - Dx - Ey - F),$$

except in the zone +64°, where the A given in the catalogue should be replaced by $A - 0.00004$. New, improved plate constants and systematic corrections due to nonlinear and magnitude terms are available in the Günther and Kox (1970a) paper.

The catalogue is arranged as follows: The measurements are given plate by plate, in zones centered at the same declination and of increasing right ascension. The "head" contains the plate number, date of exposure and a formula for the conversion of diameters to magnitudes, and the plate constants. The listing of the measures is in columns in groups of ten, containing in one line the diameter (which is printed in boldface type for reference stars), and x and y, the measured coordinates in units of 5 mm to three decimals with respect to a corner of the plate. The serial numbers (starting with 1 on every plate) are indicated at the top of the column containing the diameters and the measurements.

Schlesinger and Barney (1942) have shown that much more accurate plate constants can be derived for this zone by determining plate constants with reference star positions which had been reduced to the epoch of the exposure with proper motions obtained from a comparison of the old AGK1 positions with their more recent positions as given in the Yale photographic catalogue.

This was found to bring about a tremendous improvement in the accuracy of the plate constants. While the accuracy of positions obtained with Turner's constants corresponds to a rms. error of 0.″78 in α and 0.″72 in δ, the corresponding figures for the constants derived by Schlesinger and Barney are 0.″43 and 0.″31. Undoubtedly, these are still not the best values available. From comparing positions on overlapping plates, Gatewood (1967) estimates the rms. error of a position from one plate as 0.″39 in either coordinate. Work on the remeasurement of the original plates on modern measuring machines was started at Hamburg-Bergedorf by de Vegt. The positions derived from the newly measured coordinates have a rms. error of about 0.″17.

5.5.5.3 Catania (+54° to +47°). The introduction to this zone and the description of the procedures used in the work on it are contained in the Vols. V, Pt. 1, pp. I to XXXVI and VI, Pt. 1, pp. I to XIV by A. Ricco.

The plates were taken between 1898 and 1926, but most of them between 1900 and 1908. Ricco reports that the objective performed well only after the centering of the components was adjusted; before that, he always got "ghosts." The focus (which changed very little) was left unaltered during the entire series of photographs. Only immaculate plates were measured.

Only the brightest of the three images of each star was measured, in direct and in reverse positions, every time by the same measurer. The coordinate differences between the images of the réseau lines and the stars were measured with an eyepiece micrometer screw. If the discrepancies between the direct and reverse position measurements exceeded 8μ, the measurements were repeated. The errors of the micrometer screws and the réseau lines were determined, and the corresponding corrections applied to the published measured coordinates.

The reference stars were taken from a catalogue by G. Boccardi (Mem. Soc. Spettrosc. Ital. **30**, 138 ff.), other catalogues available at Catania, and AGK1 stars which were reobserved at Catania. All reference stars were ostensibly reduced to Newcomb's fundamental system. There was an average of about twelve reference stars per plate. The plate constants were, at the suggestion of Turner, determined by a method of averaging in each quadrant of the plate, that is, the Christie-Dyson method, because a least squares adjustment would have consumed too much time. The plate constants were obtained by the standard four constant approach (equations 2.11). For some plates new plate constants were computed from positions in the Yale catalogue. Reference stars which gave a residual larger than $1.''8$ in either coordinate were rejected. For this zone, too, new improved constants as well as corrections for higher order and magnitude terms were computed by Günther and Kox (1970a).

The catalogues are published in zones of plates which have their centers on the same parallel of declination. Every plate is treated individually. At the heading are given plate number, coordinates (1900) of center, date of exposure and various meteorological and photographic data, and furthermore, the formulas for the transfer of measured coordinates (x, y) to standard coordinates (ξ, η)—the latter are denoted by (X, Y) in this zone—the average residuals of the reference stars (about $0.^s04$, $0.''4$), and a formula for the conversion of diameters to magnitudes. Preceding the catalogue proper, there is a precession table. The serial numbers of the stars start with 1 on every plate. The fascicles issued earlier contain the measured coordinates in units of millimeters on the plate (origin at the geometrical plate center) to 0.0001 mm, the standard coordinates (denoted by X, Y) in units of minutes of arc and oriented to 1900, the diameter, photographic and BD magnitude, α_{1900} to $0.^s001$, δ_{1900} to $0.''01$ and remarks. In the later fascicles, the serial numbers are indicated at the head of the columns; (x, y) are given to 0.0001 mm each, α to $0.^s01$, δ to $0.''1$. BD magnitude and (X, Y) are left out. At the end of the measurements for every plate are found data from other catalogues (notably BD, Draper, and AGK1) and a list of reference positions and their source.

5.5.5.4 HELSINGFORS (+46° TO +40°). This is perhaps the most carefully photographed, measured, reduced and investigated of all zones. Nearly all of the plates were taken between the fall of 1892 and the spring of 1896; only a few plates that were repeated for some reason were taken as late as in 1903. Anders Donner wrote an elaborate introduction to the zone: Catalogue Photographique du Ciel, Zone de Helsingfors entre $+39°$ et $+47°$, première série, Tome I, fasc. 1: Exposé des méthodes employées.

All measurements were made with one short screw, and corrected for its known errors, the plate being turned by 90° between the measurement of the x- and the y-coordinates. Up to the spring of 1896, the 6^m (or sometimes 7^m) exposure and the 3^m exposure were measured, but only in one position. Although the personal equations of the various measures were investigated, it was found that their removal from the measurements would be of doubtful value and no corrections for them were therefore applied.

From spring 1896 on, only the brightest image of every star was measured, but by the same observer in the direct and reverse positions. Thus the personal equation was eliminated. The errors of the screw and of the réseaux were carefully determined, the former were applied to the published measures; the latter, however, were not applied but are given beside the published coordinates for the user to apply.

The stars of the Lund and Bonn (newly reduced by Mönnichmeyer) zones of the AGK1 were used as reference stars, after the Lund positions had been reduced to the Bonn system. The way in which the plate constants were determined is peculiar to this zone and shows a great deal of insight into the problems which distinguish catalogue photographic astrometry from the photographic astrometry of single unconnected plates. Originally, plate constants were determined by the conventional six constant approach following equation (2.14a). These were, however, revised and improved by subjecting them to the condition that they would render the positions of the same stars without systematic differences from overlapping plates on which they occur. This condition was imposed only on a selection of stars. This is similar to the procedure suggested by Eichhorn (1960), but lacks the completeness and rigor of Eichhorn's approach. Even so, the industry with which Donner and his collaborators tackled the enormous amount of computation involved is very impressive. If the adjustment residual of a reference star was larger than 40μ, and if this residual appeared consistent from at least two plates, the corresponding reference star was discarded. The adjustments were not made by least squares, but by the Christie-Dyson method, that is, by taking averages and computing a "normal position" in each plate quadrant. Donner favored a six constant approach (equation [2.14a]) in principle, at least for the Helsingfors zone. He points out that there is evidence that the scale value in x is different from that in y. This was confirmed (but not in this simplified form) in a later investigation by Eichhorn (1959). The catalogue was published in eight volumes, each one covering three hours in right ascension through all declinations. Vol. I covers 0^h to 3^h, II covers 3^h to 6^h and so on. The plate numbers are not chronological in sequence of their exposure, but are assigned according to a distinct scheme so that the plate coordinates of the center can be reconstructed if the plate number is known.

Serial numbers are assigned to every star on each plate, starting with 1 on every plate. The heading of the listing of the coordinates gives plate number, right ascension and declination of the center, date of exposure, meteorological data, data concerning measurement of coordinates and magnitudes, formulas for refraction, aberration and "rattachment."

Finally the formulas for the conversion of measured coordinates (x, y) to standard coordinates (ξ, η)—which are called (X, Y) throughout this zone—are also quoted in the heading. In Vols. II, III, and IV the catalogues list the number of the star, unreduced magnitude, uncorrected x and y with respect to the plate center to 0.0001 mm. In those cases where the measurements are the mean of the readings on two images, measured in only one position, the x column is headed $(x_1 + x_2)/2$; in the case where the measures are the mean of direct and reverse measurements of the same image, the x column is headed $(x_d + x_r)/2$; the y columns are headed analogously. The next columns contain the corrections for the réseau errors in x and y, in units of 0.0001 mm (note that these were not applied to the published positions), reduced magnitude, ξ, η (called X, Y, but see below), α_{1900} to $0.^s001$, δ_{1900} to $0.''01$, and remarks, mainly concerning Bonn or Lund positions. In Vols. I and V through VIII, the columns giving X and Y are not given and the remarks are at the end.

The scale value of the Helsingfors normal astrograph is about 0.998 minutes of arc per millimeter. The authors of the Helsingfors zone anticipated that the users of the zone would use the formulas given in the introduction to convert standard coordinates to spherical coordinates. For this reason, X is not the standard coordinate, but $\xi = 0.998X$, while $\eta = Y$. This is one point which must be considered when using the formulas given for the conversion of measured to standard coordinates, but better formulas are now available anyway through the efforts of Günther and Kox (1970a).

The other point is as follows: Although the Helsingfors zone is one of the best zones of the AC, even as it was left by the Helsingfors investigators, the telescope did not give a perfectly gnomonic projection; there is evidence of tangential distortion and magnitude effects. These were investigated by Eichhorn (1959) who gives formulas and diagrams for their removal.

The systematic and accidental errors of the Helsingfors zone were the subject of two previous investigations (Furuhjelm 1906, Grönstrand 1937), whose results and tables were used by Eichhorn for the derivation of his correction formulas. Actually, these investigators had already discovered the coma effect present in the measurements contained in the Helsingfors zone. Grönstrand at the end of his paper gives tables for the reduction of the Helsingfors spherical coordinates to the system of the NFK, the published coordinates supposedly being on the system of the Bonn AGK1.

The systematic errors and new plate constants were redetermined by Günther and Kox (1970a) and are, at this time, probably the best available.

The accuracy of the Helsingfors zone positions was investigated by various authors whose results are all in good agreement. The measuring accuracy derived from a comparison of direct and reverse measures and referring to their mean corresponds to a standard deviation of 0.″13, or about 2μ on the plate. (This is actually quite high, but one must consider that the personal magnitude equations of the measurers were not taken out.) The intrinsic accuracy (according to Grönstrand) of a position at the plate center is: 0.″26 in x and 0.″23 in y for stars brighter than $8.^m1$, 0.″18 in x and 0.″20 in y for stars between $8.^m1$ and $9.^m0$, and 0.″18 in x and 0.″17 in y for stars between $9.^m1$ and $10.^m0$. Grönstrand gives diagrams which allow one to obtain the weight, relative to that at the plate center, of a star position at a certain location on the plate as a function of its magnitude and its location on the plate (l.c. pp. 29, 30). These figures agree well with those recently found by the author. (Eichhorn, unpublished.)

Furuhjelm (1916, 1926) has determined relative proper motions for the stars in the Helsingfors AC zone between 6^h and 12^h right ascension from photographs taken with the same instrument with an epoch difference of about 20 years.

5.5.5.5 POTSDAM (+39° TO +32°). This zone was never finished and, at this time, no one at Potsdam has any intention of resuming the work on the catalogue. The catalogues were published containing plate centers completely at random. Work at Potsdam stopped after the end of World War I. In 1928, it was decided to reobserve at Hyderabad (+39° to +36°) and at Oxford (+35° to +32°) the entire zone originally assigned to the Potsdam Astrophysical Observatory. Hyderabad went through with the assignment, but Oxford, which at that time had completed its original assignment (+31° to +25°), published only the zones +32° and +33°. As late as 1938 the last remaining two zones, +34° and +35°, were assigned to be photographed at Uccle Observatory (near Brussels) and to be measured at Paris. These "substitute" zones will be treated below. (Sections 5.5.5.6, 5.5.5.7 and 5.5.5.8.)

Although the Potsdam zone is incomplete, whatever was published of it has now become quite valuable, since the measures were accurate and the plates exposed before 1900. The introduction is given in "Photographische Himmelskarte," Zone +31° bis +40° Declination, Band I, by Scheiner who himself was one of the pioneers of the application of photographic techniques to astronomical problems; and Band VII contains a retrospective summary by A. Biehl.

The plates for the entire zone were taken between 1893 and 1900, except for a few regions. There were two exposures on every plate of 5^m duration each. Although Scheiner gives a long argument in favor of taking two exposures of equal length on all plates, this procedure deprives one of the possibility of an *a posteriori* determination of a nonlinear magnitude equation, see Section 5.5.3.1. The exposures were taken on specially ground and polished plates of mirror glass. The hour angles were frequently large, on the average one hour, and even reached up to three hours on various plates.

The measurements were made on a Repsold-type measuring machine with two perpendicular short eyepiece micrometer screws. Both the réseau and the measuring machine were found to be practically free of errors, so that no corrections were applied to the raw measurements. These were made in only one position. Thus, the personal magnitude equation of the various measurers was not eliminated as it would have been in a direct-reverse measurement situation. The personal equations of the various measurers must be applied by the user of the catalogue. A table for the values is given on p. 27 of the seventh volume of the Photographische Himmelskarte. This volume also contains, on pages 533 through 540, a listing of plate centers for which the corresponding measurements have been published. The personal equations were determined with a "reversing prism" which flips the image in the field of the eyepiece on a vertical axis. This method has been discussed in Section 5.1.3.

The 8766 reference star positions which were intended for use in the reductions were taken from Backlund's catalogue of fundamental stars of second order (1900), 1%; from Küstner's fundamental catalogue (Bonn Publication No. 10), 21%; from the AGK1 catalogues of Leiden, Lund and Bonn, 61%; and from the BD (!), 9%. It is obvious that under these circumstances it is impossible to speak of a "system" to which the plate constants would reduce the measurements. However, Günther and Kox (1971) have derived nonlinear and magnitude dependent terms as well as constants for this zone which are vastly superior to those published with the measurements.

The original plate constants were determined from about six reference stars per plate, assuming that 1 mm on the plate corresponds exactly to 1′, so that a "three constant approach" was, in effect, taken similar to that in the Greenwich zone. The following formulas relate the published measured coordinates (x, y) to standard coordinates (ξ, η), which are called (X, Y) in the catalogue in units of minutes of arc and oriented to 1900:

$$\xi = x + x_0 + \epsilon y; \qquad \eta = y + y_0 - \epsilon x.$$

From Vol. IV to Vol. VII, the plate constants x_0, y_0 and ϵ are given at the head of each listing of coordinates for every plate. For Vols. I to III they are given by A. Biehl (1908) in the "Sonderheft" der Himmelskarte, "Ergänzungen und Berichtigungen zu den Bänden I bis VI," except for four plates in these volumes which were published by Biehl (1914) in "Berichtigungen and Bemerkungen zu den Bänden I–VII." This latter paper ought to be consulted, in any case, before using any plate of the Potsdam zone.

The catalogue gives the measurements for each plate; the plates seem to have been selected quite arbitrarily. On the top of the listings are given the plate number (from which, similar to the Helsingfors zone, the plate center can be reconstructed), date and sidereal time of exposure, meteorological data, guiding star, tangential point to which the measured coordinates refer, observer at the telescope, and measurer. Then are given (except for Vols. I–II, see above) the coordinates (α_0, δ_0) of the tangential point to which the standard coordinates refer, and the plate constants x_0, y_0 and ϵ. The listing itself contains the serial number of the star (which starts with 1 on every plate), magnitude, approximate spherical coordinates (to 1^s in α and $0.'1$ in δ)—these are intended to serve as an extension of the BD to the eleventh magnitude—the measured, uncorrected rectangular coordinates (x, y) in units of $5'$ with respect to the plate center, to $0.'0005$ in both coordinates, and remarks.

The measuring accuracy is according to Biehl (Intro. to Vol. VII) $0.''06$, (rms. error) somewhat variable for various measurers, and Scheiner (Intro. to Vol. I) found the intrinsic accuracy of the published coordinates from comparison of positions on overlapping plates to be $0.''24$ (rms. error). The author believes from his own experiences that this figure is perhaps somewhat high; application of systematic errors (tilt, distortion, personal equation, and so forth) will probably bring it down to about $0.''20$.

5.5.5.6 Hyderabad ($+39°$ to $+36°$ and $-17°$ to $-23°$). There are two zones which were observed at Hyderabad. The second (northern) zone was begun in 1928 after it was certain that the Potsdam Observatory would not finish the zone originally assigned to it. R. J. Pocock wrote the description of the work and the procedures which is found on pages iii to xxiv of Vol. I of the "Astrographic Catalogue 1900.0, Hyderabad Section."

The instrument was not a normal astrograph, but a telescope equipped with a "patent photo-visual lens," manufactured by Cooke & Sons, with a focal length of 3325 mm and an aperture of eight inches. The guiding telescope was a 10″ visual Grubb telescope, rigidly attached to it. Nothing was done to the objective system during the course of the work, except for occasional cleaning. Pocock reports, however, that there was a "curious alteration of focal length" at the end of the work on the southern part of the zone. As we shall see later, the lens used for taking the plates of this zone has serious optical defects (mainly coma, see below) and was sent to England to be refigured (Baillaud 1957). In this zone, it has thus become impossible to make strictly differential determinations of proper motions by repeating plates now. The plates for the southern part of the zone were taken between 1914 and 1929; those for the northern half between 1928 and 1936.

Two exposures were put on every plate, the first usually 12^m but occasionally up to 20^m long. The second exposure was half as long as the first. These long exposure times (in comparison to those in the other zones) were necessary to compensate for the relatively small aperture of the lens.

Only the images produced by the longer exposure were measured (in exceptional cases when these were unmeasurable, the second exposure images were measured) with a measuring machine of the Oxford type (that is, an eyepiece scale). The same person measured in the direct and the reverse positions. Whenever the difference $d - r$ exceeded four units of the last place (corresponding to 20μ) the measurement was repeated by a different measurer. The mean error of the mean of direct and reverse settings is about $0.''23$. No réseau errors were applied, and the errors of the eyepiece scales were regarded as accidental.

The originally published plate constants A, B, C, D, E, F, which relate the measured coordinates x, y and the standard coordinates ξ, η in units of minutes of arc and oriented to 1900, by use of the formulas

$$\xi = 5(x - 13 - Ax - By - C); \qquad \eta = 5(y - 13 - Dx - Ey - F),$$

($-\eta$ for the southern part of the zone) were determined from normal equations formed by an averaging process over each plate quadrant, that is, the Christie-Dyson method. For the southern part, the AGK1 Washington was the source of reference stars. Plate constants for the northern part were initially derived using stars from the AGK1 Lund as reference stars. The plate constants thus found were given for every plate before the listing of the measured coordinates of the stars. Later, constants for the plates on the northern part of the zone were redetermined (Bhaskaran 1945) taking the reference stars from Prager's (1923) catalogue whose average epoch is much closer to that of the plates of the northern Hyderabad zone than that of the Lund positions. The plate constants obtained in this way more accurate than those published originally.

At this time, the plate constants for the southern part of the zone have not yet been improved; the average mean error of a position computed with the preliminary constants (within the system of the reference stars) is estimated at 0.″60. This figure was obtained by comparing the positions of the reference stars from different plates and is, therefore, probably too low.

In order to realize the full potential accuracy of this zone, the large magnitude dependent systematic errors, which can reach about two seconds of arc and were first pointed out by Eichhorn (1957), must be removed. They are due mainly to coma, that is, to the fact that the effective focal length depends on the magnitude of the stars. Improved plate constants for the northern Hyderabad zone were published by Eichhorn and Gatewood (1967, 1967a). Although preliminary investigations had shown that there are tilt terms and possibly also radial distortion as well as color distortion, it was apparent that these could not be reliably determined for each plate from comparison with reference stars alone. At the time that these investigations were made, the size of the available computer dictated that a complete plate overlap solution could not be carried out. Therefore, Eichhorn and Gatewood gave the formulas for the reduction from measured to standard coordinates in the form $\xi = \alpha_1 x + \alpha_2 y + \alpha_3 + \alpha_4 d + \alpha_5 dx$, $\eta = \beta_1 x + \beta_2 y + \beta_3 + \beta_4 d + \beta_6 dx$, where ξ and η are in units of seconds of arc on the system of the FK3 and oriented to the coordinate system 1950. The coefficients, α_5 and β_6, of the coma terms were constrained to be similar on each plate, and the constants on five plates in each of 16 "complexes" were determined by the overlap method. There was one additional complex with 7 plates.

Unfortunately, the models chosen for the ξ and η are inadequate when the constants are determined by an overlap solution, principally because the coma effect is not a linear function of the diameter d and the distance r from the center. This was pointed out by Eichhorn (1971). For this reason, those plate constants in Eichhorn and Gatewood's (1967a) paper, which apply to plates that were in one of the overlap complexes, should be recomputed with a more complicated model before they are used. Since Günther and Kox (1971) have calculated improved constants for this zone also, Eichhorn and Gatewood's constants may be superseded, except for the fact that the latter allowed the coma coefficients to vary (with certain restrictions) from plate to plate, while Günther and Kox force all terms which are nonlinear in the measured coordinates, to be the same for all plates of the same zone. Although only the northern part of the Hyderabad zone was investigated, it can safely be assumed that the systematic errors in the southern part of the zone will be similar to those in the northern part.

Herget (1971, private communication) also points out that the rms. errors of the positions, calculated with Eichhorn and Gatewood's constants, are over 0.″5, which is a further indication that their reduction model is inadequate, especially since Standish (1962) has carefully investigated the dependence of the intrinsic accuracy of the (northern) Hyderabad measurements on image diameter and location on the plate. He found that for an image of average diameter the intrinsic accuracy corresponds to a rms. error of about 0.″27 on the inner regions of the plate, but starting about 2 cm from the edge, the accuracy drops off to an extreme rms. error of about 0.″7 in the corners.

The catalogues contain lists of the plates in the particular volume, number, epoch of exposure, plate center coordinates, hour angle at mid-exposure, duration of the exposures, observer at the telescope, measuring machine, measurer, number of stars, ratio of the number of stars to that in the BD (north or south), and the number of stars in AGK1 on the particular plate. (The conditions of the atmosphere during the exposure are not listed).

The catalogue of the coordinates contains at the heading of a plate the right ascension (1900) of the plate center, the plate number, data of exposure, the provisional constants, and a formula for the conversion of diameters to magnitudes. The stars are numbered through in every volume, but the numeration starts with an even hundred or even fifty for every new plate. Stars marked with an asterisk were used as reference stars. The next columns refer to the diameter d, and the measured coordinates x and y which are given to three decimals in units of a réseau interval. The stars are arranged in the order of x within narrow zones of the width of a réseau interval in y. In the catalogues for the northern zone, only the last three figures of the serial numbers are given (the full number of the first star on the page is given on top of the page). The last pages in every volume contain a plate by plate listing of the standard coordinates of the reference stars used for the computation of the plate constants.

5.5.5.7 Uccle—paris (+34°, +35°). An introduction by Paul Couderc is given in Vol. I (+34°) of the zone.

When the Potsdam Observatory declared its inability to finish its commitment to the Astrographic Catalogue project, the zones +35° to +32° originally assigned to Potsdam were taken over by Oxford. However, only the zones +32° to +33°

had been finished because the astrographic work there was discontinued after Turner's death. The zones +34° and +35° were reassigned to the observatories at Uccle and Paris. The plates were taken between 1940 and 1950 at Uccle, and measured and reduced at Paris. The instrument at Uccle was a normal astrograph built by Gautier, whose lens is not as good as that of the Paris astrograph, so that lower accuracy is suspected. Four exposures were put on one plate, lasting 30^s, 5^m, 5^m and 30^s, respectively. The réseau was lent by the Toulouse Observatory and had negligible errors. Both 5^m images were measured in direct and reverse orientation with the same measuring machines (short eyepiece micrometer) that were used to measure the Paris zone. The setting accuracy corresponds to a standard deviation of 0.″1. Stars of magnitude between $7.^m5$ and $9.^m5$ from the AGK1 catalogues Leyden and Lund were reobserved at Paris on a transit circle, between 1945 and 1952, so that an average density of 16 reference stars per plate is available. The reference star positions are ostensibly on the system of the FK3. The plate constants were computed (taking equation 2.14a as model) by a least squares solution from the reference star positions to which proper motions had been applied to bring them to the epoch of the plates. The standard coordinates ξ, η in units of minutes of arc and oriented to 1900, are computed from the measured coordinates and the plate constants by the formulas $\xi = x + Ax + By$; $\eta = y + A'x + B'y$ with respect to α_0, δ_0. In the catalogue, (ξ, η, x, y), respectively, are called (X_0, Y_0, X, Y).

The catalogue gives at the head of the listing of the coordinates the plate numbers, the approximate plate centers, the epoch of exposure, and A, B, A', B', α_0, δ_0. The coordinates are listed with a serial number (which starts with 1 on every plate; an asterisk indicates inaccurate measurements), magnitude (in parentheses if inaccurate), and x, y, with respect to the center of the réseau. At the end are listed the adopted positions of the reference stars, their number on the plate, their BD number and their proper motion components. The mean error of a position from one pair of measurements is given as 0.″32 in either coordinate. New plate constants and zone wide systematic corrections for nonlinear terms have become available through the efforts of Günther and Kox (1971).

5.5.5.8 OXFORD (+33° TO +25°). The two northern zones (+33° and +32°) were taken over from Potsdam, the plates for these were exposed mostly between 1932 and 1936. Thirty-two of the plates were taken at the Greenwich Observatory. No réseau was copied on them, but they were measured with a copy of the réseau clamped to them.

After 1948, the reduction of the plates was finished at the Hamburg-Bergedorf Observatory by Kox. Further differences between the original Oxford zone and the two zones taken over from Potsdam will be pointed out below.

An introduction by H. H. Turner is given in Vol. VII of the series.

The Oxford normal astrograph, with which the plates were taken, was patterned after the instrument made by the Henry brothers for the Paris Observatory. With it, the plates for the originally assigned zone, that is, with the centers at

$+31°$ to $+25°$, were taken between 1892 and 1904. Specially thin plate glass was used, and the exposures lasted 6^m, 3^m and 20^s except in dense galactic clouds where they were shorter. All photographs were taken west of the pier. The quality of the images deteriorates towards the edges. The measuring micrometer was similar to that used at Greenwich, but only one image was measured in direct and reverse positions by the same person. The mean error of measurement (mean of $d + r$) is $0.''26$.

The measurements were carried out on a very ambitious scale, with the result that the measurers almost kept up with the newly taken plates. No réseau errors were applied.

The plate constants which connect the measurements (x, y) with the standard coordinates (ξ, η) in units of minutes of arc and oriented to 1900 by the formulas

$$\xi = 5(x - Ax - By - C - 13), \qquad \eta = 5(y - Dx - Ey - F - 13),$$

were found by comparison with stars from the AGK1 (mostly Greenwich) which were precessed to 1900 with the quantities listed for this purpose in the catalogues. Six constants (following the model equations [2.14a]) were allowed for, but formulas for the computation of differential refraction and aberration are also given. The mean error of a position computed with these constants is about $0.''58$. In an unpublished investigation, Eichhorn found the intrinsic error of an Oxford AC position to be about $0.''30$.

The listings of the measured coordinates in the catalogues are preceded by a list of plates, giving plate number, epoch (to $0^y\!.001$), right ascension of center, hour angle at mid-exposure, length of exposures (the 20^s exposure was usually left out in the two northern zones), the observer, measuring machine, measurer, number of stars on the plate, ratio of the number of star images to those in the BD, number of AGK1 stars found on the plate, and the interval between exposure and measurement in months. The two northern (originally Potsdam) zones also give (where applicable) volume and page number of the Potsdam zone in which the measurements for the corresponding plates are found.

The listing of the measurements is headed by the right ascension of the plate center, plate number, date of exposure, the plate constants and a formula for the conversion of measured diameters to magnitudes. The columns carry the serial number of the star on the plate (consecutively in every volume, but starting with a number of the form 100N + 1, with N being a positive integer, on every plate, the diameter and the measured rectangular coordinates x and y to three decimals. First are listed all stars with $0.000 \leqq y \leqq 0.999$ in the order of increasing x, and so on. Numbers with asterisks correspond to AGK1 stars which were used as reference stars. The last pages in every volume contain for every plate in it the standard coordinates of the reference stars which were used for the calculation of the plate constants. Corrections for systematic errors due to nonlinear and magnitude terms, and improved plate constants, which yield standard coordinates oriented to 1950, were published by Günther and Kox (1971).

5.5.5.9 PARIS (+24° TO +18°). A description of the instruments and methods used in the work at the Paris Astrographic zone is given by Prosper Henry on pp. (11) to (37) of Vol. I (+24°). Pertinent remarks are also made in the introduction to Vol. II (+23°).

The instrument is the original normal astrograph made by the Henry brothers. All plates were taken on this instrument. Focus and tilt were carefully determined. The plates were exposed around, but mostly before, 1900.

The following data regarding the plates are given in the introduction: approximate right ascension of center, number of plate, measurer, position of the plate in the measuring machine, measuring machine and réseau used, and temperature during exposure.

The measurements were made on measuring machines which had two short eyepiece micrometer screws. The plates for the zones +24° and +23° were measured twice by different measurers in the same position. The plates for the rest of the zones were also measured twice, but in positions which were different by 180° (that is, direct and reverse). The images from the two longest exposures were measured.

Occasionally, more than one measurer measured on the same plate; réseau errors were applied.

The mean setting error is about 0.″15. The plate constants, which connect the measured coordinates x, y (in the catalogue denoted as X, Y) with the standard coordinates ξ, η in minutes of arc and oriented to 1900 with respect to the centers α_0,δ_0 by the formulas

$$\xi = \tau_x(x + i_x y) + dist, \qquad \eta = \tau_y(y - i_y x) + dist \text{ for the zone } +24° \text{ and}$$

$$\xi = \tau_x x + i_x y + dist, \qquad \eta = -i_y x + \tau_y y + dist \text{ for the other zones}$$

were determined (six constants) from reference stars compiled from many catalogues, but not by a least squares adjustment. The reference star positions were all ostensibly reduced to Newcomb's system. The term "*dist*" means the distortion correction which is given (in seconds of arc) by $bx(x^2 + y^2)$ and $by(x^2 + y^2)$, in ξ and η, respectively, where $b = +0.25(\sin^3 1'/\sin 1'')$. Later, the linear plate constants for some of the plates in the Paris Astrographic catalogue (and some for the Bordeaux zone) were recalculated by Heckmann, Dieckvoss and Kox (1949, 1954); in the first paper allowing for six, in the second for four independent plate constants. The second paper gives new, improved plate constants for all plates in the zones +21°, +22°, +23° and +24°. These new constants, which render the old ones obsolete, allow one to compute standard coordinates in units of minutes of arc (with respect to the newly computed α_0 and δ_0) and oriented to 1950.0 and on the system of the FK3 from the formulas

$$\xi = ax + by; \qquad \eta = dx + ey.$$

The papers by Heckmann, Dieckvoss and Kox give α_0, δ_0, a, b, c, d. The existence of the distortion, which was determined by the investigators at Paris, could not be confirmed by the German investigators. The improvement achieved by their constants over the old positions is spectacular. The new rms. error of a position from one coordinate is 0.″26 in ξ and 0.″20 in η, compared to 0.″45 in either coordinate (and for coordinates not on the FK3 system) computed with the original Paris constants and with distortion correction applied.

The arrangement of the catalogue proper (which was also adopted by the other French observatories—Bordeaux, Toulouse, Algiers—which published an astrographic zone) is as follows:

Every volume contains the data derived from the plates centered at a certain declination; within the volume, the plates are arranged according to the right ascension of their center. The plate heading gives the approximate right ascension of the plate center, the date of the exposure and the hour angle at mid-exposure. The columns of the catalogue contain the number of the stars on the plate (starting with 1 on every plate, in boldface type for reference stars), the magnitude, and the measured coordinates x and y in units of 1mm and to four decimal places referred to the approximate plate center in the order of increasing x first for positive, then for negative y. At the end of the listing of coordinates for every plate are given notes concerning some of the coordinates, the plate constants i_x, i_y, τ_x, τ_y, α_0, δ_0, and the adopted mean positions (1900) of the reference stars which were used for the calculation of the plate constants. Work on improved plate constants is in progress at the Strasbourg Observatory under the direction of P. Lacroute.

5.5.5.10 Bordeaux (+17° to +11°). An introduction to the procedures and methods used in the work on this zone is given on pp. 5 to 66 of Vol. I by G. Rayet.

Exposures of 6^m, 3^m and 90^s duration, respectively, were put on each plate. The réseau was exposed on the plate before the photograph of the sky was made.

The images produced by the 6^m and the 3^m exposure were measured in one position only with respect to the réseau lines, on a measuring machine equipped with two eyepiece micrometer screws. The réseau errors were applied. The published coordinates thus refer to the mean of the 3^m and the 6^m images. The setting errors are generally smaller than 0.″15 (rms. error) varying with the observer, image quality and image intensity.

Tables give for every plate (in each volume) the approximate right ascension of the center, the plate number, the measurers (all plates were measured twice), the réseau number, the number of stars on the plate, and finally barometric pressure and temperature (in centigrade) during the exposure.

The reference stars were taken from several catalogues and ostensibly reduced to Newcomb's fundamental system. The plate constants were computed by Cauchy's method, allowing for four independent unknowns, since differential aberration and refraction were considered separately. The standard coordinates in units of minutes of arc and oriented to 1900.0 are computed from the published measured coordinates (x, y) by the following formulas

$$\xi = 0.995(p_x x + r_x y) + A, \qquad \eta = 0.995(r_y x + p_y y) + A'.$$

The catalogue of the coordinates is arranged exactly as that for the Paris zone.

Heckmann, Dieckvoss and Kox (1949) also give in their paper new plate constants for a number of plates in the zones +16° and +17° of the Bordeaux zones in the same format as for the Paris zone plates. The rms. error of a coordinate derived from one Bordeaux measure is about 0.″30 according to these authors.

They also point out a systematic run with the magnitude of the differences between positions obtained from Paris measurements and the Bordeaux measurements, and between Bordeaux and Toulouse positions. According to these authors, their differences are caused by a magnitude equation in the Bordeaux zone. They therefore propose the application of corrections $\Delta\xi$ and $\Delta\eta$ to the Bordeaux positions which are given in Table V-6.

Table V-6 The Magnitude Equation of the Bordeaux Zone in the AC.

m	(Bord.) 8.ᵐ0	8.5	9.0	9.5	10.0	10.5	11.0	11.5	12.0
$\Delta\xi$	+0.″11	+.06	+.04	+.01	.00	−.01	−.02	−.02	−.02
$\Delta\eta$	+0.22	+.13	+.06	+.03	.00	−.02	−.03	−.05	−.06

These values are rather hypothetical for the stars below 10.ᵐ0. Perhaps the differences are not due to a magnitude equation in the Bordeaux zone only. Their existence, however, shows definitively that nonlinear magnitude equations may be present in the coordinates published in any of the zones of the Astrographic Catalogue. Their determination is possible, in principle, but difficult (see Section 5.5.3.1). P. Lacroute is directing the determination of new reduction parameters for this zone at the Strasbourg Observatory. See also Herget's (1973) work.

5.5.5.11 Toulouse (+10° to +5°). The procedures were patterned in principle after those used at Paris. A synoptic introduction to the methods employed, and to the use of the published coordinates is given on pp. 1–56 of Vol. I, by B. Baillaud.

Each plate was given three exposures (5^m, $2\frac{1}{2}^m$, 20^s). Mirror glass was used throughout as material for the plates. The exposures were begun in 1893.

The measurements were made with two short eyepiece micrometer screws; the images originating from the two longer exposures were measured on all plates twice, independently of each other, by different measurers; one measured in direct, the other in reverse positions. Thus, the physiological magnitude equation of the measurers is not eliminated. There were usually several months between the direct and the reverse measurements. Right ascension of the plate center, plate number, measurers, measuring machine, réseau and temperature in centigrade during the exposure are given in the introductions for all plates.

The plate constants were obtained (four constant approach) by using stars from several catalogues, but mostly from the AGK1 Leipzig, and Saint-Blancart's

Toulouse catalogue. Reference stars which gave large residuals were rejected. Standard coordinates in units of minutes of arc, oriented to 1900.0 with respect to the center α_0, δ_0 are computed from the measured coordinates x, y by

$$\xi = \tau_x[(x + \Delta x) + i_x(y + \Delta y)], \qquad \eta = \tau_y[(y + \Delta y) - i_y(x + \Delta x)],$$

where Δx and Δy were the corrections to be applied due to the réseau errors. These are given (for the réseaux 54, 79, 89) on pp. 48–53 of the introduction to Vol. I. To the plates that were exposed with reseau 54, corrections (given on pp. 54 and 55 of the same volume) were erroneously applied and must be removed, before applying the true corrections.

The internal rms. error of a position derived from one Toulouse coordinate with the constants given in the volumes is about equal to 1.″30, very considerably below their true potential accuracy. The arrangement of the catalogue of rectangular coordinates is the same as in the Paris zone catalogue. Improved plate constants for this zone are being calculated at the Strasbourg Observatory under the direction of P. Lacroute.

5.5.5.12 Algiers (+4° to −2°). An introductory volume has been written for this zone by Ch. Trépied.

The plates were all taken at the normal astrograph of Algiers Observatory between 1891 and 1903. Exposures of 5^m, $2\frac{1}{2}^m$ and 20^s were made, and a réseau was copied on all plates.

The images produced by the 5^m and $2\frac{1}{2}^m$ exposures were measured by two observers each (first one observer measured one zone, then the other observers measured the same zone in reverse order) but in only one position of the plate. Systematic differences between the observers, in particular differences depending on the magnitude of the image, were not found. The measuring machines were the usual type with a pair of micrometer screws at the eyepiece, and, as in all other zones, the coordinates were measured with respect to the square formed by the lines of the réseau. The mean measuring error is smaller than 0.″18 (that is, 3μ) for every coordinate.

This arrangement was carried out only in Vol. VII, which was published first under Trépied's directorship, also in Vol. V, fasc. I (0°, $0^h\,0^m$ to $6^h\,56^m$). When Gonnessiat took over the direction of the work after Trépied's death, the plates were rotated by 180° between the measurements. This started after $18^h 00^m$ in the zone −2°. However, the measurers were not the same persons for the direct and reverse positions. From a comparison of the direct and the reverse measurements, a mean setting error of around 0.″12 for a published coordinate is found.

Reference stars for the determination of the plate constants were taken from a total of fifty-six catalogues (reduced to the epoch of the plate and coordinate system 1900.0 listed on page XCIII of the introduction) so that an average of twenty-five reference stars was available on a plate. Four independent constants per plate were provided for (following equations [2.11]), which were computed by a rigorous least squares solution.

The connection between measured coordinates (x, y), to which neither réseau corrections nor screw corrections were applied, and the standard coordinates ξ, η in units of one minute of arc and oriented to 1900.0 with respect to the center (α_0, δ_0) is given by

$$\xi = \tau_x x + i_x y; \qquad \eta = -i_y x + \tau_y y.$$

The rms. error (within the system of the reference stars) of a position calculated from a pair of measured coordinates with the plate constants given in the catalogues is about 0.″50.

Under Gonnessiat's charge, a change in format took place in all volumes except the first fascicles of the volumes $-2°$, $0°$, and the volume for the zone $-1°$. α_0 and δ_0 were replaced by $[\xi]$ and $[\eta]$, and the standard coordinates refer to the approximate plate center. It was also discovered that, due to the noncoincidence of the approximate and the true plate center, the quantities τ_x, i_x and i_y are subject to corrections $d\tau$ and di, so that in the new format, the standard coordinates with respect to the approximate plate center are computed by

$$\xi = x(\tau_x + d\tau) + y(i_x + di) + [\xi]; \qquad \eta = y\tau_y - x(i_y + di) + [\eta],$$

where $d\tau$ and di are computed from formula (27) and the little table on p. XI of the "preambule" to Vol. VII $(-2°)$ of the zone.

In the appendix to Vol. IV $(+1°)$ constants A, B, C, A', B', C' are given, which permit for the whole zone the computation of standard coordinates in units of minutes of arc oriented to 1900, with respect to the approximate plate center, from the measured x, y by the formulas

$$\xi = 0.989(x + Ax + By + C), \qquad \eta = 0.989(y + A'x + B'y + C').$$

The constants $A, \ldots, C'$ were obtained by using reference stars whose positions had been newly observed at Abbadia.

The arrangement of the catalogue of the measured coordinates is the same as that in the Paris zone, with the following exceptions: the headings of the coordinate lists also contain the approximate declination of the center, the temperature in centigrade, and the barometric pressure (reduced to that temperature) for every plate.

After the provisional elements (which are found at the end of the measured coordinates), quantities pertaining to the magnitude and the initials of the measurers are given. New reduction parameters for this zone are being computed at the Strasbourg Observatory under the direction of P. Lacroute.

5.5.5.13 San fernando $(-3°$ to $-9°)$. An introduction to the catalogue volumes is given in Vol. I (Introduccion y Tablas) by Thomas de Azcarate (1921).

Originally, all plates for the catalogue were taken in 1894, but 40% of these were not considered satisfactory and were thus repeated sometime before 1903. Normally, exposures lasted 6^m, 3^m and 20^s, but when atmospheric conditions made it necessary, the first exposure was increased up to 8^m.

The plates were measured on an eyepiece micrometer screw measuring machine built by Gautier. The plates of the zones $-9°$ and $-8°$ were measured by two measurers, but in only one orientation. Discrepancies of up to 0.005 réseau intervals (25μ !) were tolerated. The plates within the other zones ($-3°$ to $-7°$) were measured in direct and reverse positions by the same measurer, and differences of up to 0.010 réseau intervals between direct and reverse measurements were tolerated. Both the images from the 6^m and the 3^m exposure were measured.

Optical radial distortion proved insignificant and was thus not applied. Similarly, the second order terms for aberration and refraction, which never exceed 0.″03, were neglected.

The plate constants were determined with the use of stars from the AGK1 as reference stars, allowing for six independent plate constants according to equation (2.14a). If there were not enough AGK1 stars on a plate, the positions of BD stars were computed from neighboring plates and used as reference positions. If X_0, Y_0 are rough standard coordinates which were obtained by multiplying the measured coordinates x, y by an approximate scale factor, the standard coordinates ξ, η in minutes of arc oriented to 1900 are given by

$$\xi = X_0 + c_x + \rho_x X_0 - i_x Y_0, \qquad \eta = -(Y_0 + c_y + \rho_y Y_0 + i_y X_0).$$

The catalogues are published in volumes containing one declination zone each. The listing of the coordinates is preceded by tables giving the following information about the plates: plate number, plate brand, epoch (to 1^m) of exposure, length of exposures, right ascension of plate center, hour angle, zenith distance, parallactic angle at mid-exposure, air pressure and temperature in centigrade during exposure, réseau number, number of stars on the plate, and measurer.

The listing of the rectangular coordinates is preceded by the plate number and the coordinates of the plate center, date, hour and minute of exposure, the plate constants, and the adopted positions of the reference stars.

The catalogue of coordinates has columns for the serial numbers. These start with 1 on every plate and are in boldface type for reference stars. Given in addition, are the magnitude, the measured coordinates x, y (these only in Vols. I and II), and the approximate standard coordinates X_0 and Y_0 in minutes of arc, referred to the plate center; Y_0 increases toward south. The rms. error of a position of a field star, computed with the plate constants given in the catalogue, is given as 0.″61 in the system of the reference stars.

5.5.5.14 Tacubaya ($-10°$ to $-16°$). Up to 1899 600 plates were taken, but defects in the images were noted. A second series of plates was therefore taken after adjustment of the objective, and was (except for a few plates) finished in 1912. The plates were exposed only once, usually for six minutes, but occasionally the exposure lasted as long as 10 minutes.

The measuring of the plates began in 1897. An Oxford-type measuring machine was used. The images were measured in direct and reverse positions by the same observer. There is a deviation from the standard Oxford procedure. Only two réseau lines were measured in each position, those to the west and the south of the star in the direct, and those to the north and the east in the reverse plate position. No corrections for réseau errors were applied.

The standard coordinates (ξ, η) in minutes of arc and oriented to 1900 are obtained from the measured coordinates (x, y) by the formulas

$$\xi = 5(x + ax + by + c), \qquad \eta = 5(y + dx + ey + f).$$

The plate constants a, b, c, d, e, f were calculated by comparison with the AGK1 Washington from normal positions formed for every plate quadrant, that is, essentially by the Christie-Dyson method.

Originally it was planned to publish not only the measured coordinates, but also the standard and the spherical coordinates, and number and magnitude from the BD where applicable, but this scheme was followed only in the first fascicle of Vol. I ($-15°$, 0^h to 6^h). There, plate number, coordinates of plate center, date and sidereal time of exposure, length of exposure, seeing, observer and measuring date as well as the formulas for the conversion of measured to standard coordinates are given, along with a list of the reference stars including their BD numbers, magnitudes, and their spherical coordinates.

In the rest of the catalogue, the measured coordinates, listed plate by plate, are preceded by a list of the plates giving plate numbers, coordinates of the center, date and sidereal time of mid-exposure, date of measurement (only in some early fascicles) and the plate constants. In these, the headings give only plate number, coordinates of the plate center and the plate constants. The coordinates are listed (with numbers and magnitudes) in units of a réseau interval in the order of increasing x to four decimal places with respect to the plate center. The serial numbers of the reference stars are marked with an asterisk.

From the investigation of plate overlaps, it appears that the intrinsic rms. error of a single position is close to $0.''30$, but the rms. error of a position computed with the presently available constants is $0.''48$.

A comparison of part of the Tacubaya positions with those in other catalogues (Paris, Radcliffe and Bordeaux) is given on pp. 62 through 79 of the introduction to Vol. I ($-15°$) and shows that the system of the AGK1 Washington is apparently well realized by the plate constants.

5.5.5.15 CORDOBA ($-24°$ TO $-31°$). Perinne wrote the introduction to the volume containing the plate constants: Resultados del Observatorio Nacional Argentino, Vol. **34**, entrega 1, and to the zones in general: $-24°$, Resultados Vol. **26**. When he took over the directorship in 1908, he almost completely revised the procedures used for the establishment of the catalogue. All plates were retaken between 1909 and 1913. The images showed strong spherical aberration which became larger and distorted towards the edges. The causes of the bad images were removed in two steps. A triangular distortion of the images was caused by the fact that the objective was squeezed in its cell; this was corrected in 1911. After Aug. 9, 1910, the opening of the photographic objective was stopped down from thirteen to eleven inches. This improved the quality of the images considerably. The plates taken after this date are marked in the catalogues with a "D."

The four exposures on every plate lasted 5^m, 5^m, $1\frac{1}{4}^m$ to $1\frac{1}{2}^m$, 5^s to 8^s, depending on which of two brands of plates was used. The table on the bottom of p. XV of the introduction to Vol. **26** ($-24°$) gives the details.

12% of the plates were measured on measuring machines with eyepiece micrometer screws, the remaining 88% on eyepiece scale measuring machines of the same type as at Oxford. Both images produced by the long (5^m or 6^m) exposures were measured in direct and reverse positions by the same observer. Perinne expresses the fear that two of the measurers might have cheated and manufactured the reverse measurements from the values of the direct ones; however, he goes on to say that nothing in the measurements confirmed this suspicion. Some plates in the zones $-25°$ and $-26°$, whose number is preceded by an asterisk in the list of plate constants do not have the correction for the run of the screw applied. Tables for the application of the run correction are given on the bottom of p. XI of the introduction (Res. **34**, 1), and pp. 313 and 325 of the same volume.

A catalogue of 6429 reference stars for the zones was observed between Feb. 1917 and Jan. 1918 on the 190 mm Repsold transit circle. This catalogue constitutes entrega 2 of Vol. **34** of the "Resultados." The positions are on the system of the PGC.

The plate constants were determined in the same way as at Oxford and at other places, that is, by the Christie-Dyson method. "Normal positions" were formed for each plate quadrant, and six independent plate constants were computed from these by an adjustment (but not a least squares adjustment). The following formulas connect the standard coordinates (in minutes of arc, oriented to 1900) with the measured coordinates x and y:

$$\xi = 5(x - Ax - By - C - 14), \quad \eta = 5(y - Dx - Ey - F - 14).$$

The plate constants are found in the introduction (Res. **34**, entr. 1) and also in the volumes with the measurements. The introduction also contains data concerning the plates, as date of exposure and measurement, observers and measurers, type of measurement (micrometer or eyepiece scale), hour angle and atmospheric data, including seeing.

The measurements are arranged in the same format as at Oxford, regardless of whether they were made with an eyepiece micrometer or an eyepiece scale, and $2.000 \leqq x, y \leqq 26.000$. No corrections for irregularities of either micrometer screw or scales were applied.

Apparently the measuring team was not very conscientious, and Perinne complains about its members in several passages. This may have been reflected in the setting accuracy. The rms. error of a position computed from a pair of measurements with the available plate constants is $0.''56$ in right ascension and $0.''48$ in declination, which indicates that the intrinsic accuracy is higher, probably in the neighborhood of $0.''30$ per coordinate.

5.5.5.16 PERTH ($-32°$ TO $-40°$). The plates were taken (between about 1901 and 1915) at the Grubb normal astrograph at Perth Observatory west of the pier, and usually within 1^h of the meridian. Originally, the exposures were 6^m, 3^m and 20^s, but commencing with plate 2063, which was taken in November 1910 when it was found necessary to retake some plates in the zone $-32°$, the exposure times were cut to 4^m, 2^m, and 13^s. Except for four plates, the same réseau, made by Gautier, was used throughout.

On some occasion between the exposure of plates 2002 and 2062 (probably before 2009), the focus was altered without the knowledge of the director (probably by visitors who played with the focusing screw) so that afterwards the central portions of the plates, instead of the middle regions, were in sharp focus.

The measurements were carried out at Perth only for the zones $-32°$ through $-37°$, but the plates for the zones $-38°$ through $-40°$ were sent to Edinburgh to be measured there. At Perth, the measuring procedures adopted at Oxford were followed, that is, the measurements were made on an eyepiece scale machine. Neither at Perth nor at Edinburgh were corrections for irregularities of the réseau lines applied. At Edinburgh, the 6^m images were measured in direct, and the 3^m images in reverse position. This procedure was intended to provide a safeguard against the introduction of bogus faint stars into the catalogue but it defeats, of course, the original purpose of direct and reverse measurements which is the elimination of the personal equation (setting bias). Both at Perth and Edinburgh, a star was remeasured when the difference between the direct and reverse measurements exceeded 0.003 réseau intervals ($0.''9$).

The reference stars for the determination of the plate constants were taken from the Perth meridian observations (Vol. IV). At both places, a first approximation for the plate constants was obtained by a solution with six independent constants. Second approximations were obtained by what was essentially a three (or four) constant approach. The effects of differential aberration and refraction were removed from the measurements, which were then subjected to a second approximation adjustment. For the purpose of the reductions at Edinburgh, the scale values were assumed to have remained constant over certain periods of time (this assumption was not made at Perth). However, at both places the difference between the orientation constants was kept constant over certain epochs. Thus, the orthogonality of the measurements was not enforced. A similar procedure had been followed at Greenwich, see Section 5.5.5.1.

The *format of the catalogues is different for the zones measured at Edinburgh and at Perth.* For the zones measured at *Perth*, the standard coordinates (ξ, η) in minutes of arc and oriented to 1900 are connected with the measurements (x, y) by the formulas

$$\xi = 5(x - 13 - Ax - By - C), \qquad \eta = 5(y - 13 - Dx - Ey - F).$$

The heads of the coordinate listings contain the right ascension of the plate center, plate number, exposure date and the constants A, B, C, D, E, F. The listings give serial numbers (with references to remarks), magnitude (also CPD magnitude where available), and x and y (there called X and Y), in units of a réseau interval to three decimal places. x increases with increasing right ascension, y with increasing declination. $0.000 \leqq x, y \leqq 26.000$.

The measurements x, y made at *Edinburgh* are related to the standard coordinates in minutes of arc and oriented to 1900 by the equations

$$\xi = 5(x - 14 + ax + by + c), \qquad \eta = -5(y - 14 + dx + ey + f).$$

Tables with details of plates are given, containing plate number, right ascension of center, date of exposure, hour angle at exposure, date of measurement, number of stars, approximate number of stars in the CPD and measurers. The listings of the coordinates carry at the head the right ascension of the center, plate number and exposure date, and the plate constants a, b, c, d, e, f. The coordinates, which are given in units of a réseau interval to four decimals, (x increasing with right ascension, y decreasing with increasing declination, $1.000 \leqq x, y \leqq 27.0000$) are preceded by serial numbers which start with 1 on every plate, and followed by the measured diameter, magnitude, and number from the CPD for stars in the CPD. The coordinates are listed in the order of increasing y first for all x with $1.0000 \leqq x \leqq 1.9999$, then for $2.0000 \leqq x \leqq 2.9999$, and so forth.

From plate overlaps, the intrinsic rms. error of a coordinate is estimated to be $0.''3$ or slightly less.

5.5.5.17 Cape of Good Hope ($-41°$ to $-51°$). The plates were taken on a normal astrograph built by Grubb. Initially, the lens did not produce measurable images. The fault was found to be a serious error on the outer surface of the crown lens "which Sir Howard (Grubb) was sure had not been present when he applied the final tests." Grubb attributed the fault to "intentional damage done by an evil-minded employee" and refigured the lens, after which it gave good images. A series of catalogue plates was taken between 1892 and 1896, but a second series was taken between 1897 and 1905 (with some plates as late as 1910) "to bring the epoch of photographic observation nearer to 1900," and thus make them more contemporaneous with reference stars that were observed at that time on the meridian circle.

The usual precautions (plate squared on carefully, position well defined, réseau position well defined, accurate focus, and so on were all very punctiliously observed. Four different réseaux were used in the course of the work. The focus was set so that there were sharp images 40 mm from the center. The introduction to the zone and description of the work by S. S. Hough is given on pp. vii to xliii of Vol. **1**.

The measuring machines were specially designed by D. S. Gill, and are of the eyepiece screw micrometer type. In the field of view, there appeared a square of double threads which was adjusted so as to exactly contain one réseau square. Two persons worked at the measuring machine all the time; one measuring, the other recording. Settings were made, as a rule, only on the 6^m image; when this was imperfect, the 3^m images were measured. Measurements in direct and reverse positions were made by the same observer. Reference stars were, however, measured by both the observer and the recorder at the measuring machine. This may mean that there is the danger that the system of the reference stars is different from that of the field stars. No errors for irregulatities of either the réseau or the micrometer screws were applied.

The plate constants were determined through a least squares solution (four constant approach, following equations [2.11]) by comparison with the catalogue of reference stars which was also observed at the meridian circle at the Cape. Proper motions were applied to the meridian positions to bring them to the epoch of the plates.

The listing of the coordinates is preceded by a list of plates, giving their serial numbers, right ascension of center, mean sidereal time and hour angle of exposure, definition of images, réseau used, date of measurement and measuring team, micrometer used, plate constants, and number of reference stars (about 11 on the average).

The coordinates are listed as usual, in sequential order on each plate in 2° zones, in the order of right ascension of the plate center. The head of the listing consists of the serial number of the plate, the epoch of exposure and the limiting diameter. The columns of the listing proper are: notes; the standard coordinates in minutes of arc oriented to 1900, in the order of increasing x on every plate, to three decimal places with respect to the plate center as origin; the diameter; and CPD number if a star is a CPD star. The serial numbers of the stars, which begin with 1 on every plate, are indicated on top of the column, and on the bottom of the pages are the star numbers at which observer and recorder changed place at the measuring machine, which happened frequently and was usually done when a reference star was measured. The measured coordinates themselves are not given, but can be reconstructed from the published plate constants, aberration and refraction corrections and standard coordinates.

From the adjustment residuals of the reference stars, an rms. error of 0.″38 of a position from one plate is indicated. (This may be too high, since Hough may have overestimated the accuracy of the Cape meridian positions.) From plate overlaps it follows that the intrinsic rms. accidental error of a single coordinate is 0.″41. This looks like a very high figure for a zone that was measured with micrometers. However, there are positive indications that the effective focal length of the Cape instrument depends on the image intensity, that is, there is a (still uninvestigated) coma effect in the Cape measurements. The application of corrections for this effect will probably decrease the figure for the error of a coordinate given above.

5.5.5.18 SYDNEY (−52° TO −64°). The work on this zone was started under the directorship of H. C. Russell in 1891. In 1899, the telescope (a standard normal astrograph) was moved to "Redhill," which was apparently an unfavorable site. The plates were sent to Melbourne to be measured there under the supervision of P. Baracchi. Apparently, the quality of the plates left much to be desired; the exposure time for the principal image varied between 4^m and 12^m, perhaps even more. When the new director assumed his duties, he wanted to start the work all over, but was overruled by his board of visitors. The introduction to the zone is given as introduction to Vol. III, which was the first one to be published. However, the standards for exposure of the "new" series of plates (after 1912) were very rigorous. The exposures were 4^m, 2^m and 13^s uniformly, and plates were taken only on excellent nights. The réseaux were copied on the plates with a special apparatus.

Before 1912, the plates were sent to Melbourne to be measured in the way the measures were executed in Melbourne; after that, the plates were measured at Sydney with an eyepiece scale machine in direct and reverse positions. The mean of the two positions is the published coordinate.

The plate constants were determined by comparison with meridian observations made at Perth. Six independent linear plate constants (equation [2.14a]) were allowed for. Standard coordinates ξ, η in minutes of arc and oriented to 1900 and measured coordinates x, y are connected by the formulas:

$$\xi = -5[x - 14 - A(x - 1) - B(y - 30) - C]$$

$$\eta = 5[y - 43 - D(x - 1) - E(y - 30) - F].$$

The coordinates are published in volumes comprising six hours each of right ascension within one zone. The headings consist of the right ascension of the plate center, the plate number (plates of the new series [after 1912] are indicated by "N"), the exposure date, the machine on which the measurements were made, and the constants $A, \ldots, F$. Sometimes the plate constants were not available when the catalogues were printed. In these cases the corresponding spaces are left free in the headings, and the catalogue proper is preceded by a sheet which lists the plate constants. The catalogues list (plate by plate) the serial numbers of the stars (starting with 1 on every plate), their diameters (which are coded by letters of the alphabet for plates of the new series, and the measured coordinates in units of a reseau interval to three decimal places ($1.000 \leqq x < 27.000; 30.000 \leqq y < 56.000$). Note that the x-coordinate increases with *decreasing* right ascension.

5.5.5.19 MELBOURNE ($-65°$ TO $-90°$). The introduction and description of the work on this zone is given in Vol. I (zones $-65°$ and $-66°$) by J. M. Baldwin.

The instrument used for taking the plates was a normal astrograph made by Grubb. (Baldwin complains about the lens of the 10″ guiding telescope not having been perfect when it was first delivered.) The position of the plates in the plate holder during the exposures (5^m, $2\frac{1}{2}^m$, 20^s) was very carefully defined. The emulsion was on plates of special glass with two ground edges. Setting errors larger than 1″ were not tolerated. Since the images are not very sensitive to changes of the focusing, not much attention was paid to focusing until 1909. The first plates were taken in 1891. The réseaux were copied on the plate after the exposure. The original réseaux by Gautier became progressively worse, developing many pinholes. Therefore, réseaux were ruled at Melbourne in asphalt varnish. The distortion due to film shifts during processing was found to be $2\frac{1}{2}$ microns on the average, and 10μ at maximum.

The measuring machines were all of the eyepiece micrometer type; the first was made in the workshop of the Melbourne Observatory; two more were made by Repsold and were of the type designed by Gill for the Cape zone. They had, however, only one micrometer screw, so that the micrometer had to be turned by 90° to measure in two coordinates.

The plates were measured in direct and in reverse positions. Different measurers measured the northern and southern halves of each plate. Reference stars were measured by both measurers and the mean was adopted.

The reference stars were observed at the transit circle at Melbourne, and were originally taken from the catalogues Melbourne 3 and Melbourne 4. However, since the comité permanent in 1896 had passed a resolution that there should be at least ten reference stars on every plate, 2603 supplementary reference stars were observed at the transit circle at Melbourne; these constitute the catalogue Melbourne 5. The positions were, as in all the other zones, referred to equator and equinox 1900.0. Where available, proper motions were applied to the reference star positions to reduce them to the epoch of the plates. Until 1914, the plate constants (following the model given by equations [2.14a]), which connect the standard coordinates ξ, η in minutes of arc and oriented to 1900.0 with the measured coordinates x, y by the formulas $\xi = x + ax + by + c$, $\eta = y + dx + cy + f$, were determined by a rigorous least squares solution; afterwards (at the suggestion of A. S. Eddington) from normal positions formed for every plate quadrant, that is, by the Christie-Dyson method.

The catalogues contain at the head of every listing of the measured coordinates the coordinates of the plate center, plate number and exposure date, the plate constants and the formulas for the computation of ξ and η from x and y.

5.5.6 REOBSERVATION AND DETERMINATION OF PROPER MOTIONS. The great value of the AC positions is twofold: their high intrinsic potential accuracy, and their age. The combination of these two factors permits them to be used for the determination of accurate proper motions. But, obviously, the determination of the proper motions requires the reobservation of the positions at a recent epoch. Somewhat timid and, on the whole, unorganized attempts at second epoch observations for the determination of proper motion have been made at a few of the participating observatories.

The very age which makes the AC valuable for the determination of proper motions detracts from its usefulness as a source of reference stars for the determination of positions of faint objects, as for instance, other stars, asteroids, comets or man-made bodies. Whatever has been said above about the contemporaneousness of plates, and the reference stars used for their reduction, applies here also. Even if the systematic proper motions were applied, the accuracy of the positions would still have suffered through the random components of the proper motions, that is, the so-called peculiar motions.

The obvious solution would be to repeat the Astrographic Catalogues, and thus get new accurate positions that could be reduced to any epoch with the proper motions to be determined from a comparison with the (properly reduced) positions from the original Astrographic Catalogue.

How should a task like this be undertaken?

In reflecting, it is apparent that one of the main criticisms of the Astrographic Catalogue could be directed towards its inhomogeneity and difference of format of the various zones. This is only partially due to the lack of specific directives; there are significant differences even between the two "zones" of the AGK2, which were observed and reduced at Hamburg and at Bonn, respectively, although it was the intention of the original sponsors that the various parts of the AGK2 should be treated in absolutely the same way. If the Astrographic Catalogue is repeated, the final authority for the execution of the work should probably lie with only one individual or agency.

It would furthermore be desirable to use as few telescopes for taking the plates as possible. If, as before, a format of $2° \times 2°$ is chosen, approximately 22000 plates will have to be taken in order to cover the sky at least twice. One might even use the old plate centers. With modern high speed plates, a five minute exposure should reach the limiting magnitude of the Astrographic Catalogue, and provide measurable first-order grating images for most of the stars.

It is rather obvious that one will want to plan very carefully the features of the instrument to be used. Up to a focal length of about ten meters (about three times that of a normal astrograph) the rms. error of a position is inversely proportional to the square of the focal length. A ten meter camera would thus be capable of giving positions with a weight of ten times those one would obtain on a normal astrograph.

The reason why the Astrographic Catalogue was not observed with long focus instruments, but with normal astrographs instead, is apparently merely historical. For one thing, a prototype for the normal astrograph existed. The scale of one minute per millimeter may have had a magic-like fascination, and may have been chosen with the simplicity of the reductions in mind. One must not forget that long focus refractors were used for photographic astrometric purposes by Schlesinger only about a couple of decades after work on the Astrographic Catalogue was begun. Even when long focus refractors were generally used for astrometric purposes, their fields covered mostly an area of less than $1° \times 1°$. The double coverage of the entire sky with such instruments would have required at least five times as much photographic work, and furthermore, there would not have been enough reference stars on a plate to establish the plate constants with an accuracy comparable to the accidental accuracy of the measured positions. Thus, the accuracy of the latter could never have been properly utilized.

Neither argument against long focus cameras is still valid. Large fields, or rather a large number of reference stars, are no longer necessary for the establishment of accurate plate constants if the conditions given by the plate overlaps are properly utilized. Since the plate overlapping technique requires an electronic computer, it could not possibly have been applied in the days when the original adjustments of the Astrographic Catalogue plates were made.

Secondly, long focus telescopes have been designed which have a focal length of ten meters and give images of astrometric quality over a field of more than $2° \times 2°$. To continue the use of short focus instruments for photographic astrometric purposes in the interest of obtaining large fields with many reference stars would probably be a waste of potential accuracy.

An instrument, eminently suited for the repetition of the Astrographic Catalogue, was designed and is being constructed by James G. Baker. It is a catadioptric system with about 10 meters focal length, a field of more than $2° \times 2°$, color effects below the measuring accuracy, and small magnitude effects and distortion. A duplicate and, if necessary, triplicate of Baker's instrument could probably readily be made. The problem of instrument design and construction has therefore apparently already been solved. The photographs would, of course, have to be made with gratings in front of the objective in order to cope with those magnitude effects which are produced by the circumstances incidental to the exposure (guiding, and so on) and which do not originate in the optical system. These procedures would yield positions with accidental rms. errors of not more than $0.''04$ at the epoch.

There is no serious problem with regard to reference stars. On the northern hemisphere, the AGK3 is finished and provides an excellent source of reference positions which can be reduced to any reasonable epoch. On the southern hemisphere there are the Yale and Cape Catalogues and several meridian catalogues, which could be combined to provide reference positions and proper motions, apart from the complete coverage of the southern sky that is presently planned at the U.S. Naval Observatory.

By far the most time-consuming job would be the measuring, but at this time, when automatic measuring machines are becoming quite common, this should also no longer be the bottleneck it was when all measuring was carried out by hand.

The reductions would, at the most, take half a year on a large electronic computer.

Altogether, a staff of no more than ten people, aided by modern electronic equipment, could finish the reobservation measurement and re-reduction of the Astrographic Catalogue in about, say, ten years. The reductions ought to be carried out in such a way that the original AC is at the same time re-reduced and proper motions obtained for all stars.

Bibliography

Anguita, C., Carrasco, G., Loyola, P., Shishkina, V. N. and Zverev, M. S. 1968. Meridian observations at Santiago, Chile. In *Highlights of Astronomy*, ed. L. Perek, p. 292. Dordrecht: Reidel. (Internat. Astron. Union.)

Aoki, S. 1967. On Oort's constant B. *Publ. Astron. Soc. Japan* **19,** 585; (also *Tokyo Astron. Obs. Reprint No.* 326).

Astronomisches Rechen-Institut. 1957. *Individuelle Verbesserungen des FK3 nördlich von −30° Deklination.* Veröff. des Astron. Rechen-Instituts Heidelberg No. 6.

———. 1957. *Individuelle Verbesserungen des FK3 südlich von −30° Deklination.* Veröff. des Astron. Rechen-Instituts Heidelberg No. 7.

———. 1960. *Individuelle Verbesserungen FK3R–FK3 für die Jahre 1960 bis 1962.* Veröff. des Astron. Rechen-Instituts Heidelberg No. 8.

———. 1963. *Preliminary supplement to the FK4 (FK4 SUP).* Veröff. des Astron. Rechen-Instituts Heidelberg No. 11.

Atkinson, R. d'E. 1955. On the flexure of meridian circles. *Month. Not. Roy. Astron. Soc.* **115,** 427.

Auwers, A. 1879. Fundamental-Catalog für die Zonen-Beobachtungen am nördlichen Himmel. *Publ. d. Astron. Gesellschaft* **14.**

———. 1883. Mittlere Örter von 83 südlichen Sternen für 1875.0 zur Fortsetzung des Fundamental-Catalogs für die Zonen-Beobachtungen der Astronomischen Gesellschaft, nebst Untersuchungen über die Relationen zwischen einigen neueren Sterncatalogen, insbesondere für den in Europa sichtbaren Theil des südlichen Himmels. *Publ. d. Astron. Gesellschaft* **17.**

———. 1889. Vorläufiger Fundamental-Catalog fur die südlichen Zonen der Astronomischen Gesellschaft. *Astron. Nachr.* **121,** 145.

———. 1894. Tafeln zur Reduction von Sternörtern auf das System des Fundamentalcatalogs für die Zonenbeobachtungen der Astronomischen Gesellschaft (A.G. Publ. XIV and XVII). *Astron. Nachr.* **134,** 33.

———. 1897. Fundamental-Catalog für Zonenbeobachtungen am Südhimmel und südlicher Polar-Catalog für die Epoche 1900. *Astron. Nachr.* **143,** 361.

———. 1898. Vorläufige Verbesserung des Fundamental-Catalogs für die Zonenbeobachtungen der Astronomischen Gesellschaft and seiner südlichen Fortsetzung (Publ. XIV u. XVII der Astr. Ges.). *Astron. Nachr.* **147,** 47.

———. 1900. Gewichtsfafeln für Sterncataloge. *Astron. Nachr.* **151,** 225.

———. 1903. Nachträge zu den Tafeln zur Reduktion von Sterncatalogen auf das System des Fundamentalcatalogs der AG und zu den Gewichtstafeln für Sterncataloge. *Astron. Nachr.* **162,** 357.

———. 1904. Tafeln zur Reduction von Sterncatalogen auf das System des Berliner Jahrbuchs. *Astron. Abhandlungen (Erg. Hefte zu den Astron. Nachr.)* **7.**

———. 1904a. Ergebnisse der Beobachtungen 1750–1900 für die Verbesserung des Fundamentalcatalogs des Berliner Jahrbuchs, Publ. A.G. XIV and XVII. *Astron. Nachr.* **164,** 225.

———. 1905. Weitere Nachweise der Grundlagen für die neuen Stern-Ephemeriden des Berliner Jahrbuchs. *Astron. Nachr.* **168,** 161.

———. 1907. Abgekürzte Bezeichnungen für Sterncataloge. *Astron. Nachr.* **174,** 369.

———. 1914. Ergebnisse aus Vergleichungen des Küstnerschen Catalogs von 10663 Sternen für 1900 mit anderen Sternverzeichnissen. *Astron. Nachr.* **198,** 273.

Baillaud, J. 1950. *Catalogue des 11755 étoiles de la zone +17° a + 25° et de magnitudes 9,5 à 10,5.* Paris: Observatoire de Paris.

———. 1957. Report of the IAU Commission 23, Carte du Ciel. *Trans. Internat. Astron. Union* **9,** 325.

Bakulin, P. I. 1949. *Fundamentalnye Katalogy Zvyesd.* (Fundamental Star Catalogues). Moscow and Leningrad: Gossudarstvennoye Izdatelstvo Techniko-Teoreticheskoy Literatury.

Barney, I. 1945. *Catalogue of the positions and proper motions of 8248 stars. Re-obs. by phot. of the A.G. zone between dec's. 6° and 10°, red'd. to 1950.0 w.o. appl. p.m's.* Trans. Astr. Obs. Yale Univ. **16.**

———. 1945a. *Catalogue of the positions and proper motions of 8108 stars. Re-obs. by phot. of the A.G. zone between −2° and −6°, red'd. to 1950. w.o. appl. p.m's.* Trans. Astron. Obs. Yale Univ. **17.**

———. 1947. *Catalogue of the positions and proper motions of 9092 stars. Re-obs. by phot. of the A.G. zone between dec's. +15° and +20°, red'd. to 1950.0 w.o. appl. p.m's.* Trans. Astron. Obs. Yale Univ. **18.**

———. 1948. *Catalogue of the positions and proper motions of 8967 stars. Re-obs. by phot. of the A.G. zone between dec's. +10° and +15°, red'd. to 1950.0 w.o. appl. p.m's*. Trans. Astron. Obs. Yale Univ. **19.**

———. 1949. *Catalogue of the positions and proper motions of 7996 stars. Re-obs. by phot. of the A.G. zone between dec's. +1° and +5° red'd. to 1950.0 w.o. appl. p.m's*. Trans. Astron. Obs. Yale Univ. **20.**

———. 1950. *Catalogue of the positions and proper motions of 5583 stars. Re-obs. by phot. of the A.G. zone between dec's. −2° and +1° red'd. to 1950.0 w.o. appl. p.m's*. Trans. Astron. Obs. Yale Univ. **21.**

———. 1951. *Supplementary Volume to the Yale Zone Catalogues −30° to +30°*. Trans. Astron. Obs. Yale Univ. **23.**

———. 1953. *Revised catalogue of 10358 stars, +25° to +30°*. Trans. Astron. Obs. Yale Univ. **24.**

———. 1954. *Revised catalogue of 8703 stars, +20° to +25°*. Trans. Astron. Obs. Yale Univ. **25.**

Barney, I. and van Woerkom, A. J. J. 1954. *Catalogue of the positions and proper motions of 1031 stars. Btw. dec's. +85° and +90°, red'd. w.o. appl. p.m's. to the equ. 1950.0.* Trans. Astron. Obs. Yale Univ. **26**/I.

Barney, I., Hoffeit, D., and Jones, R. B. 1959. *Catalogue of the positions and proper motions of 8380 stars. Between dec's. +50° and +55° red'd. w.o. appl. p.m's. to the equ. 1950.0.* Trans. Astron. Obs. Yale Univ. **26**/II.

———. 1959a. *Catalogue of the positions and proper motions of 8164 stars. Between dec's. +55° and +60° red'd. w.o. appl. p.m's. to the equ. 1950.0.* Trans. Astron. Obs. Yale Univ. **27.**

Bauschinger, J. 1922. Ergänzungen zum Katalog der Astronomischen Gesellschaft, Zone Leipzig II, nach Zonenbeobachtungen in der Jahren 1879–1882. *Veröff. d. Univ. Sternw. zu Leipzig* Heft 1.

Bauschinger, J., Courvoisier, L., Donner, A., Küstner, F. and Schorr, R. 1923. Vorläufige Vorschläge der A. G. Kommission für die Wiederholung der A. G. Kataloge. *Astron. Nachr.* **218,** 45.

Belyavsky, S. I. 1947. *Astrographic Catalogue of 11322 stars between 70° north declination and the north celestial pole* (in Russian). Pulkovo Publications Ser. 2, **60.**

Bemporad, A. 1907. Besondere Behandlung des Einflusses der Atmosphäre (Refraktion and Extinktion). In *Enzykl. d. Math. Wiss.* **VI-2A,** eds. K. Schwarzschild and S. Oppenheim, p. 287. Leipzig: B. G. Teubner.

Bessel, F. W. Nachricht von einer auf der Königsberger Sternwarte angefangenen allgemeinen Beobachtung des Himmels. *Astron. Nachr.* **1,** 257.

Bhaskaran, T. P. 1945. *Revised values of the constants of plates taken for the*

Astrographic Catalogue, Zones +39° to +36°. Publ. of the Nizamiah Obs. Hyderabad **13,** pt. 1.

Biehl, A. 1908. *Photographische Himmelskarte. Katalog. Ergänzungen und Berichtigungen zu den Bänden I-IV*. Potsdam: Publ. Astrophys. Obs. Potsdam. Phot. Katalog.

———. 1914. *Photographische Himmelskarte, Zone +31° bis +40° Deklination. Katalog. Berichtigungen and Bemerkungen zu den Bänden I-VIII*. Potsdam: Publ. Astrophys. Obs. Potsdam. Phot. Katalog.

Blaauw, A. 1968. The place of accurate proper motions in galactic research. In *Highlights of Astronomy*, ed. L. Perek, p. 316. Dordrecht: Reidel. (Internat. Astron. Union).

Böhme, S. and Fricke, W. 1965. Astronomical Constants, a survey of determined values. *Bull. Astron.* **25,** 269; (also *Mitt. d. Astron. Rechen-Instituts Heidelberg Ser. A*. No. 26.)

Boss, B. 1927. On the variable rotation of the Earth. *Astron. Journ.* **38,** 1.

———. 1937. *General Catalogue of 33342 Stars for the Epoch 1950, prepared at the Dudley Observatory, Albany, New York. With the collaboration of Sebastian Albrecht, Heroy Jenkins, Harry Raymond, Arthur J. Roy, William B. Varnum, Ralph E. Wilson*. 5 vols. Washington: Carnegie Institution of Washington. Publication No. 486.

Boss, B. and Jenkins, H. 1927. Comparison of the Albany and San Luis declination systems with the PGC. *Astron. Journ.* **37,** 173.

Boss, L. 1877. *The declinations of the stars employed in latitude work with the zenith telescope, embracing systematic corrections in declination deduced from various authorities, and a catalogue of five hundred stars for the mean epoch 1875*. United States Northern Boundary Commission, Rept. of the Chief Astronomer. Appendix H.

———. 1898. Standard stars south of declination −20°. *Astron. Journ.* **19,** 121.

———. 1903. Position and motion of 627 standard stars. *Astron. Journ.* **23,** 17.

———. 1903a. Method of forming the right ascensions of the Catalogue of 627 Principal Stars. *Astron. Journ.* **23,** 59.

———. 1903b. Determination of absolute magnitude equation for the Catalogue of 627 Standard Stars. *Astron. Journ.* **23,** 83.

———. 1903c. Method of forming the system of declination for the Catalogue of 627 Standard Stars. *Astron. Journ.* **23,** 117.

———. 1903d. On the systematic differences in declination between Bradley (Auwers) and the Catalogue of 627 Standard Stars. *Astron. Journ.* **23,** 157.

———. 1903e. Weights and systematic corrections of meridian observations in right ascension and declination. *Astron. Journ.* **23,** 191.

———. 1910. *Preliminary General Catalogue of 6188 Stars for the Epoch 1900, including those visible to the naked eye and other well-determined stars.* Washington: Carnegie Institution of Washington.

Brosche, P. 1966. *Representation of systematic differences in positions and proper motions of stars by spherical harmonics.* Veröff. des Astron. Rechen-Instituts Heidelberg, Nr. 17.

———. 1970. *Program for the determination of systematic differences and application to* $W3_{50}$*–FK4.* Veröff. des Astron. Rechen-Instituts Heidelberg, Nr. 23.

Brosche, P., Nowacki, H. and Strobel, W. 1964. *Systematic Differences FK4–GC and FK4–N30 for 1950.0.* Veröff. des Astron. Rechen-Instituts Heidelberg, Nr. 15.

Brouwer, D. 1934. On the determination of systematic errors in star positions from observations of minor planets. *Pub. Amer. Astron. Soc.* **8,** 44.

———. 1935. On the determination of systematic corrections to star positions from observations of minor planets. *Astron. Journ.* **44,** 57.

———. 1960. A method of constructing a revised general catalogue. *Astron. Journ.* **65,** 167.

Brown, D. C. 1955. *A matrix treatment of the general problem of least squares considering correlated observations.* Aberdeen Proving Grounds, Md.: Ballistic Research Laboratories Report No. 937.

———. 1965. *An advanced reduction and calibration for photogrammetric cameras.* Eau Gallie, Fla.: Instr. Corp. of Florida, Subcontr. D. Brown Associates.

———. 1966. Decentering distortion of lenses. *Photogrammetric Engineering* **32,** 444.

———. 1967. A unified lunar control network. *Photogrammetric Engineering,* **34,** 1272.

———. 1968. *Advanced methods for the calibration of metric cameras.* Final Rept. pt. 1, contract No. DA–44–009–AMC–1457(X) U.S. Army Engineer Topographic Lab's. Melbourne, Fla.: DBA Systems, Inc.

Brown, E. W. 1926. *The evidence for changes in the rate of rotation of the Earth and their geophysical consequence with a summary and discussion of the deviations of the Moon and the Sun from their gravitational orbits.* Trans. Astron. Obs. Yale Univ. **3**/4, 205.

Bucerius, H. 1932. Theorie des Objektivgitters. *Astron. Nachr.* **246,** 33.

Carrington, R. C. 1857. *A catalogue of 3735 circumpolar stars observed at Redhill in the years 1854, 1855 and 1856 and reduced to mean positions for 1855.0.* London: Longman, Brown, Green and Longman's.

Chandrasekhar, S. 1942. *Principles of stellar dynamics.* Chicago: Univ. of Chicago Press.

Chevalier, S. 1921. *Tour de l'equateur sur 12 plaques photographiques 24 × 30 cm portant 300 clichés photographiquement reliés.* (1er fasc.) 0h à 6h. Ann. Obs. Astron. Zô-Sè **12.**

——. 1927. *Tour de l'equateur sur 12 plaques photographiques 24 × 30 cm portant 300 clichés photographiquement reliés.* (2e fasc.) Ann. Obs. Astron. Zô-Sè **14,** C1.

——. 1928. *Catalogue de la zone $-0°50'$ a $+0°50'$ (Equin. 1920) d'après les Photographies du Tour de l'Équateur.* Ann. Obs. Astron. Zô-Sè, **15.**

Christie, W. H. M. and Dyson, F. W. 1894. Account of the measurement and comparison of a set of four astrographic plates, made at the Royal Observatory, Greenwich. *Month. Not. Roy. Astr. Soc.* **55,** 60.

Clemence, G. M. 1948. On the system of astronomical constants. *Astron. Journ.* **53,** 169.

——. 1948a. The value of minor planet observations in meridian astronomy. *Astron. Journ.* **54,** 10.

——. 1966. Inertial frames of reference. *Quart. Journ. Roy. Astron. Soc.* **7,** 10.

Clube, S. V. M. 1966. An absolute system of proper motions. *Quart. Journ. Roy. Astron. Soc.* **7,** 257.

——. 1968. Relative star positions from overlapping photographic plates. In *Highlights of Astronomy* ed. L. Perek, p. 347. Dordrecht: G. Reidel. (Internat. Astron. Union).

Cohn, F. 1899. *Bearbeitung von Bessel's Beobachtungen am Dollond'schen Mittagsfernrohre in des Jahren 1813–1819.* Astron. Beob. a.d. Kgl. Univ.-Sternw. zu Königsberg, Abt. 39.

——. 1907. Reduktion der astronomischen Beobachtungen. In *Enz. d. Math. Wiss.* **VI-2A,** eds. K. Schwarzschild and S. Oppenheim, p. 16. Leipzig: B. G. Teubner.

——. 1907a. Theorie der astronomischen Winkelmessinstrumente. In *Enz. d. Math. Wiss.* **VI-2A,** eds. K. Schwarzschild and S. Oppenheim, p. 195. Leipzig: B. G. Teubner.

——. 1913. Über die periodischen Fehler der Rektaszensionen der Fundamentalkataloge. *Astron. Nachr.* **195,** 225.

Danjon, A. 1955. L'astrolabe impersonnel de l'observatoire de Paris. *Bull. Astron.* **18,** 251.

——. 1959. *Astronomie Générale.* 2d ed. Paris: J. & P. Sennac.

Débarbat, S. and Guinot, B. 1970. *La méthode des hauteurs égales en astronomie.* Paris, London and New York: Gordon & Breach.

de Sitter, W. 1927. On the secular accelerations and the fluctuations of the longitudes of the Moon, the Sun, Mercury and Venus. *Bull. Astron. Inst. Netherl.* **4,** 21.

de Vegt, Chr. 1968. Report on overlap methods in photographic astrometry. In *Highlights of Astronomy*, ed. L. Perek, p. 343. Dordrecht: G. Reidel. (Internat. Astron. Union).

Deutsch, A. N. 1970. A determination of secular parallaxes of reference stars from relative proper motions of open clusters. In *Internat. Astron. Union Coll. No. 7, Proper Motions*, ed. W. J. Luyten, p. 74. Minneapolis: Univ. of Minnesota.

Dick, J. 1963. *Praktische Astronomie an visuellen Instrumenten*. Leipzig: Joh. Ambrosius Barth.

Dieckvoss, W. 1960. Progress on the AGK3. *Astron. Journ.* **65,** 171.

———. 1962. Systematic errors in the AGK2 and final reductions in the "AGK3" program. *Astron. Journ.* **67,** 686.

———. 1970. Progress with a general catalogue of positions and absolute proper motions based on AGK2, AGK3 and the astrographic catalogue as sources of data. In *Internat. Astron. Union Colloquium No. 7. Proper Motions*, ed. W. J. Luyten, p. 116. Minneapolis: Univ. of Minnesota.

———. 1970a. Experience with a simple plate overlap reduction method in a field around alpha Persei. In *Internat. Astron. Union Coll. No. 7, Proper Motions*, ed. W. J. Luyten, p. 158. Minneapolis, Univ. of Minnesota.

———. 1971. The AGK3, a basis for a general (northern) reference catalogue of positions and proper motions. In *Conference of Photographic Astrometric Technique*, ed. H. Eichhorn, p. 161. NASA Contractor Report NASA CR-1825. Washington: NASA. (Also *Astron. Contrib. from the Univ. of South Florida at Tampa No. 34*).

Dneprovsky, N. 1932. On the methods for the improvement of the fundamental declinations of stars. *Bull. Obs. Centr. a Poulkovo,* **13,** 1.

Downing, A. M. W. 1905. Comparison of the positions of Auwers' corrected places of southern fundamental stars with the catalogue of fundamental stars of Newcomb, for the epoch 1900. *Astron. Nachr.* **169,** 138.

Dreyer, 1876. On personal error in astronomical transit observations. *Proc. Roy. Irish. Acad.* ser. 2, **2,** 484.

Eckert, W. J. and Jones, R. B. 1954. Automatic measurement of photographic star positions. *Astron. Journ.* **59,** 83.

Eichelberger, W. S. 1925. Positions and proper motions of 1504 standard stars for the equinox 1925.0. *Astron. Papers Amer. Eph. and Naut. Alm.* **10,** pt. 1.

Eichhorn, H. 1953. Ein verkürztes Verfahren zur exakten Bestimmung von Schrauben- oder Skalenfehlern und Untersuchung des Töpferschen Messapparates der Wiener Universitätssternwarte. *Mitt. Univ. Sternw. Wien,* **7,** 45 (also *Sitz Ber. d. öst. Ak. d. Wiss., math.-naturw. Kl. IIa,* **162,** 327.)

———. 1957. Correction formulae for the coma effect in the Hyderabad Astrographic Catalogue. *Astron. Journ.* **62,** 204.

———. 1959. Formulae and Diagrams for the correction of the systematic errors of the Helsingfors zone of the Astrographic Catalogue. *Astron. Journ.* **64,** 2.

———. 1960. Zur Reduktion von photographischen Sternpositionen und Eigenbewegungen. *Astron. Nachr.* **285,** 233.

———. 1963. On the relationship between standard coordinates of stars and the measured coordinates of their images. *Applied Optics* **2,** 17.

———. 1964. Modern developments and problems in photographic astrometry. *Photogrammetric Engineering* **30,** 771. (also *Astron. Contrib. Univ. South Fla. at Tampa,* No. 1).

———. 1969. Star Catalogues. In *Spherical and Practical Astronomy* by I. I. Mueller, ch. 6, pp. 179–248. New York: F. Ungar Publishing Co.

———. 1971. Is the astrometric plate reduction method rigorous? *Astron. Nachr.* **293,** 127; (also *Astron. Contrib. from the Univ. of South Florida at Tampa,* No. 53).

———. 1971a. The behavior of magnitude dependent systematic errors. In *Conference on Photographic Astrometric Technique,* ed. H. Eichhorn, p. 241. NASA Contractor Report NASA CR-1825. Washington: NASA. (Also *Astron. Contrib. from the Univ. of South Florida at Tampa,* No. 34.)

Eichhorn, H. and Gatewood, G. D. 1967. New plate constants for the Northern Hyderabad zone (+35° to +40°) of the Astrographic Catalogue, in part computed by the plate overlap method. *Astron. Journ.* **72,** 1191. (Also *Astron. Contrib. from the Univ. of South Florida at Tampa,* No. 4).

———. 1967a. Tables containing the improved plate constants for the Northern Hyderabad zone of the Astrographic Catalogue. *Astron. Contrib. from the Univ. of South Florida at Tampa,* No. 3.

Eichhorn, H., Gatewood, G. D. and Sofia, S. 1970. The determination of the proper motions of X-ray sources. In *Internat. Astron. Union Coll. No. 7, Proper Motions,* ed. W. J. Luyten, p. 164. Minneapolis: Univ. of Minnesota.

Eichhorn, H., and Googe, W. D. 1969. The improvement of Star Catalogues by the incorporation of new data. *Astron. Nachr.* **291,** 125. (Also *Astron. Contrib. from the Univ. of South Florida at Tampa,* No. 15).

Eichhorn, H., Googe, W. D., Lukac, C. and Murphy, J. K. 1969. *A catalog of positions for 502 stars in the region of the Pleiades.* TOPOCOM Tech. Rept. No. 70. Washington: U.S. Army Topographic Command.

Eichhorn, H. and Rust, A. 1970. Rigorous computation of proper motions and their effect on star positions. *Astron. Nachr.* **292,** 37.

Eichhorn, H. and Williams, C. A. 1963. On the systematic accuracy of photographic astrometric data. *Astron. Journ.* **68,** 221.

Ellery, R. L. J. and Baracchi, P. 1917. *Third Melbourne General Catalogue of 3,068 stars for the equinox 1890.* Melbourne: A. J. Mullett, Government Printer.

Fatchikhin, N. V. 1968. Preliminary results of the determination of absolute proper motions of stars referred to galaxies. In *Highlights of Astronomy* ed. L. Perek, p. 297. Dordrecht: Reidel, (Internat. Astron. Union).

Felsmann, G. 1961. Ergänzungen zu Abteilung I der Geschichte des Fixsternhimmels. *Astron. Nachr.* **286,** 144.

Freundlich, E. 1916. *Katalog von 1886 Sternen zwischen +79° und +90° Deklination, beobachtet von L. Courvoisier und E. Freundlich.* Veröff. d. Königl. Sternwarte zu Berlin-Babelsberg, **2**/1.

Fricke, W. 1966. Probleme der fundamentalen Astrometrie und deren Beziehung zur Milchstrassenforschung. *Mitt. Astron. Ges.* No. 21, p. 47. (Also *Mitt. d. Astron. Rechen.-Instituts Heidelberg, Ser. A*, No. 28).

———. 1967. Precession and galactic rotation derived from McCormick and Cape proper motions in the systems of FK3, N30 and FK4. *Astron. Journ.* **72,** 642, (Also *Mitt. d. Astron. Rechen-Instituts Heidelberg, Ser. B*, No. 15).

———. 1967a. Precession and galactic rotation derived from fundamental proper motions of distant stars. *Astron. Journ.* **72,** 1368. (also *Mitt. d. Astron. Rechen-Instituts Heidelberg, Ser. B*, No. 16).

———. 1968. Precession and galactic rotation on the basis of various proper motion systems. In *Highlights of Astronomy*, ed. L. Perek, p. 306. Dordrecht: G. Reidel (Internat. Astron. Union). (also *Mitt. d. Astron. Rechen-Instituts Heidelberg*, Ser. A, No. 34).

———. 1970. Can the fundamental system be replaced by a reference system of galaxies? In *Internat. Astron. Union Colloquium No. 7, Proper Motions*, ed. W. J. Luyten, p. 105. Minneapolis: Univ. of Minnesota.

———. 1970a. *A fundamental system of planets, stars, galaxies and radio sources.* Unpublished manuscript.

———. 1970b. *Oort's constant B and the secular change of obliquity.* Manuscript.

———. 1971. A rediscussion of Newcomb's determination of precession. *Astron. and Astrophys.* **13,** 298.

———. 1971a. Determinations of precession. *Celest. Mech.* **4,** 150. (Also *Mitt. d. Astron. Rechen-Instituts Heidelberg, Ser A*, No. 52).

———. 1972. *Fundamental systems of positions and proper motions.* Ann. Review of Astron. and Astrophys. **10,** 101, eds. L. Goldberg, D. Layzer and J. G. Phillips. Palo Alto: Annual Reviews Inc.

Fricke, W., Brouwer, D., Kovalevsky, J., Mikhailov, A. A. and Wilkins, G. A. 1965. Report to the Executive Committee of the Working Group on the System of Astronomical Constants. *Bulletin Géodésique* **75,** 59.

Fricke, W. and Gliese, W. 1968. Desiderata for FK5. In *Highlights of Astronomy*, ed. L. Perek, p. 301. Dordrecht: Reidel. (Internat. Astron. Union).

Fricke, W., Kopff, A. et al. 1963. *Fourth Fundamental Catalogue (FK4)* Veröff. des Astron. Rechen-Instituts Heidelberg, No. 10.

Furuhjelm, R. 1916. Recherches sur les mouvements propres des étoiles dans la zone photographique de Helsingfors. I.: clichés de 9^h a 12^h *Acta Soc. Scient. Fennicae* **48,** No. 1.

Furuhjelm, R. 1926. Recherches sur les mouvements propres des étoiles dans la zone photographique de Helsingfors. II.: Cliches de 6^h a 9^h. *Acta Soc. Scient. Fennicae* **50,** No. 7.

Garfinkel, B. 1944. An investigation in the theory of astronomical refraction. *Astron. Journ.* **50,** 169.

———. 1967. Astronomical refraction in a polytropic atmosphere. *Astron. Journ.* **72,** 235.

Gatewood, G. D. 1967. On the intrinsic relative accuracy of the positions in the Vatican zone of the Astrographic Catalogue. *Astron. Nachr.* **290,** 189. (Also *Astron. Contrib. from the Univ. of South Florida at Tampa* No. 9).

Gill. D. and Hough, S. S. 1923. *Zone Catalogue of 20,843 stars, equinox 1900.* Cape of Good Hope: Royal Observatory, H.M. Stationery Office.

Gliese, W. 1963. *The right ascension system of the fourth fundamental catalogue (FK4).* Veröff. des Astron. Rechen-Instituts Heidelberg No. 12.

———. 1970. Recent observations relevant to the improvement of the FK4. In *Internat. Astron. Union Coll. No. 7, Proper Motions*, ed. W. J. Luyten, p. 146. Minneapolis: Univ. of Minnesota.

Googe, W. D., Lukac, C. F. and Eichhorn, H. 1968. The overlap approach toward the derivation of photographic stellar coordinates. In *Highlights of Astronomy*, ed. L. Perek, p. 338. Dordrecht: Reidel. (Internat. Astron. Union).

Googe, W. D., Eichhorn, H. and Lukac, C. F. 1970. The overlap algorithm for the reduction of photographic star catalogues. *Month. Not. Roy. Astron. Soc.* **150,** 35. (Also *Astron. Contrib. from the Univ. of South Florida at Tampa* No. 16).

Grönstrand, H. O. 1937. Bestimmung der systematischen Korrektionen der Sternörter im photographischen Zonenkataloge der Sternwarte Helsingfors. *Acta Soc. Scient. Fennicae N.S. A* **2,** No. 7.

Guinot, B. 1955. Astrolabe impersonnel. Reduction des observations. Étude des resultats. *Bull. Astron.* **18,** 283.

———. 1961. Comparisons du catalogue d'étoiles de l'astrolabe de Paris et d'autres catalogues. *Bull. Astron.* **23,** 343.

Günther, A. and Kox, H. 1970. Definitive plate constants for the astrographic catalogue north of $+40°$ declination. *Astron. and Astrophys.* **4,** 156.

———. 1970a. Tables of definitive plate constants for the zones Greenwich, Rome-Vatican, Catania, Helsingfors of the astrographic catalogue (carte du ciel). *Astron. and Astrophys. Suppl. Series* **3,** 85.

———. 1971. Definitive plate constants for the astrographic catalogue for the zones from +32° to +39° declination. *Astron. and Astrophys.* **12,** 175.

———. 1972. Tables of definitive plate constants for the zones Potsdam, Hyderabad, Uccle, Oxford of the astrographic catalogue (carte du ciel) *Astron. and Astrophys. Suppl. Series* **6,** 201.

Gyllenberg, W. 1926. *Katalog von 11800 Sternen der Zone +35° bis +40°, A.G. Lund*. Lund Univ. Obs. Ann. **1.**

———. 1948. Systematic corrections and weights of 108 star catalogues. *Medd. fr. Lunds Astron. Obs. Ser II*, Nr. 122. (Also *Lunds Univ. Arsskrift N. F. Avd. 2,* **44,** Nr. 2, also *Kungl. Fys. Sällskapets Handl.* N.F. **59,** Nr. 2).

Haas, J. 1954. *Geschichte des Fixsternhimmels Abt. II,* **14,** pp. III and IV. Karlsruhe: G. Braun Verlag.

Haramundanis, K. 1967. Experience of the Smithsonian Astrophysical Observatory in the construction and use of star catalogues. *Astron. Journ.* **72,** 588.

Heckmann, O., Dieckvoss, W. and Kox, H. 1949. Die Ableitung der Eigenbewegungen von Sternen der AG Kataloge mit Benutzung der PHK, insbesondere in der Zone +20° bis +25° (Berlin B). *Sitz. Ber. d. Deutschen Ak. d. Wiss., math naturw. Kl.*, Jg. **1948,** No. VII.

———. 1954. New plate constants in the system of the FK3 for the declination zones +21°, +22°, +23°, +24° of the Astrographic Catalogue Paris. *Astron. Journ.* **59,** 143.

Heckmann, O. and Dieckvoss, W. 1958. Questions of method and technique concerning proper motions. *Astron. Journ.* **63,** 156.

Heinemann, K. 1964. *Verzeichnis von Sternkatalogen 1900–1962*. Veröff. des Astr. Rechen-Instituts Heidelberg, No. 16.

Herget, P. 1970. The physical distortion of Schmidt Plates. In *Internat. Astron. Union Coll. No. 7, Proper Motions*, ed. W. J. Luyten, p. 93. Minneapolis: Univ. of Minnesota.

Herget, P. 1973. *Plate constants for the Bordeaux zone of the Astrographic Catalogue.* Publ. of the Cincinnati Obs., No. 24.

Hlavaty, W. 1939. *Differentialgeometrie der Kurven und Flächen und Tensorrechnung*. Groningen: P. Noordhoff.

Hoffleit, D. 1967. *Catalogue of the positions and proper motions of stars between declinations −30° and −35° reduced to the equ. of 1950. without appl. proper motions*. Trans. Astron. Obs. Yale Univ. **28.**

———. 1967a. Current work on southern zone catalogues at the Yale Observatory *Astron. Journ.* **67,** 696.

———. 1968. *Catalogue of the positions and proper motions of stars between declinations −35° and −40° reduced to the equin. of 1950 without appl. proper motions*. Trans. Astron. Obs. Yale Univ. **29.**

Hoffleit, D. with Eckert, D., Lü, P. and Paranya, K. 1970. *Catalogue of the positions and proper motions of stars between declinations −40 and −50° reduced to the equin. of 1950 without appl. proper motions*. Trans. Astron. Obs. Yale Univ. **30.**

Hoffleit, D. and Paranya, K. 1970. Common proper motion stars in a southern zone catalogue. In *Internat. Astron. Union, Coll. No. 7, Proper Motions*, ed. W. J. Luyten, p. 80. Minneapolis: Univ. of Minnesota.

Høg, E. 1961. Determination of division corrections. *Astron. Nachr.* **286,** 65, (Also *Mitt. Sternw. Hamburg-Bergedorf* **10,** No. 114).

———. 1962. The photoelectric micrometer for the Bergedorf Meridian Circle. *Sitz. Ber. Heidelb. Akad. Wiss, Jg.* 1962/63, p. 48.

———. 1968. Refraction Anomalies: The mean power spectrum of star image motion. *Zeitschrift für Astrophysik* **69,** 313.

———. 1970. A theory of a photoelectric multislit micrometer. *Astron. and Astrophys.* **4,** 89.

Hollis, H. P. 1938. The beginning of personal equation. *Observatory* **61,** 301.

Hough, P. V. C. and Powell, B. W. 1960. A method for faster analysis for bubble chamber photographs. *Nuovo Cimiento X,* **18,** 1184.

Hough, S. S. 1905. *A catalogue of 917 circumpolar stars*. Annals of the Cape Observatory **11**/II.

Jackson, E. S. 1968. Determination of the equinox and equator from meridian observation of the minor planets. *Astron. Papers Amer. Eph. and Naut. Alm.* **20,** pt. 1.

Jackson, J. and Stoy, R. H. 1954. *Cape photographic catalogue for 1950.0. Zone −30° to −35°*. Annals of the Cape Observatory **17.** London: H.M. Stationery Office.

———. 1954a *Cape photographic catalogue for 1950.0. Zone −35° to −40°*. Annals of the Cape Observatory **18.** London: H.M. Stationery Office.

———. 1954b. *Cape photographic catalogue for 1950.0. Zone −52° to −56°*. Annals of the Cape Observatory **19**. London: H.M. Stationery Office.

———. 1958. *Cape photographic catalogue for 1950.0. Zone −56° to −64°*. Annals of the Cape Observatory **20**. London: H.M. Stationery Office.

Jacoby, H. 1892. The Rutherfurd photographic measures of the group of the Pleiades. *Annals of the New York Acad. of Sci.* **6,** 239. (Also *Contr. from the Obs. of Columbia College, New York*, No. 3).

Jeffers, H. M. 1932. Meridian Observations of 1188 stars between 20° and 30° north declination. *Lick Observatory Bulletin* No. 436.

Jenkins, L. 1952. *General catalogue of trigonometric stellar parallaxes.* New Haven: Yale Univ. Observatory.

Kahrstedt, A. 1927. Über die systematischen Fehler von der From $\Delta\alpha_\alpha$ im N.F.K. *Astron. Nachr.* **229,** 400.

———. 1931. Über die Verbesserung des Äquinoctiums des N.F.K. *Astron. Nachr.* **244,** 33.

———. 1961–1966. *Index der Sternörter 1925–1960 (Index II).* Berlin: Akademie-Verlag.

Kapteyn, J. C. 1922. On the proper motions of the faint stars and the systematic errors of the Boss fundamental system. *Bull. Astron. Inst. Netherl.* **1,** 69.

Klock, B. L. 1970. *The automatic transit circle of the U.S. Naval Observatory.* Commission 8 Report, Internat. Astron. Union.

Klock, B. L. and Scott, D. K. 1970. Orientation of the FK4 catalogue from meridian observations of the Moon. *Astron. Journ.* **75,** 851.

Knobel, E. B. 1877. The chronology of star catalogues. *Mem. Roy. Astr. Soc.* **43,** 1.

Kohlschütter, A. 1957. *Zweiter Katalog der Astronomischen Gesellschaft für das Äquinoktium 1950.* Bd. **11.** Bonn: Dümmler.

König, A. 1933. Reduktion photographischer Himmelsaufnahmen. In *Handb. d. Astrophys.* vol. 1, ch. 6, p. 502. Berlin: Springer-Verlag.

———. 1962. Astrometry with astrographs. In *Astronomical Techniques (Stars and Stellar Systems vol. 2,* ed. A. Hiltner), p. 461). Chicago: U. of Chicago Press.

Kopff, A. 1928. Vorläufige Verbesserung der Sterne des Neuen Fundamentalkatalogs von Auwers nördlich von −20° für 1925.0. *Astron. Nachr.* **231,** 325.

———. 1931. Vorläufige Verbesserung der Sterne des Neuen Fundamentalkatalogs von Auwers südlich des Äquators für 1925.0. *Astron. Nachr.* **242,** 313.

———. 1933. Abgekürzte Bezeichnungen für Sternkataloge. *Astron. Nachr.* **248,** 333, (Also *Mitt. Astr. Rechen-Institut Berlin-Dahlem* **3** No. 9).

———. 1936. Star catalogues, especially those of fundamental character. *Month. Not. Roy. Astron. Soc.,* **96,** 714.

———. 1937. *Dritter Fundamentalkatalog des Berliner Astronomischen Jahrbuchs. I. Teil: Die Auwers-Sterne für die Epochen 1925 und 1950.* Veröff. des Astr. Rechen-Instituts zu Berlin-Dahlem No. 54.

———. 1937a. Die Entstehung und Weiterbildung des Fundamentalkatalogs von Auwers. *Sitz. Ber. d. Preuss. Ak. Wiss., math.-naturwiss. Kl.* **1937,** 101.

———. 1938. *Dritter Fundamentalkatalog des Berliner Astronomischen Jahrbuchs,*

II. Teil: Die Zustazsterne. Abh. d. Preuss. Ak. Wiss., math.-naturwiss. K1. **1938** No. 3.

———. 1938a. Bemerkungen zum Katalog der Zusatzsterne des FK3. *Astron. Nachr.* **267,** 301. (Also *Mitt. d. Astron. Rechen-Instituts Berlin-Dahlem* **4** No. 23).

———. 1939. *Tafeln zur Reduktion des Systems des General Catalogue auf das System des FK3.* Abh. d. Preuss. Ak. Wiss., math.-naturwiss. Kl. **1939** No. 18.

———. 1939a. Vergleich des FK3 mit dem GC von B. Boss. *Astron. Nachr.* **269,** 160. (Also *Mitt. Kop.-Inst. Berlin-Dahlem* **5** No. 4. Tab's. (Tafeln) **5** No. 5).

———. 1942. *Die Genauigkeit des Katalogs der Anhalsterne des AGK2.* Abh. d. Preuss. Ak. Wiss., math.-naturwiss. K1. **1942,** No. 5 (also *Mitt. Kop.-Inst. Berlin-Dahlem* **5,** No. 21).

———. 1949. A comparison of the systems of the General Catalogue and the Dritter Fundamentalkatalog. *Month. Not. Roy. Astron. Soc.* **109,** 580.

———. 1954. Supplement-Katalog des FK3 (FK3Supp). In *Astron.-Geod. Jahrbuch 1954.* Heidelberg: Astron. Rechen-Institut.

———. 1954a. Remarks on the revision of the FK3 and its relation to N30. *Month. Not. Roy. Astron. Soc.* **114,** 478.

Kopff, A. and Nowacki, H. 1934. Zusatzsterne des Dritten Fundamentalkatalogs des Berliner Astronomischen Jahrbuchs. *Astron. Nachr.* **252,** 185.

Kopff, A., Nowacki, H. and Strobel, W. 1964. *Individual corrections to FK3 and the declination system of the Fourth Fundamental Catalogue (FK4).* Veröff. des Astron. Rechen-Instituts Heidelberg No. 14.

Kopff, A. and Peters, J. 1943. *Katalog des Anhaltsterne für das Zonnenunternehmen der Astronomischen Gesellschaft (nach Beobachtungen an den Sternwarten Babelsberg, Bergedorf, Bonn, Breslau, Heidelberg, Leipzig and Pulkovo).* Veröff. des Astron. Rechen-Instituts Berlin-Dahlem No. 55.

Kulikov, K. A. 1964. *Fundamental constants of astronomy.* (Trans. from the orig. (1956) Russian edition for the U.S. NASA and the NSF). Washington: Office of Tech. Services, U.S. Dept. of Commerce.

Küstner, F. 1900. *Beobachtungen von 4070 Sternen. . .* Veröff. d. Kgl. Stw. zu Bonn, No. 4.

———. 1902. Beobachtete Correctionen des Fundamental-Catalogs von Auwers in AN 3508–09 und Ermittlung seiner Helligkeitsgleichung. *Astron. Nachr.* **158,** 129.

———. 1920. *Der kugelförmige Sternhaufen Messier 56.* Veröff. d. Univ.-Stw. zu Bonn. No. 14.

LaBonte, A. E. 1970. Automated proper motion survey: Data reduction and initial system performance. In *Internat. Astron. Union Colloquium No. 7, Proper Motions*, ed. W. J. Luyten, p. 26. Minneapolis: Univ. of Minnesota.

Lacroute, P. 1964. Amélioration des constants des clichés en utilisant leur recouvrement. *Ann. Obs. Strasbourg* **6,** 97.

———. 1968. Étude sur l'emploi de recouvrements de plaques pour l'établissement de catalogues photographiques. In *Highlights of Astronomy,* ed. L. Perek, p. 319–337. Dordrecht: G. Reidel. (Internat. Astron. Union).

———. 1970. Use of the overlap method for the connexion of proper motions to galaxies. Problems of the zone without nebulae. In *Internat. Astron. Union Coll. No. 7, Proper Motions,* ed. W.J. Luyten, p. 188. Minneapolis: Univ. of Minnesota.

Lacroute, P. and Valbousquet, A. 1970. Some results on AGK2–AGK3. *Internat. Astron. Union Coll. No. 7, Proper Motions,* ed. W. J. Luyten, p. 153. Minneapolis: Univ. of Minnesota.

———. 1970a. Theoretical computations on the efficiency of the overlap method in the Lick program. In *Internat. Astron. Union Coll. No. 7, Proper Motions,* ed. W. J. Luyten, p. 197. Minneapolis: Univ. of Minnesota.

Lanoë, C. 1965. Recherches des déclages radiaux systematiques des images stellaires dans la zone $22° < \delta < 24°$ du Catalogue Photographique de Paris. Notes Informations. *Publ. Obs. Paris fasc. 25, Astrométrie. No. 13.*

Larink, J. 1942. Bemerkungen zur Aufstellung eines Generalkatalogs schwacher Sterne. *Astron. Nachr.* **273,** 21.

Larink, J., Bohrmann, A., Kox, H., Groeneveld, I. and Klauder, H. 1955. *Katalog von 3356 schwachen Sternen für das Äquinoktium 1950, beobachtet an den Meridiankreisen der Sternwarten Hamburg-Bergedorf und Heidelberg-Königstuhl.* Hamburg-Bergedorf: Verlag der Sternwarte.

Laustsen, S. 1967. Development of impersonal transit circle methods. *Publ. og mindre Medd. fra Kφbenhavn's Obs. No. 190.* (Also *Mat. Fys. Medd. Dan. Vid. Selsk.* **3,** No. 3).

———. 1968. Meridian observations made in Brorfelde (Copenhagen University Observatory) 1964–1967. Positions of 972 stars brighter than 11.0 vis. mag. *Mat. Fys. Skr. Dan. Vid. Selk.* **3,** No. 6. (also *Pub. og. mindre Medd. fra Kφbenhavn's Obs.* No. 191).

Levy, J. 1955. Détermination des corrections de graduation des cercles divisées. *Bull. Astron.* **20,** 35.

Lieske, J. H. 1970. On the secular change of the obliquity of the ecliptic. *Astron. and Astrophys.* **5,** 90.

Lukac, C. F., Murphy, J. K., Googe, W. D. and Eichhorn, H. 1970. *An overlap reduction of the measurements used to produce the Cape Photographic Catalogs for* $-52°$ *to* $-56°$ *and* $-56°$ *to* $-60°$. TOPOCOM Tech. Report No. 1–2. Washington: U.S. Army Topographic Command.

Luyten, W. J. 1970. The Bruce and Palomar-Schmidt proper motion surveys. In

Internat. Astron. Union Coll. No. 7, Proper Motions, ed. W. J. Luyten, p. 59. Minneapolis: Univ. of Minnesota.

Martin, E. G. 1945. A 150th anniversary. *Observatory* **66,** 191.

Melchior, P. 1970. Precession-nutation and tidal potential. Paper presented to the I.A.U. Colloquium on Astronomical Constants, Heidelberg.

Melchior, P. and Georis, B. 1968. Earth tides, precession-nutation and the secular retardation of the earth's rotation. *Phys. Earth and Planet. Interiors* **1,** 267.

Mihalas, D. and Routly, P. M. 1968. *Galactic Astronomy*. San Francisco and London: W. H. Freeman.

Mönnichmeyer, C. 1909. *Verbesserte Örter des A.G.K. Bonn nebst gelegentlich bestimmten Örtern von weiteren 757 Sternen der Zone +40° bis +50°*. Veröff. d. Univ. Stw. zu Bonn **9**/1.

Morgan, H. R. 1932. Observed corrections to Newcomb's equinox. *Astron. Journ.* **41,** 177.

———. 1949. Corrections to the fundamental catalogs. *Astron. Journ.* **54,** 145.

———. 1952. Catalog of 5268 standard stars, based on the normal system N30. *Astron. Papers Amer. Eph. and Naut. Alm.* **13,** pt. 3.

Morgan, H. R. and Lyons, U. S. 1938. *Positions and proper motions of 1117 reference stars in declination −10° to −20°*. Pub. U.S.N.O. 2nd ser. **14,** pt. II.

Mt. Stromlo Observatory 1959. *Fourth Melbourne General Catalogue, reduced without proper motion to the equinox 1900.0*. Canberra: Mt. Stromlo Obs.

Mueller, I. I. 1969. *Spherical and Practical Astronomy as applied to Geodesy*. New York: Frederick Ungar Publishing Co.

Muhlemann, D. O., Holdridge, D. B. and Block, N. 1962. The Astronomical Unit determined by radar reflections from Venus. *Astron. Journ.* **67,** 191.

Murray, C. A. 1968. The relationships between various techniques for obtaining proper motions. In *Highlights of Astronomy*, ed L. Perek, p. 311. Dordrecht: Reidel. (Internat. Astron. Union).

Murray, C. A., Tucker, R. H. and Clements, E. D. 1969. Optical positions of radio sources. *Nature* **221,** 1229.

Murray, C. A. and Clube, S. V. M. 1970. Kinematics of faint stars, and the reduction from relative to absolute proper motions. In *Internat. Astron. Union Coll. No. 7, Proper Motions*, ed. W. J. Luyten, p. 131. Minneapolis: Univ. of Minnesota.

Nautical Almanac Offices of the United Kingdom and the United States of America. 1961. *Explanatory Supplement*. London: Her Majesty's Stationery Office.

Newcomb, J. 1970. The Luyten-Control Data stellar proper motion measuring machine. In *Internat. Astron. Union Colloquium No. 7, Proper Motions*, ed. W. J. Luyten, p. 5. Minneapolis: Univ. of Minnesota.

Newcomb, S. 1872. *On the right ascensions of the equatorial fundamental stars*. Washington Observations for 1870 (appendix). Washington: U.S. Naval Observatory.

———. 1880. Catalogue of 1,098 clock and zodiacal stars. *Astron. Papers Amer. Eph. and Naut. Alm.* **1,** pt. 3, 147.

———. 1896. On Boss' system of declinations and on that of the Astronomische Gesellschaft. *Astron. Journ.* **16,** 33.

———. 1898. A new determination of the precessional constant. *Astron. Papers Amer. Eph. and Naut. Alm.* **8,** pt. 1, p. 1.

———. 1899. Catalogue of fundamental stars for the epochs 1875 and 1900, reduced to an absolute system, *Astron. Papers Amer. Eph. and Naut. Alm.* **8,** pt. 2, p. 77.

———. 1906. *A compendium of spherical astronomy*. New York: The Macmillan Co.; reprinted 1960 by Dover Publications, N.Y.

Niethammer, Th. 1947. *Die genauen Methoden der astronomisch-geographischen Ortsbestimmung*. Basel: Birkhäuser.

Nowacki, H. 1935. Vergleich des FK3 mit den Fundamentalkatalogen von A. Auwers, L. Boss and W. S. Eichelberger für 1925. *Astron. Nachr.* **255,** 301.

Nowacki, H. and Strobel, W. 1968. *Systematic relations between 71 star catalogues and the FK3, and weights*. Veröff. des Astron. Rechen-Instituts Heidelberg No. 20.

Numerov, B. V. 1933. On the problem of determination of systematic errors of declination of fundamental stars. *Bull. de l'Inst. Astron. Lenningrad* **32,** 139.

———. 1935. On the problem of the determination of systematic errors of star positions. *Astronomicheskii Zhournal* **12,** 339.

———. 1936. Zur Schaffung einer Fundamentalsystems schwacher Sterne, *Astron. Nachr.* **260,** 305.

———. 1936. On the problem of simultaneous determination of corrections to the elements of the planets and of the Earth. *Astron. Journ.* **45,** 105.

Nyrén, M. 1876. Das Aequinoctium für 1865. *Mem. Acad. St. Petersburg, VII,* **23,** No. 3.

Oort, J. H. 1943. Some remarks on the fundamental systems of the General Catalogue and the Dritter Fundamentalkatalog. *Bull. Astr. Inst. Netherlands* **9,** 417.

———. 1943a. Tentative corrections to the FK3 and GC systems of declinations. *Bull. Astr. Inst. Netherlands* **9,** 423.

———. 1943b. The constants of precession and galactic rotation. *Bull. Astr. Inst. Netherlands* **9,** 424.

Orlov, B. A. 1956. *Refraction tables of Pulkovo observatory*. 4th ed. Pulkovo: Principal State Observatory.

Osvalds, V. V. 1954. Neue Untersuchungen über Eigenbewegungen im Gebiet der Hyaden. *Astron. Nachr.* **281,** 193. (Also *Mitt. d. Hamburger Sternw. in Bergedorf* No. 88).

Peters, J. 1907. *Neuer Fundamentalkatalog des Berliner Astronomischen Jahrbuchs nach den Grundlagen von A. Auwers für die Epochen 1875 and 1900.* Veröff d. Kön. Astron. Rechen-Instituts zu Berlin, No. 33.

Pierce, D. A. 1970. Star catalog corrections determined from photographic observations of selected minor planets (summary and results). Presented to meeting of I.A.U. Commission 8.

Podobed, V. V. 1965. *Fundamental astrometry.* Chicago: Univ. of Chicago Press. (Trans. from the Russian "Fundamentalnaya Astrometriya". 1962. Moscow: State Publishing House for Physical and Mathematical Literature).

Polozhetsev, D. D. and Zverev, M. S. 1958. *Predvatnitelniy svodinij katalog fundamentalnikh slavikh svyost so skolozhenyami ot +90° to −20° (PFKSZ).* Pulkovo Publications **72.**

Prager, R. 1923. *Katalog von 8803 Sternen zwischen 31° und 40° nördlicher Deklination. Nach gemeinschaftlich mit K. F. Bottlinger am Pistor und Martinschen Meridiankreise der Sternwarte zur Berlin-Babelsberg ausgeführten Beobachtungen.* Veröff. Univ.-Sternw. Berlin-Babelsberg **4.**

Protheroe, W. M. 1961. Stellar scintillation. *Science* **134,** 1593.

Raymond, H. 1926. The corrections to the declination system of Boss' preliminary General Catalogue. *Astron. Journ.* **36,** 129.

———. 1927. Corrections to the Boss system of declinations in the order of right ascension. *Astron. Journ.* **37,** 88.

Rees, J. K. 1906. The Rutherfurd photographic measures. *Contr. from the Obs. of Columbia Univ.* Nos. 1 and 2.

Ristenpart, F. 1901. Sterncataloge und Karten. In *Handwörterbuch der Astronomie,* herausg. v. Dr. W. Valentiner. **3-II,** p. 455. Breslau: Eduard Trewendt.

———. 1901a. *Verzeichnis von 336 Sterncatalogen.* Breslau: Eduard Trewendt.

———. 1909. Fehlerverzeichnis zu den Sternkatalogen des 18. und 19. Jahrhunderts. *Astron. Abh., Erg. Hefte zu den Astron. Nachr.* **3,** No. 16, pp. 1–500.

Schaub, W. *Vorlesungen über sphärische Astronomie.* Leipzig: Akademische Verlagsgesellschaft Geest & Porstig.

Schlesinger, F. and Barney, I. 1925. *Catalogue of the positions and proper motions of 8359 stars. Reobservation by means of photography of the Astronomische Gesellschaft zone between declinations +50° and +55°. With an appendix containing the positions of 1070 comparison stars observed by R. H. Tucker with the Lick Meridian Circle.* Trans. Astron. Obs. Yale Univ. **4.**

———. 1930. *Catalogue of the positions and proper motions of 7727 stars. Reobservation by photography of the Astronomische Gesellschaft zone between de-*

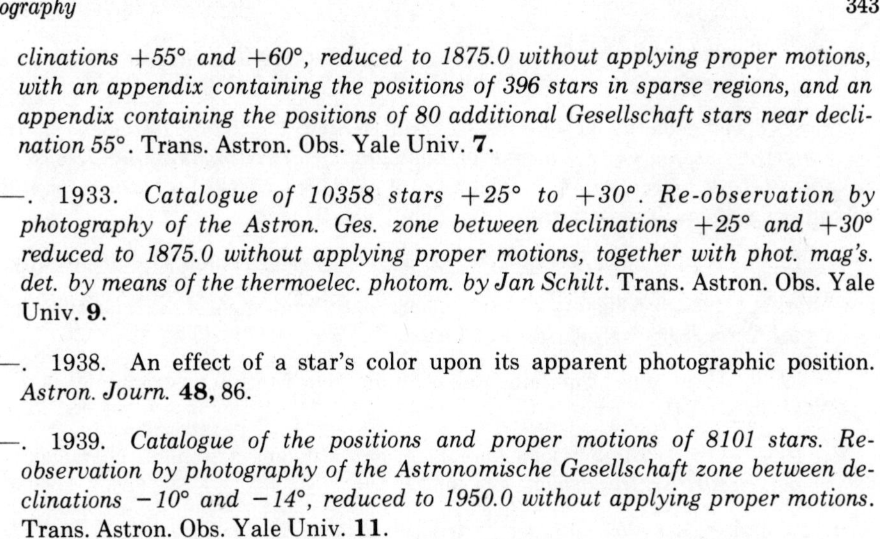

clinations +55° and +60°, reduced to 1875.0 without applying proper motions, with an appendix containing the positions of 396 stars in sparse regions, and an appendix containing the positions of 80 additional Gesellschaft stars near declination 55°. Trans. Astron. Obs. Yale Univ. **7**.

———. 1933. *Catalogue of 10358 stars +25° to +30°. Re-observation by photography of the Astron. Ges. zone between declinations +25° and +30° reduced to 1875.0 without applying proper motions, together with phot. mag's. det. by means of the thermoelec. photom. by Jan Schilt.* Trans. Astron. Obs. Yale Univ. **9**.

———. 1938. An effect of a star's color upon its apparent photographic position. *Astron. Journ.* **48,** 86.

———. 1939. *Catalogue of the positions and proper motions of 8101 stars. Re-observation by photography of the Astronomische Gesellschaft zone between declinations −10° and −14°, reduced to 1950.0 without applying proper motions.* Trans. Astron. Obs. Yale Univ. **11**.

———. 1939a. On the accuracy of the proper motions in the General Catalogue, Albany 1938. *Astron. Journ.* **48,** 51.

———. 1940. *Catalogue of the positions and proper motions of 8563 stars. Re-observation by photography of the AG zone between declinations −14° and −18°, reduced to 1950.0 without applying proper motions.* Trans. Astron. Obs. Yale Univ. **12**/1.

———. 1940a. *Catalogue of the positions and proper motions of 4553 stars. Re-observation by photography of the AG zone between declinations −18° and −20°, reduced to 1950.0 without applying proper motions.* Trans. Astron. Obs. Yale Univ. **12**/2.

———. 1942. New reductions of astrographic plates with the help of the Yale Photographic Catalogues. *Astron. Jour.* **49,** 39.

———. 1943. *Catalogue of the positions and proper motions of 4292 stars. Re-observation by photography of the AG zone between declinations −20° and −22°, reduced to 1950.0 without applying proper motions.* Trans. Astron. Obs. Yale Univ. **13**/1.

———. 1943a. *Catalogue of the positions and proper motions of 9455 stars. Re-observation by photography of the Cordoba zone between declinations −27° and −30°, reduced to 1950 without applying proper motions.* Trans. Astron. Obs. Yale Univ. **13**/2.

———. 1943b. *Catalogue of the positions and proper motions of 15110 stars. Re-observation by photography of the Cordoba zone between declinations −22° and −27°, reduced to 1950.0 without applying proper motions.* Trans. Astron. Obs. Yale Univ. **14**.

Schlesinger, F., Barney, I., and Gesler, C. 1934. *Catalogue of 8703 stars +20° to +25°*. Trans. Astron. Obs. Yale Univ. **10**.

Schlesinger, F. and Hudson, C. J. 1914. *The determination of star positions by means of a wide angle camera.* Publ. Allegheny Obs. **3,** No. 9.

Schlesinger, F., Hudson, C. J., Jenkins, L., and Barney, I. 1926. *Catalogue of 1275 stars. Re-observation by means of photography of Astronomische Gesellschaft stars between declinations +1° and +2° reduced to 1875.0 without applying proper motions.* Trans. Astron. Obs. Yale Univ. **3,** 137.

———. 1926a. *Catalogue of 5833 stars −2° to +1°.* Trans. Astron. Obs. Yale Univ. **5.**

Schmeidler, F. 1952. A history of extreme errors in fundamental declinations. *Occasional Notes Roy. Astron. Soc.* **2,** 160.

———. 1957. Über eine Fundamentalgleichung der Meridianastronomie. *Astron. Nachr.* **283,** 241.

———. 1958. Über systematische Fehler bei der Messung astronomischer Deklinationen. *Naturwiss. Rundschau* **11,** 458.

Schorr, R. and Kruse, W. 1928. *Index der Sternörter 1900–1925, Bd.* **1:** *Der nördliche Sternhimmel; Bd.* **2:** *Der südliche Sternhimmel.* Bergedorf: Verlag der Sternwarte.

Scott, F. P. 1966. Observer's manual for use with the seven-inch transit circle. Washington: U.S. Naval Observatory. Unpublished manuscript.

———. 1967. International reference star programs progress report. 1 August 1967, Washington: U.S. Naval Observatory. Mimeographed manuscript.

———. 1968. TheAGK3, SRS, and related projects. In *Highlights of Astronomy*, ed. L. Perek, p. 279. Dordrecht: G. Reidel (Internat. Astron. Union).

———. 1970. Private communication.

Scott, F. P. and Schombert, J. L. 1970. Status and problems of the international reference star programs. In *International Astron. Union Colloquium No. 7, Proper Motions*, ed. W. J. Luyten, p. 120. Minneapolis: Univ. of Minnesota.

Scott, F. P. and Smith, C. 1971. Comparison of the SAO and AGK3R star catalogues. In *Conference on Photographic Astrometric Technique*, ed. H. Eichhorn, p. 167. NASA contractor report NASA CR-1825. Washington: NASA. (Also *Astr. Contr. from the Univ. of So. Fla. at Tampa* No. 34).

Shepherd, W. M. and Radenkovič, D. 1954. The bending of a square plate on a spherical former with applications to the photographic plate of a Schmidt camera. *Month. Not. Roy. Astron. Soc.* **114,** 211.

Smithsonian Astrophysical Observatory (Staff of). 1966. *Star Catalog. Positions and proper motions of 258.997 stars for the epoch and equinox of 1950.0.* Publications of the Smithsonian Institution of Washington, D.C. No. 4562. Washington: Smithsonian Institution.

Sofia, S., Eichhorn, H., and Gatewood, G. 1969. A study of the positions and proper motions of 83 stars in the region of Sco X-1, *Astron. Journ.* **74,** 20. (Also *Astron. Contr. from the Univ. of South Fla. at Tampa* No. 12).

Spencer Jones, H. 1939. Some problems of meridian astronomy. *Month. Not. Roy. Astron. Soc.* **99,** 424.

Spencer Jones, H. and Jackson, J. 1936. *Proper motions of stars in the zone catalogue of 20,843 stars, 1900.* Cape of Good Hope: Royal Observatory, H. M. Stationery Office.

Stoy, R. H. 1954. Astrometric work at the Cape Observatory and the KSZ. *Astron. Journ.* **59,** 35.

———. 1966. *Cape Photographic Catalogue for 1950.0, zones −64° to −80°.* Annals of the Cape Observatory **21**. London: Her Majesty's Stationery Office.

———. 1968. *Cape Photographic Catalogue for 1950.0, zones −80° to −90°.* Annals of the Cape Observatory **22**. London: Her Majesty's Stationery Office.

———. 1970. The initial performance of the GALAXY machine. In *Internat. Astron. Union Colloquium No. 7, Proper Motions*, ed. W. J. Luyten, p. 48. Minneapolis: Univ. of Minnesota.

Standish, E. M. 1962. *The accuracy of the coordinate measurements in the Northern Hyderabad zone of the Astrographic Catalogue.* Master's Thesis. Middletown, Conn.: Wesleyan University.

Strand, K. A. 1960. The 60″ astrometric reflector of the U.S. Naval Observatory. *Astron. Journ.* **65,** 502.

Tucker, R. H. 1922. Comparison of standard star systems. *Astron. Journ.* **34,** 143.

van Altena, W. F. and Jones, B. F. 1970. The absolute proper motion of the Pleiades cluster, In *Internat. Astron. Union Coll. No. 7, Proper Motions*, ed. W. J. Luyten, p. 103 Minneapolis: Univ. of Minnesota.

van de Kamp. P. 1951. Long focus photographic astrometry. *Popular Astronomy* **59,** No. 2.

———. 1967. *Principles of astrometry.* San Francisco and London: W. H. Freeman and Co.

———. 1968. Parallax, proper motion, acceleration and orbital motion of Barnard's star. *Astron. Journ.* **73,** S121.

———. 1970. Secular perspective acceleration. In *Internat. Astron. Union Coll. No. 7, Proper Motions*, ed. W. J. Luyten, p. 77. Minneapolis: Univ. of Minnesota.

van Herk, G. 1952. The influence of the spectral type of a star on its declination as derived from meridian-circle observations. *Bull. Astron. Inst. Netherlands* **11,** 489.

van Rhijn, P. J. and Bok, B. J. 1931. The secular parallax of the stars of different apparent magnitude and galactic latitude. *Publ. Kapteyn Astron. Lab. Groningen* No. 45.

Vasilevskis, S. and Klemola, A. R. 1970. First results of the Lick proper motion program. In *Internat. Astron. Union Coll. No. 7, Proper Motions*, ed. W. J. Luyten, p. 167. Minneapolis: Univ. of Minnesota.

von der Heide, J. 1937. Untersuchung der auf der Hamburger Sternwarte in Bergedorf für das Zonen-Unternehmen der Astronomischen Gesellschaft benutzten Aufnahme- und Mess-Instrumente. *Astron. Abh. der Sternw. in Bergedorf,* **4,** No. 9.

———. 1952. Untersuchungen an Zeitbestimmungen. *Astron. Nachr.* **281,** 31. (Also *Mitt. d. Hamburger Stw. in Bergedorf* No. 85).

Wahl, E. 1937. Systematische Beziehungen einiger Kataloge zum System des FK3. *Astron. Nachr.* **263,** 349.

Walther, M. E. 1946. Determination of proper motions using the Astrographic Catalogue. *Popular Astronomy* **54,** 499.

Watts, C. B., Scott, F. P. and Adams, A. N. 1952. Results of observations made with the six-inch transit circle 1941–1949. (Obsn's. of the Sun, Moon, and planets, Catalog of 5216 stars for 1950.0. Corr'ns. to GC and FK3). *Pub. U.S.N.O.* 2nd ser. **16,** pt. III.

Watts, C. B. 1960. The transit circle. In *Telescopes*, ed. J. Kuiper and B. Middlehurst, ch. 6. Chicago: Univ. of Chicago Press.

Wayman, P. A. 1966. Determination of the inertial frame of reference. *Quart. Journ. Roy. Astron. Soc.* **7,** 138.

Wesselink, A. J. 1970. The Yale-Columbia southern proper motion program. In *Internat. Astron. Union Coll. No. 7, Proper Motions*, ed. W. J. Luyten, p. 184. Minneapolis: Univ. of Minnesota.

Williams, C. A. 1967. *An analysis of the Performance of the Loomis polar telescope.* Doctoral Dissertation. New Haven: Yale University.

Williams, E. T. R. 1947. Reduction of the Cape photographic proper motions to the FK3 system. *Astron. Journ.* **53,** 59. (Also *Pub. Leander McCormick Obs.* **10,** 12).

Williams, E. T. R. and Vyssotsky, A. N. 1948. The constants of solar motion, precession, and galactic rotation as derived from McCormick and Cape proper motions. *Pub. Leander McCormick Obs.* **10,** 17.

———. 1948a. The constants of galactic rotation and precession. *Pub. Leander McCormick Obs.* **10,** 27.

Wood, H. 1970. The Astrographic Catalogue as a source of old epoch plates. In *Internat. Astron. Union, Coll. No. 7, Proper Motions*, ed. W. J. Luyten, p. 67. Minneapolis: Univ. of Minnesota.

Woolard, E. W. 1953. Theory of the rotation of the Earth around its center of mass. *Astron. Papers Amer. Eph. and Naut. Alm.* **15**/1, (see also *Astron. Journ.* **58,** 2).

Woolard, E. W. and Clemence, G. M. 1966. *Spherical Astronomy.* New York and London: Academic Press.

Zurhellen, W. 1904. *Darlegung und Kritik der zur Reduktion photographischer Him-*

melsaufnahmen aufgestellten Formeln und Methoden. Bonn: Inaugural-Dissertation.

Zverev, M. S. 1951. *Catalogue of faint stars. List of 5120 stars with declinations from +90° to +30°* (1st part; in Russian). Moscow: State Astron. Sternberg Institute.

———. 1951a. *Catalogue of faint stars. List of 10235 stars with declinations from +30° to −30°* (2nd part; in Russian). Moscow: State Astron. Sternberg Institute.

———. 1951b. *Catalogue of faint stars. List of those 645 stars of the Fundamental Catalogue of 931 faint stars with declinations north of −30°* (in Russian). Moscow: State Astronomical Sternberg Institute.

———. 1960. An investigation of the PFKSZ. *Astron. Journ.* **65.** 223.

———. 1963. *Fundamental astrometry*. Translated for the U.S. Naval Observatory, Washington, D.C. by the U.S. Dept. of Commerce, Office of Technical Services, Translation from *Uspekhi Astronomicheskich Nauk* **5,** (1950) and **6,** (1954).

———. 1968. Recent work on fundamental astrometry in the U.S.S.R. In *Highlights of Astronomy*, ed. L. Perek, p. 286. Dordrecht: G. Reidel (Internat. Astron. Union).

Index